W9-AAA-754

ESSENTIALS OF GEOLOGY

Second Edition

Frederick K. Lutgens □ **Edward J. Tarbuck**

Illinois Central College

Charles E. Merrill Publishing Company
A Bell & Howell Company
Columbus   Toronto   London   Sydney

Published by
Charles E. Merrill Publishing Co.
A Bell & Howell Company
Columbus, Ohio 43216

This book was set in Frutiger.
Production Editor: Rex Davidson.
Text Design by Cynthia Brunk.
Cover Design by Cathy Watterson.
Cover Photograph: Deer Creek Gorge, Grand Canyon National Park, Arizona.
(Photo by David Muench)
Illustrations by Dennis Tasa, Tasa Graphic Arts, Inc.

Copyright © 1986, 1982, by Bell & Howell Company. All rights reserved. No
part of this book may be reproduced in any form, electronic or mechanical,
including photocopy, recording, or any information storage and retrieval
system, without permission in writing from the publisher. ''Bell+Howell'' is
a trademark of the Bell & Howell Company. ''Charles E. Merrill Publishing Co.''
and ''Merrill'' are registered trademarks of the Bell & Howell Company.

Library of Congress Catalog Card Number: 85–61319
International Standard Book Number: 0–675–20464–X
Printed in the United States of America
2 3 4 5 6 7 8 9 10 — 90 89 88 87 86

QE
26.2
L87
1986
c.4

PREFACE

In recent years, media reports have made us increasingly aware of the geological forces at work in our physical environment. News stories graphically portray the violent force of a volcanic eruption, the devastation generated by a strong earthquake, and the large numbers left homeless by mudflows and flooding. Such events, and many others as well, are destructive to life and property, and we must be better able to understand and deal with them. However, our natural environment has an even greater importance, for the earth is our home. The earth provides the mineral resources so basic to modern society, as well as most of the ingredients necessary to support life. Therefore, as many members of society as possible should acquire an understanding of how the earth works.

With this in mind, we have written a text to help people increase their understanding of the physical environment. We hope this new knowledge will encourage some to actively participate in the preservation of the environment, while others may be sufficiently stimulated to pursue a career in the earth sciences. Equally important, however, is our belief that a basic understanding of earth will greatly enhance appreciation of our planet and thereby enrich the reader's life.

The Second Edition of *Essentials of Geology*, like its predecessor, is a nonquantitative text intended for students taking their first course in geology. We have attempted to write a text that is not only informative and timely, but one that is highly usable as well. The language is straightforward and written to be understood by a student with little or no college-level science experience. We have deliberately refrained from using excessive jargon and when new terms are introduced, they are placed in boldface type and defined. A glossary is also included at the conclusion of the text for easy reference to important terms. Further, review questions conclude each chapter to help the reader prepare for exams and quizzes. Useful information on metric conversions, common minerals, and topographic maps is also provided in the appendices.

The Second Edition of *Essentials of Geology* has been extensively rewritten to make the material more timely and more readable. For example, Chapter Thirteen, Plate Tectonics, has been totally restructured using an historical approach. Here, we begin with the development of the concept of continental drift and end with the evidence that led to the unfolding of the plate tectonics theory. In addition to thoroughly revising the manuscript, we have included a new chapter entitled Earth History: A Brief Summary. This addition provides the reader with an overview of the major events which have shaped our physical and biological environment through geologic time.

Perhaps the most obvious change in this edition is the use of a full-color format. This has allowed us to include many color photographs and to place them in the text where they can best aid the reader. Color is also used in the body of the text to emphasize headings and is used in much of the line art to highlight selected information. Moreover, the more complex diagrams and maps have been produced in full color to help clarify important concepts. Despite all of these changes, the focus and structure of the Second Edition remain the same as its predecessor— to foster a basic understanding of the earth sciences.

The authors wish to express their thanks to the many individuals, institutions, and state and federal agencies that provided us with information, photographs, and illustrations. Special gratitude goes to those colleagues who were kind enough to offer their suggestions for improving the manuscript. Their critical comments have greatly aided us. We thank William R. Shirk, Shippensburg State College; Frederick Goldstein, Trenton State College; Lawrence McAdam, Seminole Community College; Harold E. Featherman, Monroe Community College; Louis J. Pinto, Monroe Community College; Stuart J. Inglis, Chabot College; Stanley E. Karp, Bakersfield College; Sr. Michael Ann Durrer, St. Francis College; and Douglas L. Smith, University of Florida.

Our thanks also go to Diane Weber who typed much of the manuscript, Dennis Tasa for his imaginative production of the line art, and to Rex Davidson, our production editor, who along with the many other fine people at Charles E. Merrill Publishing skillfully transformed our manuscript into a finished product.

Frederick K. Lutgens
Edward J. Tarbuck

CONTENTS

INTRODUCTION

The spectacular eruption of a volcano, the terror brought by an earthquake, the magnificent scenery of a mountain valley, the destruction created by a landslide—all are subjects for the geologist. The study of geology deals with many fascinating and practical questions about our physical environment. What forces produce mountains? What was the Ice Age like? Will there be another? What created this cave and the stone icicles hanging from its ceiling? Should we look for water here? Is strip mining practical in this area? Will oil be found if a well is drilled at this location? What will result if the landfill is located in the old quarry?

The subject of this text is **geology**, a word that literally means "the study of the earth." To understand the earth is not an easy task because our planet is not an unchanging mass of rock, but rather a dynamic body possessing a long and complex history.

The science of geology is traditionally divided into two broad areas—physical and historical. *Physical geology*, which is the primary focus of this book, examines the materials composing the earth and seeks to understand the many processes that operate beneath and upon its surface. The aim of *historical geology*, on the other hand, is to understand the origin of the earth and its development through time. Thus, it strives to establish an orderly chronological arrangement of the multitude of physical and biological changes that have occurred in the geologic past. The study of physical geology logically precedes the study of earth history, because we must first understand how the earth works before we attempt to unravel its past.

SOME HISTORICAL NOTES ABOUT GEOLOGY

The nature of our earth—its materials and processes—has been a focus of study for centuries. Writings about such topics as fossils, gems, earthquakes, and volcanoes date back to the Greeks, more than 2300 years ago. Certainly the most influential of the Greek philosophers was Aristotle. Because Aristotle was a philosopher, his explanations were not always based on observations and experiments but often were arbitrary pronouncements. He believed that rocks were created under the "influence" of the stars and that earthquakes occurred when air crowded into the ground was heated by central fires and escaped explosively. When confronted with a fossil fish, he explained that, "a great many fishes live in the earth motionless and are found when excavations are made." Although Aristotle's explanations may have been adequate for his day, they unfortunately continued to be expounded for many centuries, thus thwarting the acceptance of ideas that were more closely in accord with observations. Frank D. Adams states in *The Birth and Development of the Geological Sciences* (New York: Dover, 1938) that, "throughout the Middle Ages Aristotle was regarded as the head and chief of all philosophers; one whose opinion on any subject was authoritative and final."

Catastrophism

During the seventeenth and eighteenth centuries the doctrine of **catastrophism** strongly influenced the formulation of explanations about the dynamics of the earth. Briefly stated, catastrophists believed that the earth's landscape had been developed primarily by great catastrophes. Features such as mountains and canyons, which today we know take great periods of time to form, were explained as having been produced by sudden and often worldwide disasters produced by unknowable causes that no longer operate. This philosophy was an attempt to fit the rate of earth processes to the then-current ideas on the age of the earth. In 1654, Archbishop

James Ussher, a scholar of the Bible, concluded that the earth was approximately 6000 years old, having been created in 4004 B.C. Later, another biblical scholar named Lightfoot was even more specific, declaring that the earth had been created at 9:00 A.M. on October 26, 4004 B.C.

The relationship between catastrophism and the age of the earth has been summarized nicely as follows:

> That the earth had been through tremendous adventures and had seen mighty changes during its obscure past was plainly evident to every inquiring eye; but to concentrate these changes into a few brief millenniums required a tailor-made philosophy, a philosophy whose basis was sudden and violent change.*

The Birth of Modern Geology

The late eighteenth century is generally regarded as the beginning of modern geology, for it was during this time that James Hutton, a Scottish physician and gentleman farmer, published his *Theory of the Earth* in which he put forth a principle that came to be known as the doctrine of **uniformitarianism** (Figure I.1). Uniformitarianism is a fundamental concept in modern geology. It simply states that the physical, chemical, and biological laws that operate today have also operated in the geologic past. That is to say that the forces and processes that we observe presently shaping our planet have been at work for a very long time. Thus, to understand ancient rocks, we must first understand present-day processes and their results. This idea is commonly stated as "the present is the key to the past."

Prior to Hutton's *Theory of the Earth,* no one had effectively demonstrated that geology had to deal with extremely long periods of time. However, Hutton persuasively argued that processes which appear weak and slow-acting could, over long spans of time, produce effects that were just as great as those resulting from sudden catastrophic events. Unlike his predecessors, Hutton cited verifiable observations to support his ideas.

Since Hutton's literary style was cumbersome and difficult, his work was not widely read nor easily

*H. E. Brown, V. E. Monnett, and J. W. Stovall. *Introduction to Geology* (New York: Blaisdell, 1958).

FIGURE I.1 James Hutton, the 18th century Scottish geologist who is often called the "father of modern geology." (Photo courtesy of the British Museum)

understood. It is the English geologist Charles Lyell who is given the most credit for advancing the basic principles of modern geology. Between 1830 and 1872 Lyell produced eleven editions of his great work, *Principles of Geology.* As was customary, Lyell's book had a rather lengthy subtitle that outlined the main theme of the work: *Being an Attempt to Explain the Former Changes of the Earth's Surface, by Reference to Causes now in Operation.* In the text, he painstakingly illustrated the concept of the uniformity of nature through time. He was able to show more convincingly than his predecessors that those geologic processes observed today can be assumed to have operated in the past. Although the doctrine of uniformitarianism did not originate with Lyell, he is the person who was most successful in interpreting and publicizing it for society at large.

Despite its importance in modern geology, the doctrine of uniformitarianism should not be taken too literally. To say that geologic processes in the past were the same as those occurring today is not to suggest that they always operated at precisely the same rate. Although the processes have remained

A.

B.

FIGURE I.2 Geologic processes often act so slowly that changes may not be visible during an entire lifetime. These two photographs were taken from the same vantage point in the Grand Canyon, nearly one hundred years apart. Photograph A was taken by J. K. Hillers in 1872 and photograph B was taken in 1968 by E. M. Shoemaker. The photos reveal practically no visible signs of erosion. (Photos courtesy of U.S. Geological Survey)

essentially the same, their rates have undoubtedly varied during geologic time.

The acceptance of the concept of uniformitarianism, however, meant the acceptance of a very long history for the earth, for although processes vary in their intensity, they still take a very long time to create or destroy major landscape features.

For example, rocks containing fossils of organisms that lived in the sea more than 15 million years ago are now part of mountains that stand 3000 meters (9800 feet) above sea level. This means that the mountains were uplifted 3000 meters in about 15 million years—a rate of only 0.2 millimeter per year! Rates of erosion are equally slow (Figure I.2). Estimates indicate that the North American continent is being lowered at a rate of just 3 centimeters per 1000 years. Thus, as you can see, it takes tens of millions of years for nature to build mountains and

wear them down again. But even these time spans are relatively short on the time scale of earth history, for the rock record contains evidence that shows the earth has experienced many cycles of mountain building and erosion. Concerning the everchanging nature of the earth through great expanses of geologic time, Hutton stated: "We find no vestige of a beginning, no prospect of an end." A quote from William L. Stokes sums up the significance of Hutton's basic concept:

> In the sense that uniformitarianism implies the operation of timeless, changeless laws or principles, we can say that nothing in our incomplete but extensive knowledge disagrees with it.*

Essentials of Earth History (Englewood Cliffs, New Jersey: Prentice-Hall, 1966), p. 34.

FIGURE I.3 View of the earth that greeted the *Apollo 8* astronauts as their spacecraft came from behind the moon. (Courtesy of NASA)

It is important to remember that although many features of our physical landscape may seem to be unchanging in terms of the tens of years we might observe them, they are nevertheless changing, but on time scales of hundreds, thousands, or even many millions of years.

A VIEW OF THE EARTH

Figure I.3 shows the earth as seen by *Apollo 8* astronauts as their spacecraft came from behind the moon after circling it for the first time in December, 1968. A view such as this from a distance of 160,000 kilometers (100,000 miles) provided the astronauts, as well as those of us back on earth, a unique perspective of our planet. For the first time we were able to see the earth from the depths of space as a small, fragile-appearing sphere surrounded by the blackness of an infinite universe. Such views were not only spectacular and exciting, but also humbling, for they showed us as never before what a tiny part of the universe our planet occupies.

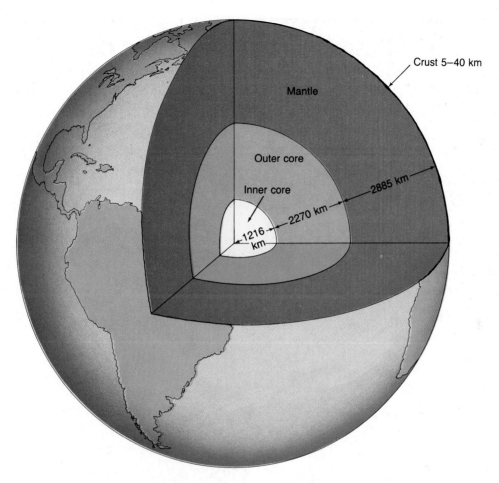

FIGURE I.4 View of the earth's layered structure. The inner core, outer core, and mantle are drawn to scale, but the thickness of the crust is exaggerated by about five times.

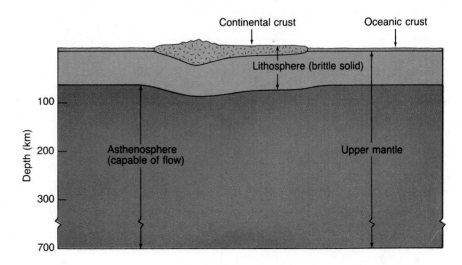

FIGURE I.5 Schematic of the respective positions of the asthenosphere and lithosphere.

As we look more closely at our planet from space, the most conspicuous features are not the continents, but the swirling clouds suspended above the surface and the vast global ocean. From such a vantage point, we can appreciate why the earth's physical environment is traditionally divided into three major parts: the envelope of air called the **atmosphere**; the **hydrosphere**, or water portion of our planet; and, of course, the solid earth. Ours is a dynamic planet that is not dominated by rock, water, or air alone. Rather, it is characterized by continuous interactions as air comes into contact with rock, rock with water, and water with air.

The atmosphere, the earth's life-giving blanket of air that is hundreds of kilometers thick, is an integral part of the planet. It not only provides the air that we breathe, but also acts to protect us from the sun's intense heat and dangerous radiation. The energy exchanges that continually occur between the atmosphere and the earth's surface, and between the atmosphere and space produce the effects we call weather and climate.

The hydrosphere is a dynamic mass of liquid that is continually on the move, from the oceans to the air, to the land, and back again. The global ocean is obviously the most prominent feature of the hydrosphere, blanketing 71 percent of the earth's surface and accounting for about 97 percent of the earth's water. However, the hydrosphere also includes the fresh water found in streams, lakes, and glaciers, as well as that found in the ground. Although these latter sources constitute just a tiny fraction of the total, they are much more important than their meager percentage indicates, because they are responsible for sculpturing and creating many of our planet's varied landforms.

Lying beneath the atmosphere and the ocean is the solid earth. Rather than being a homogeneous body, the earth's interior consists of shells or spheres composed of materials having different properties. The principle divisions of the earth include: (1) The **inner core**, a solid iron-rich zone having a radius of 1216 kilometers (756 miles); (2) The **outer core**, a molten metallic layer some 2270 kilometers (1410 miles) thick; (3) The **mantle**, a solid rocky layer having a maximum thickness of 2885 kilometers (1789 miles); and (4) The **crust**, a relatively light outer skin that ranges from 5 to 40 kilometers (3 to 25 miles) thick (Figure I.4, page 5).

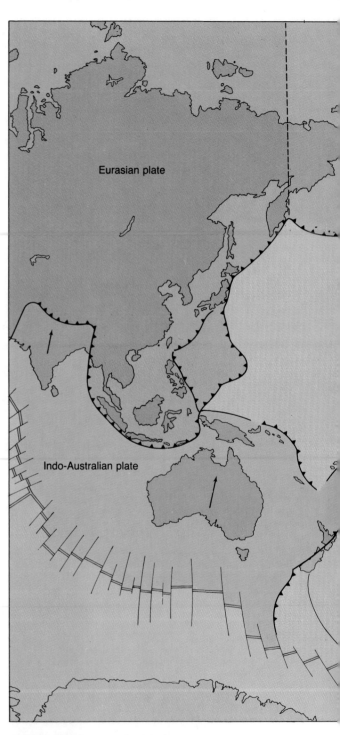

FIGURE I.6 Mosaic of rigid plates that constitute the earth's outer shell. **A.** Divergent boundary. **B.** Convergent boundary. **C.** Transform fault boundary. (After W. B. Hamilton, U.S. Geological Survey)

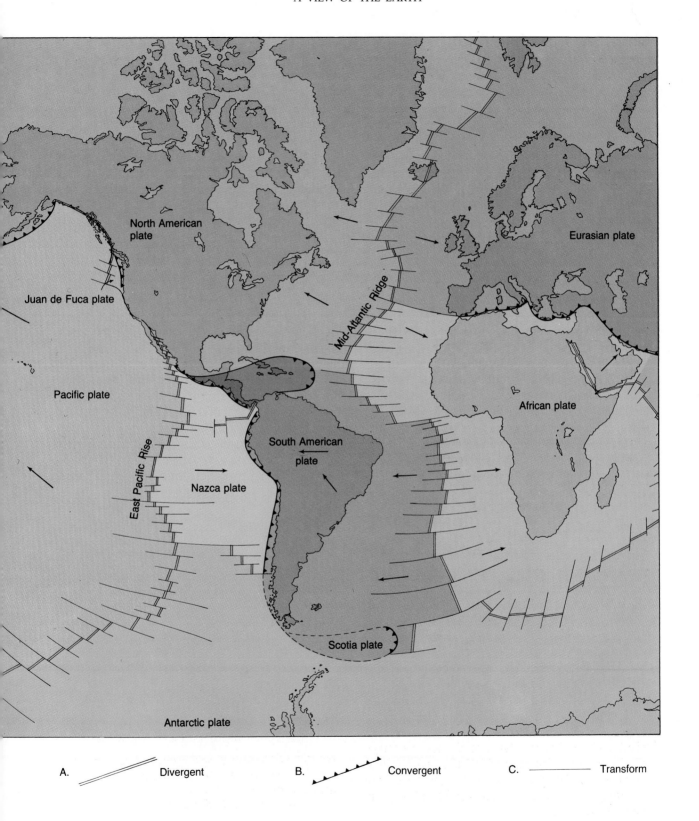

A. ——— Divergent   B. ▲▲▲▲ Convergent   C. ——— Transform

A very important zone exists within the mantle and deserves special mention. This region, called the **asthenosphere**, is located between the depths of approximately 100 and 700 kilometers. The asthenosphere is a hot, weak zone that is capable of gradual flow. Situated above the asthenosphere, geologists recognize a zone called the **lithosphere** ("sphere of rock"), which includes the crust and uppermost mantle (Figure I.5, page 5). In contrast to the asthenosphere upon which it rests, the lithosphere can be considered to be cool and rigid.

THE DYNAMIC EARTH

The earth is a dynamic planet. If we could go back in time a billion years or more, we would find a planet whose surface was dramatically different than it is today. Such prominent features as the Grand Canyon, the Rocky Mountains, and even the much older Appalachian Mountains did not exist. Moreover, we would find that the continents would have different shapes and be located in different positions than today. On the other hand, a billion years ago the moon's surface was almost the same as we now find it. In fact, if viewed telescopically from the earth, perhaps only a few craters would be missing. Thus, when compared to the earth, the moon is a lifeless body wandering through space and time.

The processes that alter the earth's surface can be divided into two categories. Those forces which wear away the land include weathering and erosion. Unlike the moon, where weathering and erosion progress at infinitesimally slow rates, these processes are continually altering the landscape of the earth. In fact, these destructive forces would have long ago leveled the continents had it not been for opposing constructional processes. Included among the constructional processes are volcanism and mountain building, which increase the average elevation of the land in opposition to gravity. As we shall see, these forces depend upon the earth's internal heat for their source of energy.

Within the last few decades, a great deal has been learned about the workings of our dynamic planet. In fact, many have called this period a revolution in our knowledge about the earth which has been unequalled at any other time. This revolution began in the early part of the twentieth century with the radical proposal that the continents had drifted about the face of the earth. This idea contradicted the established view that the continents and ocean basins are permanent and stationary features on the face of the earth. For that reason, it was received with great skepticism. More than 50 years passed before enough data was gathered to transform this relatively simple hypothesis into a working theory

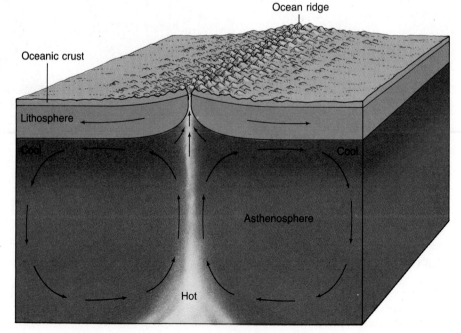

FIGURE I.7 Unequal temperature distribution in the earth's interior is thought to produce convection currents that move the earth's rigid outer shell.

which weaved together the basic processes known to operate on the earth. The theory that finally emerged, called **plate tectonics**,* provided geologists with a comprehensive model of the earth's internal workings.

According to the plate tectonics model, the earth's rigid outer shell, the lithosphere, is broken into several individual pieces called **plates** (Figure I.6, pages 6 and 7). It is further thought that these rigid plates are slowly, but nevertheless continually, in motion. This motion is believed to be driven by a thermal engine, the result of an unequal distribution of heat within the earth. As hot material wells up from deep within the earth and spreads laterally, the plates are set in motion (Figure I.7). Ultimately, this movement of the earth's lithospheric plates generates earthquakes, volcanic activity, and the deformation of large masses of rock into mountains.

Because each plate moves as a distinct unit, all interaction among individual plates occurs along their boundaries. The first approximations of plate boundaries were made on the basis of earthquake

---

*Tectonics is the study of large scale deformation of the earth's lithosphere that results in the formation of major structural features such as those associated with mountains.

and volcanic activity. Later work indicated the existence of three distinct types of plate boundaries, which are differentiated by the movement they exhibit (Figure I.8). These are:

1 **Divergent boundaries**—zones where plates move apart, leaving a gap between them.
2 **Convergent boundaries**—zones where plates move together, causing one to go beneath the other, as happens when oceanic crust is involved; or where plates collide, which occurs when the leading edges are made of continental crust.
3 **Transform fault boundaries**—zones where plates slide past each other, scraping and deforming as they pass.

Each plate is bounded by a combination of these zones (Figure I.6). Movement along one boundary requires that adjustments be made at the others.

Plate spreading (divergence) is believed to occur at the oceanic ridges. As the plates separate, the gap created is immediately filled with molten rock that wells up from the hot asthenosphere (Figure I.9). This material slowly cools to produce a new sliver of

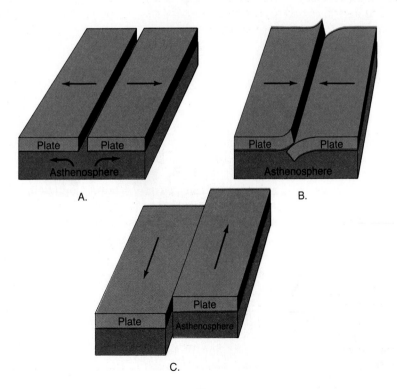

FIGURE I.8 Schematic of plate boundaries.
A. Divergent boundary.
B. Convergent boundary.
C. Transform fault boundary.

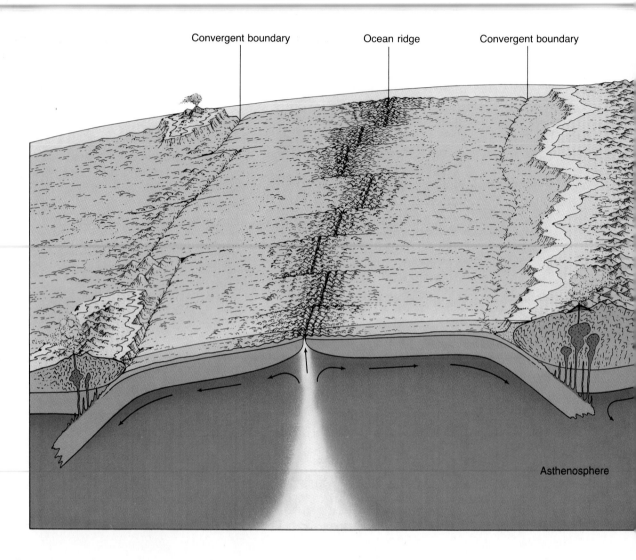

Convergent boundary   Ocean ridge   Convergent boundary

Asthenosphere

sea floor. Successive separations and fillings continue to add new oceanic lithosphere between the diverging plates. This mechanism, which has produced the floor of the Atlantic Ocean during the past 200 million years, is appropriately called **sea-floor spreading**. The typical rate of sea-floor spreading is estimated to be 5 centimeters (2 inches) per year, although it varies considerably from one location to another. This seemingly slow rate of movement is nevertheless rapid enough so that all of the existing ocean basins could have been generated within the last 5 percent of geologic time.

Although the earth's rigid outer layer is constantly being generated at the oceanic ridges, the total surface area of the earth remains constant. Therefore, lithosphere must be destroyed at the same rate that it is created. The zone of plate convergence is the site of this destruction. As two plates move together, the leading edge of one of the slabs is bent downward, allowing it to slide beneath the other. Whenever continental and oceanic lithosphere collide, it is always the denser oceanic material that plunges into the weak asthenosphere below (Figure I.9).

The regions where oceanic lithosphere is being consumed are called **subduction zones**. Here, as the solid plates move downward, they enter high-pressure, high-temperature environments. Some subducted material is thought to melt and migrate upward into the overriding plate. Occasionally this molten rock may reach the surface, where it gives rise to volcanic eruptions such as those of Mount St. Helens.

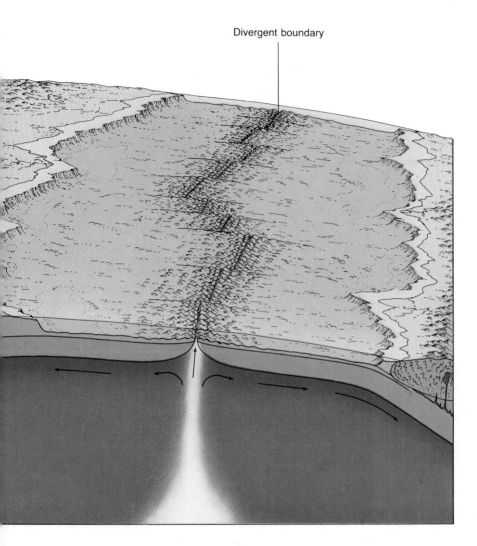

Divergent boundary

FIGURE I.9 View of the earth showing the relationship between divergent and convergent plate boundaries.

Other boundaries, represented by transform faults, are located where plates slip past each other without producing or destroying crust. These faults form in the direction of plate movement and were first discovered in association with offsets in the oceanic ridges (Figure I.9). Although most transform faults are located within the ocean basins, a few slice through the continents. The San Andreas fault of California is a famous example. Along this fault the Pacific plate is moving toward the northwest, past the North American plate. The movement along this boundary does not go unnoticed. As these plates pass, strain builds in the rocks on opposite sides of the fault and is occasionally released in the form of a great earthquake of the type that devastated San Francisco in 1906.

A recent discovery is that the interaction of plates along their boundaries initiates most of our planet's volcanism, earthquakes, and mountain building. Further, these boundaries do not remain constant through time. For example, a divergent boundary which runs through eastern Africa appears to have developed in the relatively recent past. If spreading continues there, Africa will split into two continents separated by a new ocean basin. At other locations continents are presently moving toward each other and may eventually join into a "supercontinent." When continents collide, the thick accumulations of rocks and sediments along their margins are gradually thrust into majestic mountain ranges.

As long as the temperatures deep within the earth remain significantly higher than those near the surface, the material within the earth will continue to move. This internal flow, in turn, will keep the rigid outer shell of the earth in motion. Thus, as long as the earth's internal heat engine operates, the positions and shapes of the continents and ocean basins will change, and the earth will remain a dynamic planet.

In the remaining chapters we will examine in more detail the workings of our dynamic planet in light of the plate tectonics model.

THE ROCK CYCLE

The **rock cycle** is one means of viewing many of the interrelationships of geology. By studying the rock cycle we may ascertain the origin of the three basic rock types and gain some insight into the role of various geologic processes in transforming one rock type into another. The concept of the rock cycle, which may be considered as a basic outline of physical geology, was initially proposed by James Hutton. The rock cycle shown in Figure I.10 indicates processes by arrows and materials in boxes.

The first rock type, **igneous rock**, originates when molten material called **magma** cools and so-

lidifies. This process, called **crystallization**, may occur either beneath the earth's surface or, following a volcanic eruption, at the surface. Initially, or shortly after forming, the earth's outer shell is believed to have been molten. As this molten material gradually cooled and crystallized, it generated a primitive crust that consisted entirely of igneous rocks.

If igneous rocks are exposed at the surface of the earth, they will undergo **weathering**, in which the day-in-and-day-out influences of the atmosphere slowly disintegrate and decompose rocks. The materials that result will be picked up, transported, and deposited by any of a number of erosional agents— gravity, running water, glaciers, wind, or waves. Once these particles and dissolved substances, called **sediment**, are deposited, usually as horizontal beds in the ocean, they will undergo **lithification**, a term meaning "conversion into rock." Sediment is lithified when compacted by the weight of overlying layers or when cemented as percolating water fills the pores with mineral matter. If the resulting **sedimentary rock** is buried deep within the earth or involved in the dynamics of mountain building, it will be subjected to great pressures and heat. The sedimentary

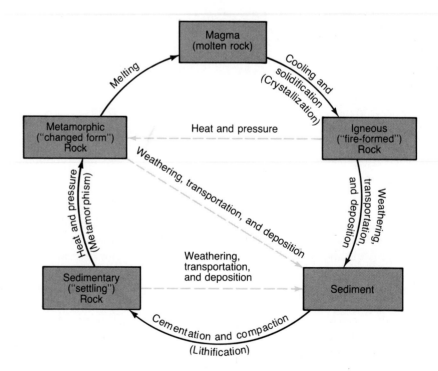

FIGURE I.10 The rock cycle. Originally proposed by James Hutton, the rock cycle illustrates the role of the various geologic processes which act to transform one rock type into another.

rock will react to the changing environment and turn into the third rock type, **metamorphic rock**. When metamorphic rock is subjected to still greater heat and pressure, it will melt, creating magma, which will eventually solidify as igneous rock.

The full cycle just described does not always take place. "Shortcuts" in the cycle are indicated by dashed lines in Figure I.10. Igneous rock, for example, rather than being exposed to weathering and erosion at the earth's surface, may be subjected to the heat and pressure found far below and change to metamorphic rock. On the other hand, metamorphic and sedimentary rocks, as well as sediment, may be exposed at the surface and turned into new raw materials for sedimentary rock.

When the rock cycle was first proposed by James Hutton, very little was actually known about the processes by which one rock was transformed into another; only evidence for the transformation existed. In fact, it was not until very recently with the development of the theory of plate tectonics that a complete picture of the rock cycle became clear.

According to this model, weathered material from elevated landmasses is transported to the continental margins where it is deposited in layers that collectively are thousands of meters thick. Once lithified, these sediments create a thick wedge of sedimentary rocks flanking the continents.

Eventually the relatively quiescent activity of sedimentation along a continental margin may be interrupted if the region becomes a convergent plate boundary (see Figure I.9). When this occurs, the oceanic lithosphere adjacent to the continent begins to inch downward into the asthenosphere beneath the continent. Along active continental margins such as this, the converging plate deforms the margin's sedimentary rocks and transforms them into linear belts of metamorphic rocks. Further, as the oceanic plate descends, some of the overlying sediments that were not crumpled into mountains are carried downward into the hot asthenosphere where they too undergo metamorphism. Eventually some of this metamorphic material will be transported to depths where the temperatures and pressures are sufficiently great to initiate melting. This newly formed magma will then migrate upward and occasionally erupt at the surface. Crystallization of this magma generates igneous rocks that are immediately attacked by the processes of weathering. Thus, the rock cycle is ready to begin anew.

CHAPTER ONE

MATTER AND MINERALS

The earth's outer layer, which we call the crust, is only as thick when compared to the remainder of the earth as a peach skin is to a peach, yet it is of supreme importance to us. We depend on it for fossil fuels and as a source of such diverse minerals as talc for baby powder, salt to flavor food, and gold for world trade. In fact, on occasion, the availability or absence of certain earth materials has altered the course of history.

In addition to the economic uses of rocks and minerals, all of the processes studied by geologists are in some way dependent upon the properties of these basic earth materials. Events such as volcanic eruptions, mountain building, weathering and erosion, and even earthquakes involve rocks and minerals. Consequently, a basic knowledge of earth materials is essential to the understanding of all geologic phenomena.

ROCKS VERSUS MINERALS

Many people consider rocks to be rather nondescript objects that are hard and often dirty. Minerals are considered by many to be dietary supplements, or possibly rare ores or precious gems that are mined for their economic value. However, these common perceptions are far from the actual situation.

A **rock** can be defined simply as an aggregate of one or more minerals. Here, the term *aggregate* implies that the minerals are found together as a *mixture* in which the properties of the individual minerals are retained. Although most rocks are composed of more than one mineral, certain minerals are commonly found by themselves in large quantities. In these instances they are considered to be both a mineral and a rock. A common example is the mineral calcite, which frequently is the dominant

constituent in large rock units, where it is given the name *limestone*.

By contrast, **minerals** are defined as naturally occurring inorganic solids, which possess a definite internal structure and a specific chemical composition. Although this definition is quite precise, it is not without shortcomings. For example, this definition excludes organic compounds; however, many geologists would classify coal, and occasionally even petroleum, as minerals. Further, the chemical composition of many minerals actually varies over a wide range.

This chapter deals primarily with the nature of minerals. However, keep in mind that rocks are simply aggregates of minerals. Thus, the properties of rocks are determined solely by the chemical composition and internal structure of those minerals which compose them.

THE COMPOSITION OF MATTER

Minerals, like all matter, are made of **elements**. At present, over 100 elements are known, a dozen and a half of which have been produced only in the laboratory. Some minerals such as gold and sulfur are made entirely of one element, but most are a combination of two or more elements joined to form a chemically stable **compound**. In order to better understand how elements combine to form compounds, we must first consider the **atom**, the smallest part of matter that still retains the characteristics of an element, because it is this extremely small particle that does the combining.

Atomic Structure

Individual atoms are far too small to be observed directly; therefore, our concept of atomic structure has come from experimental evidence and mathe-

---

Pyrite crystals. (Courtesy of JLM Visuals)

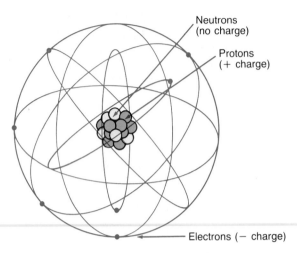

Neutrons
(no charge)

Protons
(+ charge)

Electrons (− charge)

FIGURE 1.1 Simplified model of an atom. Atoms consist of a central nucleus composed of protons and neutrons that is encircled by electrons.

matical models. A simplified model of the structure of an atom is shown in Figure 1.1. Each atom has a central region, called the **nucleus**, which contains very dense positively charged **protons** and equally dense neutral particles called **neutrons**. Orbiting the nucleus are negatively charged particles known as **electrons**. Unlike the orderly orbiting of the planets around the sun, electrons move so rapidly that their positions cannot be pinpointed. Hence, a more realistic picture of the positions of electrons can be obtained by envisioning a cloud of electrons surrounding the nucleus. It is also known that individual electrons are located at given distances from the nucleus in regions called **energy-level shells**. As we shall see, an important fact about these shells is that each can hold only a specific number of electrons.

The number of protons in the nucleus determines the **atomic number** and name of the element. For example, all atoms with six protons are carbon atoms, all those with eight protons are oxygen atoms, and so forth. Since all atoms have the same number of electrons as protons, the atomic number also equals the number of electrons surrounding the nucleus. Moreover, since neutrons have no charge, the positive charge of the protons is exactly balanced by the negative charge of the electrons. Consequently, every atom is an electrically neutral particle. Elements can be considered to be a large collection of electrically neutral atoms, all having the same atomic number.

The simplest element, hydrogen, is composed of atoms that have only one proton in the nucleus and one electron surrounding the nucleus. Each successively heavier atom has one more proton and one more electron, in addition to a certain number of neutrons (Table 1.1). Studies of electron configurations have shown that each electron is added in a systematic fashion to a particular energy level, or shell. In general, electrons enter higher energy levels only after lower energy levels have been filled to capacity. The first principal shell holds a maximum of two electrons, while each of the higher shells holds eight or more electrons. However, any shell that is an outermost shell (other than the first shell which is filled with two electrons) will contain a maximum of only eight electrons. As we shall see, it is the electrons found in outermost shells which are generally involved in chemical bonding.

Bonding

Chemical bonds occur when atoms of two or more elements join to form a compound. When the atoms separate, the bonds are broken and the compound is destroyed. Through experimentation it has been learned that the forces bonding the atoms together are electrical in nature. Further, it is known that chemical bonding results in a change in the electronic structures of the bonded atoms. Hence, the electron configurations of the atoms involved are important in determining the strength and nature of the chemical bonds that are produced.

As we noted earlier, the outermost electrons are generally involved in chemical bonding. Further, the atoms of most elements have less than the maximum number of electrons in their outermost shell. Only the noble gases such as neon and argon have a complete outer shell, which accounts for their chemical stability and the fact they do not readily react with other elements. However, every atom seeks a full outer shell to become chemically stable like the noble gases. The octet rule, literally meaning "a set of eight," refers to the concept of a completely filled outermost energy level. Simply, the **octet rule** states that atoms combine in order that each may have the electron arrangement of a noble gas, with the outer energy level containing eight electrons.

In order to satisfy the octet rule, an atom can either gain, lose, or share electrons with one or more

TABLE 1.1 Atomic number and distribution of electrons in the main shells.

Element	Symbol	Atomic Number	Number of Electrons in Each Shell			
			1	2	3	4
Hydrogen	H	1	1			
Helium	He	2	2			
Lithium	Li	3	2	1		
Beryllium	Be	4	2	2		
Boron	B	5	2	3		
Carbon	C	6	2	4		
Nitrogen	N	7	2	5		
Oxygen	O	8	2	6		
Fluorine	F	9	2	7		
Neon	Ne	10	2	8		
Sodium	Na	11	2	8	1	
Magnesium	Mg	12	2	8	2	
Aluminum	Al	13	2	8	3	
Silicon	Si	14	2	8	4	
Phosphorus	P	15	2	8	5	
Sulfur	S	16	2	8	6	
Chlorine	Cl	17	2	8	7	
Argon	Ar	18	2	8	8	
Potassium	K	19	2	8	8	1
Calcium	Ca	20	2	8	8	2

atoms. The result of this process is the formation of an electrical ''glue'' that bonds the atoms. The electrons involved in the bonding process are commonly called **valence electrons**. The number of valence electrons that an element has determines the number of bonds it will form. For example, the element silicon has four valence electrons and forms four bonds in the process of completing its outer shell. On the other hand, oxygen forms two bonds, and hydrogen forms only one.

Ionic Bonds. Perhaps the easiest type of bond to visualize is an **ionic bond**. In ionic bonding, one or more valence electrons are transferred from one

atom to another. One atom becomes stable by giving up its valence electrons and the other uses them to complete its outer shell. An example of ionic bonding using sodium (Na) and chlorine (Cl) to produce sodium chloride (common table salt) is shown in Figure 1.2. Notice that sodium loses its single outer electron to chlorine. As a result, sodium acquires the electron configuration of the noble gas neon, which has two electrons in the first shell and eight in its outermost shell. By adding an electron, a chlorine atom fills its outermost shell to acquire the arrangement of argon. However, these atoms are no longer electrically neutral because neither contains an equal number of protons and electrons. Atoms

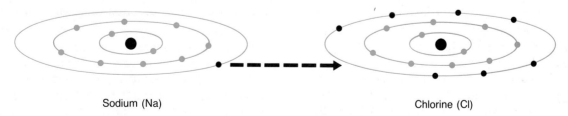

Sodium (Na)   Chlorine (Cl)

FIGURE 1.2 Chemical bonding of sodium and chlorine to produce sodium chloride. Through the transfer of one electron from sodium to chlorine, sodium becomes a positive ion and chlorine a negative ion.

such as these, which have an unequal charge because of a gain or loss of electrons, are called **ions**. Sodium becomes a positively charged ion and chlorine becomes a negatively charged ion. An ionic bond results from the attraction of these oppositely charged particles. Restated simply, an ionic bond is one in which oppositely charged ions attract one another to produce a neutral chemical compound. Figure 1.3 illustrates the arrangement of sodium and chloride ions in ordinary table salt. Notice that salt consists of alternating sodium and chloride ions, positioned such that each positive ion is attracted to and surrounded on all sides by negative ions, and vice versa. Ionic compounds therefore consist of an orderly arrangement of oppositely charged ions assembled in a definite ratio that provides overall electrical neutrality.

This is an appropriate place in our discussion to point out that the properties of a chemical compound are dramatically different from the properties of the elements composing it. For example, chlorine is a green, poisonous gas that is so toxic it was used as a weapon during World War I. Sodium is a soft, silvery metal that reacts vigorously with water and, if held in your hand, could burn it severely. Together, however, these atoms produce the compound so-dium chloride (table salt), which is a clear crystalline solid essential for human life. This example also illustrates an important difference between a rock and a mineral. A *mineral* is a *chemical compound* with unique properties that are very different from the elements which make it up. A *rock,* on the other hand, is a *mixture* of minerals, with each mineral retaining its own identity.

Covalent Bonds. Not all atoms combine by forming ions. For example, the gaseous elements oxygen (O_2), hydrogen (H_2), and chlorine (Cl_2) exist as stable molecules consisting of two atoms bonded together without the complete transfer of electrons. This is necessary because even if one of the atoms in each pair did accept one or more electrons to form a stable octet, the other atom would move farther away from such a stable condition. Instead, a stable octet is obtained when some of the outer electrons of both atoms are shared. Figure 1.4 illustrates the sharing of a pair of electrons between two chlorine atoms to form a molecule of chlorine gas. By overlapping the outer shells, one electron in each chlorine atom, which has seven electrons in its outer shell, has acquired through cooperative action, the needed electron to complete the octet. The bond

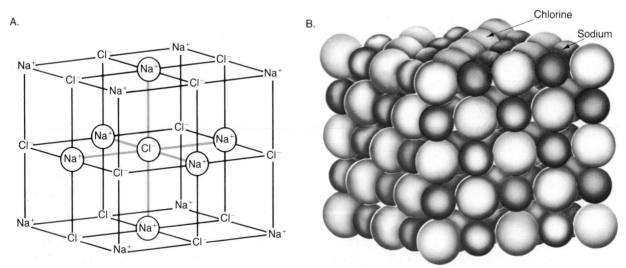

FIGURE 1.3 Schematic illustrating the arrangement of sodium and chloride ions in table salt. **A.** Structure has been opened up to show arrangement of ions. **B.** Actual ions are closely packed.

FIGURE 1.4 Schematic drawing showing the sharing of a pair of electrons between two chlorine nuclei to form a chlorine molecule.

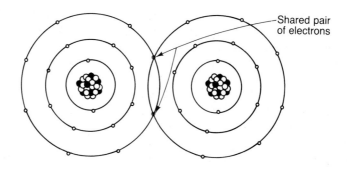

Shared pair of electrons

produced by the sharing of electrons to acquire the stable noble gas arrangement is called a **covalent bond**. The most common mineral group, the silicates, contains the element silicon that readily forms covalent bonds with oxygen.

A common analogy may help you visualize a covalent bond. Imagine two people at opposite ends of a dimly lit room, both reading under a lamp. By moving the lamps to the center of the room, they are able to combine their light sources so each can see better. Just as it is impossible to determine the source of the overlapping light, shared electrons are indistinguishable from each other.

It should be pointed out that most chemical bonds are actually a blend, consisting to some degree of electron sharing, as in covalent bonding, and to some degree of electron transfer, as in ionic bonding. In addition, there also exists an extreme type of electron sharing, in which the electrons move freely from atom to atom. This type of bonding is found in metals such as copper, gold, aluminum, and silver. The term **metallic bonding** is applied to this type of electron sharing. Metallic bonding accounts for the high electrical conductivity of metals, the ease with which metals are reshaped, and numerous other special properties of metals.

Atomic Mass

Subatomic particles, such as protons, are so incredibly small that a special unit was devised to express their mass. A proton or a neutron has a mass just slightly more than one **atomic mass unit**, whereas an electron is only about one two-thousandth of an atomic mass unit. Thus, although electrons play an active role in chemical reactions, they do not contribute significantly to the mass of an atom. Because of this, the **mass number** of an atom is obtained simply by totaling the number of neutrons and the number of protons in the nucleus. Atoms of the same element commonly have varying numbers of neutrons and, therefore, different mass numbers. Such atoms are called **isotopes** of that element. For example, carbon has two well-known isotopes, one having a mass number of 12 (carbon-12), the other a mass number of 14 (carbon-14). Recall that all atoms of the same element must have the same number of protons (atomic number) and that carbon always has six. Hence, carbon-12 must have six neutrons to give it a mass number of 12, whereas carbon-14 must have eight neutrons to give it a mass number of 14. The term commonly used to express the average of the atomic masses of isotopes for a given element is **atomic weight**. The atomic weight of carbon is much closer to 12 than 14, because carbon-12 is the more common isotope. Note that in a chemical sense all isotopes of the same element are nearly identical. To distinguish among them would be like trying to differentiate individual members from a group of similar objects, all having the same shape, size, and color, with only some being slightly heavier.

Although the vast majority of atoms are stable, many elements do have isotopes that are unstable. Unstable isotopes such as carbon-14 go through a process of natural disintegration called **radioactivity**, which occurs when the forces that bind the nucleus are not strong enough. The rate at which unstable nuclei break apart (decay) is measurable and makes such elements useful ''clocks'' in dating the events of earth history. A discussion of radioactivity and its application in dating events of the geologic past can be found in Chapter Fifteen.

PHYSICAL PROPERTIES OF MINERALS

Minerals are solids formed by inorganic processes. Each mineral has an orderly arrangement of atoms (crystalline structure) and a definite chemical composition which give it a unique set of physical properties. Since the internal structure and chemical composition of a mineral are difficult to determine without the aid of sophisticated tests and apparatus, the more easily recognized physical properties are frequently used in identification. A discussion of some diagnostic physical properties follows.

Crystal Form

Most people think of a crystal as a rare commodity, when in fact most inorganic solid objects are composed of crystals. The reason for this misconception is that most crystals do not exhibit their crystal form. The **crystal form** is the external expression of a mineral that reflects the orderly internal arrangement of atoms. Any time a mineral is permitted to form without space restrictions, it will develop individual crystals with well-formed crystal faces. Some crystals such as those of the mineral quartz have a very distinctive crystal form that can be helpful in identification (Figure 1.5). However, most of the time crystal growth is interrupted because of competition for space, resulting in an intergrown mass of crystals, none of which exhibits its crystal form.

Luster

Luster is the appearance or quality of light reflected from the surface of a mineral. Minerals that have the appearance of metals, regardless of color, are said to have a *metallic luster*. Minerals with a *nonmetallic luster* are described by various adjectives, including vitreous (glassy), pearly, silky, resinous, and earthy (dull). Some minerals appear partially metallic in luster and are said to be *submetallic*.

Color

Although **color** is the most obvious feature of a mineral, it is often an unreliable diagnostic property. Slight impurities in the common mineral quartz, for example, give it a variety of colors, including pink, purple (amethyst), white, and even black. When a mineral, such as quartz, exhibits a variety of colors, it is said to possess *exotic coloration*. Other minerals, for example, sulfur, which is generally yellow, and malachite, which is bright green, are said to have *inherent coloration* because their color does not vary significantly.

Streak

Streak is the color of a mineral in its powdered form and is obtained by rubbing the mineral across a piece of unglazed porcelain termed a *streak plate*. Although the color of a mineral may vary from sample to sample, the streak usually does not, and is therefore the more reliable property. Streak can also be an aid in distinguishing minerals with metallic lusters from those having nonmetallic lusters. Metallic minerals generally have a dense, dark streak, whereas minerals with nonmetallic lusters do not.

FIGURE 1.5 Quartz crystals exhibit a characteristic external form. (Photo by E. J. Tarbuck)

Hardness

One of the more useful diagnostic properties is **hardness**, the resistance of a mineral to abrasion or scratching. This is a relative property that is determined by rubbing a mineral of unknown hardness against one of known hardness, or vice versa. A numerical value can be obtained by using **Mohs scale** of hardness, which consists of ten minerals arranged in order from 1 (softest) to 10 (hardest) as follows:

Hardness	Mineral
1	Talc
2	Gypsum
3	Calcite
4	Fluorite
5	Apatite
6	Orthoclase
7	Quartz
8	Topaz
9	Corundum
10	Diamond

Any mineral of unknown hardness can be compared to these or to other objects of known hardness. For example, a fingernail has a hardness of 2.5, a copper penny 3, and a piece of glass 5.5. The mineral gypsum, which has a hardness of 2, can be easily scratched with your fingernail. On the other hand, the mineral calcite, which has a hardness of 3, will scratch your fingernail but will not scratch glass. Quartz, the hardest of the common minerals, will scratch a glass plate with ease.

Cleavage

Cleavage is the tendency of a mineral to break along planes of weak bonding. Minerals that possess cleavage are identified by the smooth surfaces which are produced when the mineral is broken. The simplest type of cleavage is exhibited by the micas (Figure 1.6). Because micas have excellent cleavage in one direction, they break to form thin, flat sheets. Some minerals have several cleavage planes which produce smooth surfaces when broken, while others exhibit poor cleavage, and still others have no cleavage at all. When minerals break evenly in more than one direction, cleavage is described by the number of planes exhibited and the angles at which they meet (Figure 1.7).

FIGURE 1.6 Sheet-type cleavage common to the micas. (Courtesy of Ward's Natural Science Establishment, Inc., Rochester, N.Y.)

FIGURE 1.7 Smooth surfaces produced when a mineral with cleavage is broken. These samples exhibit three planes of cleavage (six sides). The mineral on the left has cleavage planes which meet at 90 degree angles, whereas the mineral on the right has cleavage planes which meet at 75 degree angles. (Photo by E. J. Tarbuck)

Cleavage should not be confused with crystal form. When a mineral exhibits cleavage, it will break into pieces that have the same configuration as the original sample. By contrast, the quartz crystals shown in Figure 1.5 do not have cleavage, and if broken, would shatter into shapes that do not resemble each other or the original crystals.

Fracture

Such minerals as the quartz just described do not exhibit cleavage and are therefore said to **fracture**

FIGURE 1.8 Conchoidal fracture. The smooth curved surfaces result when minerals break in a glasslike manner. (Photo by E. J. Tarbuck)

when broken. Those that break into smooth curved surfaces resembling broken glass have a *conchoidal fracture* (Figure 1.8). Others break into splinters or fibers, but most minerals fracture irregularly.

Specific Gravity

Specific gravity is a number representing the ratio of the weight of a mineral to the weight of an equal volume of water. For example, if a mineral weighs three times as much as an equal volume of water, its specific gravity is 3. With a little practice, you can estimate the specific gravity of minerals by hefting them in your hand. For example, if a mineral feels as heavy as the common rocks you have handled, its specific gravity will probably be somewhere between 2.5 and 3. Some metallic minerals have a specific gravity two or three times the average. Galena, which is an ore of lead, has a specific gravity of roughly 7.5, while the specific gravity of 24-carat gold is approximately 20.

MINERAL GROUPS

Over two thousand minerals are presently known to exist and new ones are still being discovered. Fortunately for those who study minerals, no more than two dozen are abundant. Collectively, these few make up most of the rocks of the earth's crust and as such, are classified as the *rock-forming minerals*. It is also interesting to note that only eight elements compose the bulk of these minerals and represent over 98 percent (by weight) of the continental crust (Table 1.2). The two most abundant elements are silicon and oxygen, which combine to form the framework of the most common mineral group, the **silicates**. Every silicate mineral contains oxygen and silicon, and except for quartz, one or more additional elements are needed to acquire electrical neutrality. Perhaps the next most common mineral group is the carbonates, of which calcite is the most prominent member. Other common rock-forming minerals include gypsum and halite (table salt).

In addition to the rock-forming minerals, a number of minerals are prized for their economic value. Included in this group are the ores of metals such as hematite (iron), sphalerite (zinc), and galena (lead); the native elements including gold, silver, and carbon (diamonds); and a host of others such as fluorite, corundum, and uraninite. Note that the rock-forming minerals themselves are not without economic value. For instance, quartz is used in the production of glass, calcite is the main constituent in

TABLE 1.2 Relative abundance of the more common elements in the earth's crust.

Element	Approximate Percentage by Weight
Oxygen (O)	46.6
Silicon (Si)	27.7
Aluminum (Al)	8.1
Iron (Fe)	5.0
Calcium (Ca)	3.6
Sodium (Na)	2.8
Potassium (K)	2.6
Magnesium (Mg)	2.1
All others	1.5
Total	100

Source: Data from Brian Mason.

portland cement, and plaster is composed of the mineral gypsum.

SILICATE STRUCTURES

All silicate minerals have the same fundamental building block, the **silicon-oxygen tetrahedron**. This structure consists of four oxygen atoms surrounding a much smaller silicon atom positioned in the space between them (Figure 1.9). The silicon-oxygen tetrahedron is not, however, a stable compound; rather, it is a complex ion with a charge of -4. This excess negative charge results because each of the four oxygen atoms contributes a charge of -2, whereas the one silicon atom has a charge of $+4$. In nature, one simple means by which these tetrahedra are neutralized is through the addition of positively charged ions. In this way a chemically stable structure consisting of individual tetrahedra linked together by positively charged ions is produced.

In addition to positive ions acting as the "glue" to bind the tetrahedra, the tetrahedra themselves may be linked in a variety of configurations. For example, the tetrahedra may join to form single chains, double chains, or sheet structures as shown in Figure 1.10. The joining of tetrahedra in each of these configurations results from the sharing of oxygen atoms by pairs of silicon atoms. In order to better understand how this sharing takes place, select one

of the silicon atoms (small spheres) near the middle of the single chain structure shown in Figure 1.10. Notice that this silicon atom is completely surrounded by four larger oxygen atoms. Also notice that two of the four oxygen atoms are joined to two silicon atoms, whereas the other two are not shared in this manner. It is the linkage across the shared oxygen atoms that joins the tetrahedra into a chain structure. Now, examine a silicon atom near the middle of the sheet structure and count the number of shared and unshared oxygen atoms surrounding it. The increase in the degree of sharing accounts for the sheet structure. Although they are not shown, other silicate structures exist. The most common silicate structure has all of the oxygen atoms shared to produce a complex three-dimensional framework.

By now we can see that the ratio of oxygen atoms to silicon atoms differs in each of the silicate structures. In the isolated tetrahedron there are 4 oxygen atoms for every silicon atom, in the single chain the oxygen-to-silicon ratio is 3 to 1, and in the three-dimensional framework this ratio is 2 to 1. Consequently, as more of the oxygen atoms are shared, the percentage of silicon in the structure increases. The silicate minerals are therefore described as having a high or low silicon content based on their ratio of oxygen to silicon. This difference in silicon content is quite important as we shall see later when we consider the formation of igneous rocks.

These silicate structures, with the exception of the three-dimensional framework, are not neutral chemical compounds themselves. Thus, like the individual tetrahedra, they all are neutralized by the inclusion of positively charged metallic ions that bond them together into a variety of crystalline configurations. The ions that most often link silicate structures are those of the elements iron (Fe), magnesium (Mg), potassium (K), sodium (Na), aluminum (Al), and calcium (Ca). Notice in Figure 1.11 that each of these positive ions has a particular atomic size and a particular charge. Generally, ions of approximately the same size are able to freely substitute for one another. For instance, the ions of iron (Fe^{2+}) and magnesium (Mg^{2+}) are nearly the same size, and substitute for each other without altering the mineral structure. This also holds true for calcium and sodium, which can occupy the same site in a crystalline structure, and aluminum (Al^{3+}), which substitutes for silicon in the silicon-oxygen tetrahedron.

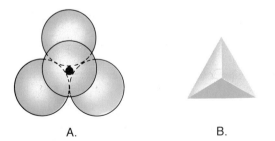

A.　　　　　　　　B.

FIGURE 1.9 Top view of the silicon-oxygen tetrahedron. **A.** The four large spheres represent oxygen atoms, and the one dark sphere represents a silicon atom. **B.** Diagrammatic representation of the tetrahedron using four points to represent the positions of the oxygen atoms.

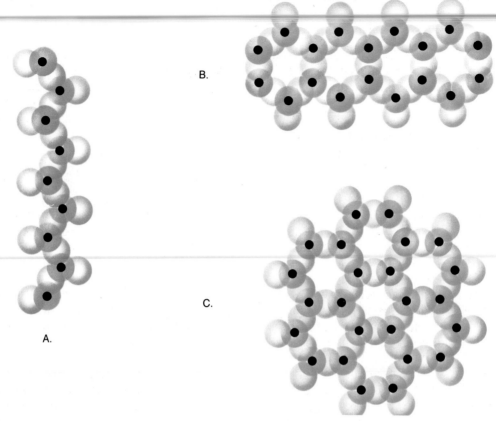

FIGURE 1.10 Three types of silicate structures. **A.** Single chains. **B.** Double chains. **C.** Sheet structures.

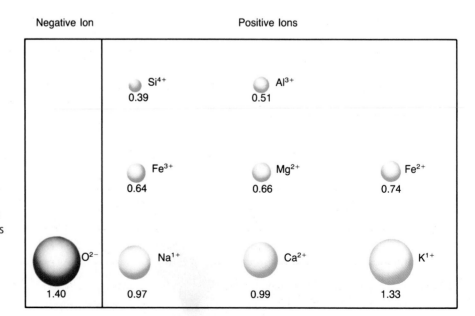

Negative Ion

Positive Ions

Si^{4+}
0.39

Al^{3+}
0.51

Fe^{3+}
0.64

Mg^{2+}
0.66

Fe^{2+}
0.74

O^{2-}
1.40

Na^{1+}
0.97

Ca^{2+}
0.99

K^{1+}
1.33

FIGURE 1.11 Relative sizes and electrical charges of ions commonly found in rock-forming minerals. Ionic radii are expressed in Angstroms (one Angstrom equals 10^{-8} cm).

Due to the ability of mineral structures to readily accommodate different ions at a given bonding site, individual specimens of a particular mineral may contain varying amounts of certain elements. A mineral of this type is often expressed by a chemical formula that uses parentheses to set apart the variable component. A good example is the mineral olivine, $(Fe,Mg)_2SiO_4$. As we can see from the formula, it is the iron (Fe^{2+}) and magnesium (Mg^{2+}) ions in olivine that freely substitute for one another. At one extreme, olivine may contain iron without any magnesium (Fe_2SiO_4) and at the other, iron is totally lacking (Mg_2SiO_4). Between these end members any ratio of iron to magnesium is possible. Thus olivine, as well as many other silicate minerals, is actually a family of minerals that has a range of composition between the two end members.

In certain substitutions the ions that interchange do not have the same electrical charge. For instance, when calcium (Ca^{2+}) substitutes for sodium (Na^{1+}), the structure gains a positive charge. In nature, one way in which this substitution is accomplished, while still maintaining overall electrical neutrality, is that a simultaneous substitution of aluminum (Al^{3+}) for sil-icon (Si^{4+}) takes place. This particular double substitution occurs in the feldspar group, which is the most abundant family of minerals found in the crust. The end members of this particular feldspar series are anorthite, $CaAl_2Si_2O_8$, and albite, $NaAlSi_3O_8$.

SILICATE MINERALS

The main groups of silicate minerals and common examples of each are given in Figure 1.12. The feldspars are by far the most abundant group, comprising over 50 percent of the earth's crust. Quartz, the second most common mineral in the continental crust, is the only one made completely of silicon and oxygen.

Notice in Figure 1.12 that each group has a particular silicate structure. A relationship exists between the internal structure of a mineral and the cleavage it exhibits. Because the silicon-oxygen bonds are strong, silicate minerals tend to cleave between the silicon-oxygen structures rather than across them. For example, the micas have a sheet structure and tend to cleave into flat plates (Figure 1.6). Quartz, which has equally strong silicon-

Mineral		Idealized Formula	Cleavage	Silicate Structure	
Olivine		$(Mg,Fe)_2SiO_4$	None	Single tetrahedron	
Pyroxene group		$(Mg,Fe)SiO_3$	Two planes at right angles	Chains 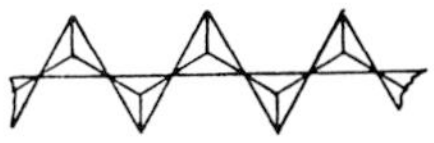	
Amphibole group		$(Ca_2Mg_5)Si_8O_{22}(OH)_2$	Two planes at 60° and 120°	Double chains 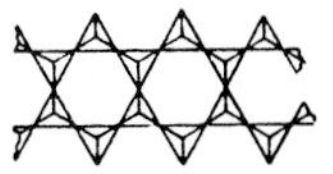	
Micas	Muscovite	$KAl_3Si_3O_{10}(OH)_2$	One plane	Sheets 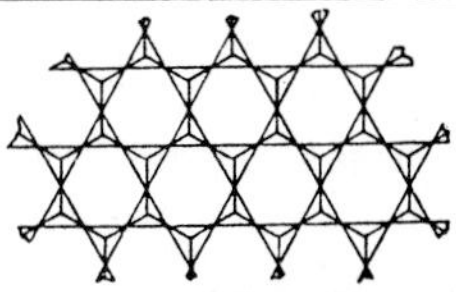	
	Biotite	$K(Mg,Fe)_3Si_3O_{10}(OH)_2$			
Feld-spars	Orthoclase	$KAlSi_3O_8$	Two planes at 90°	Three-dimensional networks 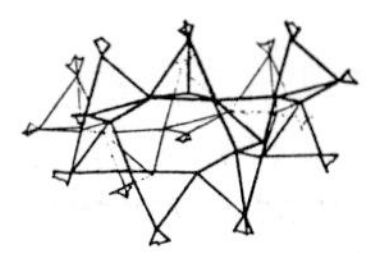	
	Plagioclase	$(Ca,Na)AlSi_3O_8$			
Quartz		SiO_2	None		

FIGURE 1.12 Common silicate minerals. Note that the complexity of the silicate structure increases down the chart.

oxygen bonds in all directions, has no cleavage.

Most silicate minerals form when molten rock cools. This cooling can occur at or near the earth's surface, or at great depths where temperatures and pressures are very high. The environment during crystallization and the chemical composition of the molten rock to a large degree determine the minerals that are produced. For example, the silicate mineral olivine crystallizes at high temperatures and possesses a chemical structure that is stable at high temperatures. Quartz, on the other hand, crystallizes at much lower temperatures. In addition, some silicate minerals are stable at the earth's surface and represent the weathered products of pre-existing silicate minerals. Still other silicate minerals are formed under the extreme pressures associated with metamorphism. Each silicate mineral therefore has a structure and a chemical composition that indicate the conditions under which it formed.

The various silicate minerals can be divided on the basis of chemical makeup. The *ferromagnesian silicates* are those minerals containing ions of iron and/or magnesium in their structure. Those minerals that do not contain these ions are simply called *nonferromagnesians.* Ferromagnesian minerals are typically dark in color and have a specific gravity between 3.2 and 3.6. By comparison, nonferromagnesian silicates are generally light in color and have an average specific gravity of 2.7. These observed differences are mainly attributable to the presence or absence of iron.

Ferromagnesian Silicates

Olivine is a high-temperature silicate mineral that is black to olive green in color, has a glassy luster, and a conchoidal fracture. Rather than developing large crystals, olivine commonly forms small, rounded crystals that give the mineral a granular appearance. Olivine is composed of individual tetrahedra which are bonded together by a mixture of iron and magnesium ions positioned so as to link the oxygen atoms together. Since the three-dimensional network generated in this fashion does not have its weak bonds aligned, olivine does not possess cleavage.

Pyroxene is a black, opaque mineral with two planes of cleavage that meet at nearly a 90 degree angle. Its crystalline structure consists of single chains of tetrahedra bonded together by ions of iron and magnesium. Since the silicon-oxygen bonds are stronger than the bonds joining the silicate structures, pyroxene cleaves parallel to the silicate chains. Pyroxene is one of the dominant minerals in basalt, a common igneous rock of the oceanic crust that is also prevalent in volcanic areas on the continents.

Hornblende is the most common member of a chemically complex group of minerals called *amphiboles.* Hornblende is usually dark green to black in color and except for its cleavage angles, which are about 60 degrees and 120 degrees, it is very similar in appearance to pyroxene (Figure 1.13). The double chains of tetrahedra in the hornblende structure account for its particular cleavage. In a rock, hornblende often forms elongated crystals. This helps distinguish it from pyroxene, which forms rather blocky crystals. Hornblende is found predominantly in continental rocks, where it often makes up the dark portion of an otherwise light-colored rock.

Biotite is the dark iron-rich member of the mica family. Like other micas, biotite possesses a sheet structure which gives it excellent cleavage in one direction. Biotite also has a very shiny black appearance that helps distinguish it from the other dark ferromagnesian minerals. Like hornblende, biotite is a common constituent of continental rocks, including the igneous rock granite.

Garnet is similar to olivine in that its structure is composed of individual tetrahedra linked by metallic ions. Also like olivine, garnet has a glassy luster, lacks cleavage, and possesses conchoidal fracture. Although the colors of garnet are varied, this mineral is most often brown to deep red. Garnet readily forms equidimensional crystals that are most commonly found in metamorphic rocks (see Figure 5.8). When garnets are transparent, they may be used as gemstones.

Nonferromagnesian Silicates

Muscovite is a common member of the mica family. It is light in color and has a pearly luster. Like other micas, muscovite has excellent cleavage in one direction. In thin sheets muscovite is clear, a property which accounts for its use as window "glass" during the Middle Ages. Since muscovite is very shiny, it can

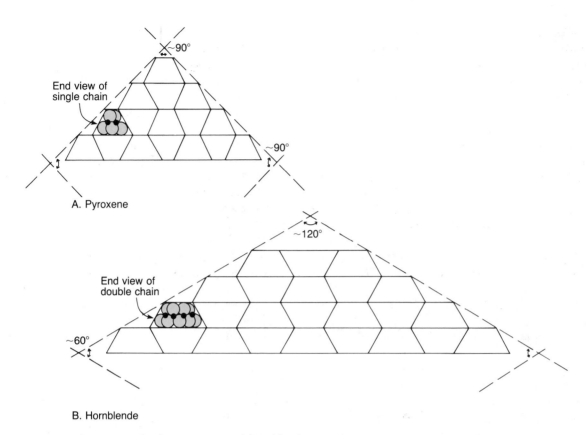

A. Pyroxene

B. Hornblende

FIGURE 1.13 Cleavage angles for pyroxene and hornblende.

often be identified by the sparkle it gives a rock. If you have ever looked closely at beach sand, you may have seen the glimmering brilliance of the mica flakes scattered among the other sand grains.

Feldspar, the most common mineral group, can form under a very wide range of temperatures and pressures, a fact that partially accounts for its abundance. All feldspars have similar physical properties. They have two planes of cleavage meeting at or near 90 degree angles, are relatively hard (6 on Mohs scale), and have a luster which ranges from glassy to pearly. As one component in a rock, feldspar crystals can be identified by their rectangular shape and rather smooth shiny faces (Figure 1.14).

The structure of the feldspar minerals is a three-dimensional framework formed when oxygen atoms are shared by adjacent silicon atoms. In addition, one-fourth to one-half of the silicon atoms in the

FIGURE 1.14 Samples of the mineral feldspar. (Photo by E. J. Tarbuck)

feldspar structure are replaced by aluminum atoms. The difference in charge between aluminum (+3) and silicon (+4) is made up by the inclusion of one or more of the following ions into the crystal lattice: potassium (+1), sodium (+1), and calcium (+2). Due to the large size of the potassium ion as compared to the size of the sodium and calcium ions, two different feldspar structures exist. *Orthoclase feldspar* is a common member of a group of feldspar minerals that contains potassium ions in its structure. The other group, called *plagioclase feldspar,* contains both sodium and calcium ions that freely substitute for one another depending upon the environment during crystallization.

Orthoclase feldspar is usually light cream to salmon pink in color. The plagioclase feldspars, on the other hand, range in color from white to medium gray. However, color should not be used to distinguish these groups. The only sure way to physically distinguish the feldspars is to look for a multitude of fine parallel lines, called *striations.* Striations are found on some cleavage faces of plagioclase feldspar, but are not present on orthoclase feldspar (Figure 1.15).

Quartz is the only silicate mineral consisting entirely of silicon and oxygen. As such, the term *silica* is applied to quartz, which has the chemical formula SiO_2. Since the structure of quartz contains a ratio of two oxygen ions (O^{2-}) for every one silicon ion (Si^{4+}), no other positive ions are needed to attain neutrality. In quartz, a three-dimensional framework is developed through the complete sharing of oxygen by adjacent silicon atoms. Thus, all of the bonds in quartz are of the strong silicon-oxygen type. Consequently, quartz is hard, very resistant to weathering, and does not have cleavage. When broken, quartz generally exhibits conchoidal fracture. In a pure form, quartz is clear and if allowed to solidify without interference, will form hexagonal crystals which develop pyramidal-shaped ends. However, like most other clear minerals, quartz is often colored by the inclusion of various ions (impurities) and forms without developing good crystal faces. The most common varieties of quartz are milky (white), smoky (gray), rose (pink), amethyst (purple), and rock crystal (clear).

Clay is a term used to describe a variety of complex minerals which, like the micas, have a sheet structure. The clay minerals are generally very fine grained and can only be studied microscopically. Most clay minerals originate as products of the chemical weathering of the other silicate minerals. Thus, clay minerals make up a large percentage of the surface material we call soil. Because of the importance of soil in agriculture, and because of its role as a supporting material for buildings, clay minerals are extremely important to humans. One of the more common clay minerals is *kaolinite,* which is used in the manufacture of fine chinaware and occasionally, pottery.

NONSILICATE MINERALS

Although many are important from an economic standpoint, other mineral groups can be considered to be scarce when compared to the silicates. Table 1.3 lists examples of oxides, sulfides, sulfates, halides, and native elements of economic value. A discussion of a few of the more common nonsilicate, rock-forming minerals follows.

The carbonate minerals are much simpler structurally than the silicates. This mineral group is composed of the complex carbonate ion, (CO_3^{2-}), and one or more positive ions. The two most common carbonate minerals are *calcite,* $CaCO_3$, and *dolomite,* $CaMg(CO_3)_2$. Because these minerals are quite similar both physically and chemically, they are difficult to distinguish from one another. Both have a vitreous luster, a hardness between 3 and 4, and

FIGURE 1.15 These parallel lines, called striations, are a distinguishing characteristic of the plagioclase feldspars. (Photo by E. J. Tarbuck)

TABLE 1.3 Common nonsilicate mineral groups.

Group	Member	Formula	Economic Use
Oxides	Hematite	Fe_2O_3	Ore of iron
	Magnetite	Fe_3O_4	Ore of iron
	Corundum	Al_2O_3	Used as an abrasive
	Ice	H_2O	Solid form of water
Sulfides	Galena	PbS	Ore of lead
	Sphalerite	ZnS	Ore of zinc
	Pyrite	FeS_2	Fool's gold
	Chalcopyrite	$CuFeS_2$	Ore of copper
Sulfates	Gypsum	$CaSO_4 \cdot 2H_2O$	Used for plaster
	Anhydrite	$CaSO_4$	Used for plaster
Halides	Halite	$NaCl$	Common salt
	Fluorite	CaF_2	Used in steel making, chemicals, ceramics
Carbonates	Calcite	$CaCO_3$	Portland cement
	Dolomite	$CaMg(CO_3)_2$	Portland cement
	Malachite	$Cu_2(OH)_2CO_3$	Ore of copper
Native elements	Gold	Au	Used for trade
	Copper	Cu	Used as an electrical conductor
	Diamond	C	Gemstone
	Sulfur	S	Used in numerous chemicals
	Graphite	C	Pencil lead and dry lubricant

nearly perfect rhombic cleavage. They can, however, be distinguished by using dilute hydrochloric acid. Calcite reacts vigorously with this acid, whereas dolomite will react only when powdered. Calcite and dolomite are usually found together as the primary constituents in the sedimentary rocks limestone and dolostone. When calcite is the dominant mineral, the rock is called limestone, whereas dolostone results from a predominance of dolomite. Limestone has numerous economic uses, including road aggregate, building stone, and as the main ingredient in portland cement.

Two other nonsilicate minerals frequently found in sedimentary rocks are *halite* and *gypsum*. Both minerals are commonly found in thick layers, which are the last vestiges of ancient seas that have long since evaporated. Halite is the mineral name for common table salt ($NaCl$). Gypsum ($CaSO_4 \cdot 2H_2O$) is the mineral from which plaster and other similar building materials are composed.

REVIEW QUESTIONS

1   Define the term *rock*.

2   List the three main particles of an atom and explain how they differ from one another.

3   If the number of electrons in an atom is 35 and its mass number is 80, calculate the following:
    a   The number of protons.
    b   The atomic number.
    c   The number of neutrons.

4   What is the octet rule? What is the significance of valence electrons?

5   Briefly distinguish between ionic and covalent bonding.

6   What occurs in an atom to produce an ion?

7   What is an isotope?

8   Although all minerals have an orderly internal arrangement of atoms (crystalline structure), most mineral samples do not demonstrate their crystal form. Why?

9   Why might it be difficult to identify a mineral by its color?

10 If you found a glassy-appearing mineral while rock hunting and had hopes that it was a diamond, what simple test might help you make a determination?

11 Explain the use of corundum as given in Table 1.3 in terms of Mohs hardness scale.

12 Gold has a specific gravity of almost 20. If a 25-liter pail of water weighs about 25 kilograms, how much would a 25-liter pail of gold weigh?

13 Explain the difference between the terms *silicon* and *silicate*.

14 What do ferromagnesian minerals have in common? List examples of ferromagnesian minerals.

15 What do muscovite and biotite have in common? How do they differ?

16 Should color be used to distinguish between orthoclase and plagioclase feldspar? What is the best means of distinguishing between the two types of feldspar?

17 Each of the following statements describes a silicate mineral or mineral group. In each case, provide the appropriate name.
    a   The most common member of the amphibole group.
    b   The most common nonferromagnesian member of the mica family.
    c   The only silicate mineral made entirely of silicon and oxygen.
    d   A high-temperature silicate with a name that is based on its color.
    e   Characterized by striations.
    f   Originates as a product of chemical weathering.

18 What simple test can be used to distinguish calcite from dolomite?

CHAPTER TWO

IGNEOUS ROCKS

I n our discussion of the rock cycle, it was pointed out that igneous rocks form when **magma** cools and crystallizes. This molten rock, which originates at depths as great as 200 kilometers within the earth, consists primarily of the elements found in silicate minerals, along with some gases, particularly water vapor, which are confined within the magma by the pressure of the surrounding rocks. Because the magma body is lighter than the surrounding rocks, it works its way toward the surface, and on occasion breaks through, producing a volcanic eruption (Figure 2.1). The spectacular explosions that sometimes accompany an eruption are produced by the gases (volatiles) escaping as the confining pressure lessens near the surface. Sometimes blockage of the vent coupled with surface water seepage into the magma chamber can produce catastrophic explosions. Along with ejected rock fragments, a volcanic eruption often generates extensive lava flows. **Lava** is similar to magma, except that most of the gaseous component has escaped. The rocks which result when lava solidifies are classified as **extrusive**, or **volcanic**. The magma not able to reach the surface eventually crystallizes at depth. Igneous rocks produced in this manner are termed **intrusive**, or **plutonic**, and would never be observed if not for the processes of erosion stripping away the overlying rocks.

CRYSTALLIZATION OF MAGMA

Because magma is a hot liquid, the ions that compose it move about freely and are said to be unordered. However, as magma cools, the random movements of the ions slow and the ions begin to arrange themselves into orderly patterns. This process is called **crystallization**. Before we look at crystallization in more detail, let us first examine how a simple crystalline solid melts. In a crystalline solid, the ions are arranged in a closely packed regular pattern. However, they are not without some motion. They exhibit a sort of restricted vibration about a fixed point. As the temperature rises, the ions vibrate more and more rapidly, and consequently collide with ever-increasing vigor with their neighbors. Continued heating causes the ions to occupy additional space, which results in expansion of the solid and greater distance between ions. When the melting point is reached, the ions are far enough apart and are vibrating rapidly enough to overcome the force of the chemical bonds which had joined them. At this stage, the ions are able to slide past one another, destroying their orderly crystalline structure. Thus, what was once a solid has become a liquid composed of unordered ions moving randomly about.

In the process of crystallization, cooling reverses the events of melting. As the temperature of the liquid drops, the ions pack closer together and begin to lose their freedom of movement. When cooling is sufficient, the force of the chemical bonds will again confine the atoms to an orderly crystalline arrangement. Usually, all of the molten material does not solidify at the same time. Rather, as it cools, numerous embryo crystals develop. In a systematic fashion, ions are added to these centers of crystal growth. When the crystals grow large enough that their edges meet, their growth ceases and crystallization continues elsewhere. Eventually, all of the liquid is transformed into a solid mass of interlocking crystals (Figure 2.2).

The rate of cooling strongly influences the crystallization process, in particular the size of the crystals. When a magma cools very slowly, relatively few

Igneous rocks exposed in Yosemite National Park, California. (Photograph used by permission of Dennis Tasa)

FIGURE 2.1 Parícutin Volcano in eruption at night, 1943. (Photo by K. Segerstrom, U.S. Geological Survey)

centers of crystal growth develop. Slow cooling also allows ions to migrate over relatively great distances. Consequently, slow cooling results in the formation of rather large crystals. On the other hand, when cooling occurs quite rapidly, the ions quickly lose their motion and readily combine. This results in the development of large numbers of nuclei which all compete for the available ions. The outcome is the formation of a solid mass formed of very small intergrown crystals.

When molten material is quenched instantly, there is not sufficient time for the ions to arrange themselves into a crystalline network. Therefore, the solids produced in this manner consist of randomly distributed ions. Rocks that consist of unordered atoms are referred to as **glass** and are quite similar to ordinary manmade glass.

The crystallization of a magma, although more complex, occurs in a manner similar to that just described. Rather than being composed of only one or

FIGURE 2.2 Photomicrograph of interlocking crystals in a coarse-grained igneous rock. (Photo by A. H. Koschmann, U.S. Geological Survey)

two different elements, most magma consists of the eight elements that are the primary constituents of the silicate minerals. These include silicon, oxygen, aluminum, sodium, potassium, calcium, iron, and magnesium. In addition, trace amounts of many other elements, as well as volatiles, particularly water and carbon dioxide, are also found in magma. A *volatile* is a material that is commonly a gas at temperatures and pressures existing at the earth's surface.

When magma cools, it is generally the silicon and oxygen atoms that link together first to form silicon-oxygen tetrahedra. As cooling continues, the tetrahedra join with each other and with other ions to form crystal nuclei of the various silicate minerals. Each crystal nucleus grows as ion after ion is added to the crystalline network in an unchanging pattern. However, the minerals which compose a magma do not all form at the same time or under the same conditions. As we shall see, certain minerals crystallize at much higher temperatures than others. Consequently, magmas often consist of solid crystals surrounded by molten material.

In addition to the rate of cooling, the mineral composition of a magma and the amount of volatile material influence the crystallization process. Since magmas differ in each of these aspects, the physical appearance and mineral composition of igneous rocks vary widely. Nevertheless, it is possible to clas-

sify igneous rocks based on their mode of origin and mineral constituents. The environment during crystallization can be inferred from the size and arrangement of the mineral grains, a property called texture. Consequently, igneous rocks are most often classified by their texture and mineral composition. We will consider both of these rock characteristics in the following sections.

IGNEOUS TEXTURES

The term *texture,* when applied to an igneous rock, is used to describe the overall appearance of the rock based on the size and arrangement of its interlocking crystals (Figure 2.3). Texture is a very important characteristic since it reveals a great deal about the environment in which the rock formed. This fact allows geologists to make inferences about a rock's origin while working in the field where sophisticated equipment is not available.

The most important factor affecting the texture of a rock is the rate at which the magma cooled. From our discussion of crystallization, we learned that rapid cooling produces small crystals, whereas very slow cooling results in the formation of much larger crystals. As we might expect, the rate of cooling is quite slow in magma chambers lying deep within the crust, whereas a thin layer of lava extruded upon the earth's surface may chill in a matter of hours, and small molten blobs ejected into the air during a violent eruption can solidify almost instantly.

Igneous rocks that form at the earth's surface or as small masses within the upper crust possess a very fine-grained texture termed **aphanitic**. By definition, the grains of aphanitic rocks are too small for individual minerals to be distinguished with the unaided eye (Figure 2.3A). Although mineral identification is not possible, fine-grained rocks are commonly characterized as being light, intermediate, or dark in color. Using this system of grouping, light-colored aphanitic rocks are those composed primarily of light-colored nonferromagnesian silicate minerals, and so forth.

A common feature in many aphanitic rocks are the voids left by escaping gases (Figure 2.4). These spherical or elongated openings are called **vesicles** and are limited to the outer portion of lava flows

A.

B.

C.

D.

FIGURE 2.3 Igneous rock textures. **A.** Aphanitic. **B.** Phaneritic. **C.** Porphyritic. **D.** Glassy. (Photos by E. J. Tarbuck)

(Figure 2.5). It is in the outer zone of a lava flow that cooling occurs rapidly enough to "freeze" the lava, thereby preserving the openings produced by the escaping gas.

When large masses of magma solidify far below the surface, they form igneous rocks that exhibit a coarse-grained texture described as **phaneritic**. These coarse-grained rocks have the appearance of a mass of intergrown crystals, which are roughly equal in size and large enough so that the individual minerals can be identified with the unaided eye (Figure 2.3B). Due to the fact that phaneritic rocks form deep within the crust, their exposure at the surface results only after erosion removes the overlying rocks that once surrounded the magma chamber.

A large mass of magma located at depth may require tens of thousands, even millions, of years to solidify. Since all minerals within a magma do not crystallize at the same rate or at the same time during cooling, it is possible for some to become quite large before others even start to form. If magma containing some large crystals should change environments, by erupting at the surface, for example, the molten portion of the lava would cool quickly. The resulting rock, which has large crystals embedded in a matrix of smaller crystals, is said to have a **porphyritic texture** (Figure 2.3C). The large crystals in such a rock are referred to as **phenocrysts**, while the matrix of smaller crystals is called **groundmass**. A rock which has such a texture is called a **porphyry**.

FIGURE 2.4 Vesicular texture. Vesicles form as gas bubbles escape near the top of a lava flow. (Photo by E. J. Tarbuck)

During some volcanic eruptions, molten rock is ejected into the atmosphere, where it is quenched very quickly. Rapid cooling of this type may generate rock with a **glassy texture**. As was indicated earlier, glass results when the ions have not been permitted the time to unite into an orderly crystalline structure. *Obsidian,* a common type of natural glass, is similar in appearance to a dark chunk of manmade glass (Figure 2.3D).

Although the rate of cooling is the major factor determining the texture of an igneous rock, other factors are also important. In particular, the composition of the magma influences the resulting texture. For example, basaltic magma, which is very fluid, will usually generate crystalline rocks when cooled quickly in a thin lava flow. Under the same conditions, granitic magma, which is quite viscous (resists flow), is much more likely to produce a rock with a glassy texture. Consequently, most of the lava flows that are composed of volcanic glass are granitic in composition. However, when basaltic lava flows into the sea, its surface may be quenched rapidly enough to form a thin, glassy skin. Moreover, small ash fragments of basaltic composition are usually cooled rapidly enough to produce a glassy texture.

Some igneous rocks are formed from the consolidation of individual rock fragments that are ejected during a violent eruption. The ejected particles may be very fine ash, molten blobs, or large angular blocks which are torn from the walls of the vent during the eruption. Igneous rocks composed of these rock fragments are said to have a **pyroclastic texture**.

A common type of pyroclastic rock is composed of glass shards (thin strands) which remained hot enough during their flight to fuse together upon impact. Other pyroclastic rocks are composed of

FIGURE 2.5 A vesicular lava which flowed from the base of Sunset Crater, Arizona. (Photo by E. J. Tarbuck)

fragments that solidified before impact and became cemented together at some later time. Because pyroclastic rocks are made of individual rock fragments rather than interlocking crystals, their overall textures are often more similar to sedimentary rocks than to igneous rocks.

MINERAL COMPOSITION

The mineral makeup of an igneous rock is ultimately determined by the chemical composition of the magma from which it crystallized. Such a large variety of igneous rocks exists that it is logical to assume an equally large variety of magmas must also exist. However, geologists have found that various eruptive stages of the same volcano often extrude lavas exhibiting somewhat different mineral compositions, particularly if an extensive period of time separated the eruptions. Evidence of this type led them to look into the possibility that a single magma might produce rocks of varying mineral content.

A pioneering investigation into the crystallization of magma was carried out by N. L. Bowen in the first quarter of this century. Bowen discovered that as magma cools in the laboratory, certain minerals crystallize first. At successively lower temperatures, other minerals begin to crystallize as shown in Figure 2.6. As the crystallization process continues, the composition of the melt (liquid portion of a magma, excluding any solid material) continually changes. For example, at the stage when about 50 percent of the magma has solidified, the melt will be greatly depleted in iron, magnesium, and calcium, because these elements are found in the earliest-formed minerals. But at the same time, it will be enriched in the elements contained in the later-forming minerals, namely sodium and potassium. Further, the silicon content of the melt becomes enriched toward the latter stages of crystallization.

Bowen also demonstrated that if a mineral remained in the melt after it had crystallized, it would react with the remaining melt and produce the next mineral in the sequence shown in Figure 2.6. For this

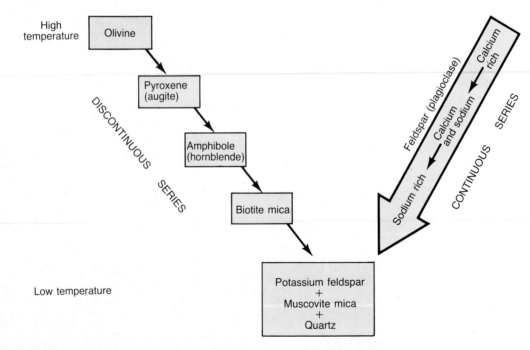

FIGURE 2.6 Bowen's reaction series shows the sequence in which minerals crystallize from a magma. Compare this figure to the mineral composition of the rock groups in Table 2.1. Note that each rock group consists of minerals that crystallize at approximately the same time.

reason, this arrangement of minerals became known as **Bowen's reaction series**. On the upper left branch of this reaction series, olivine, the first mineral to form, will react with the remaining melt to become pyroxene. This reaction will continue until the last mineral in the series, biotite, is formed. This left branch is called a *discontinuous reaction series* because each mineral has a different crystalline structure. Recall that olivine is composed of single tetrahedra and that the other minerals in this sequence are composed of single chains, double chains, and sheet structures, respectively. Ordinarily, these reactions are not complete so that various amounts of each of these minerals may exist at any given time.

The right branch of the reaction series is a continuum in which the earliest-formed calcium-rich feldspar crystals react with the sodium ions contained in the melt to become progressively more sodium rich. Oftentimes the rate of cooling occurs rapidly enough to prohibit the complete transformation of calcium-rich feldspar into sodium-rich feldspar. In these instances, the feldspar crystals will have calcium-rich interiors surrounded by zones that are progressively richer in sodium.

During the last stage of crystallization, after most of the magma has solidified, the remaining melt will form the minerals quartz, muscovite, and potassium feldspar. Although these minerals crystallize in the order shown, this sequence is not a true reaction series.

Bowen demonstrated that minerals crystallize from magma in a systematic fashion. But how does Bowen's reaction series account for the great diver-

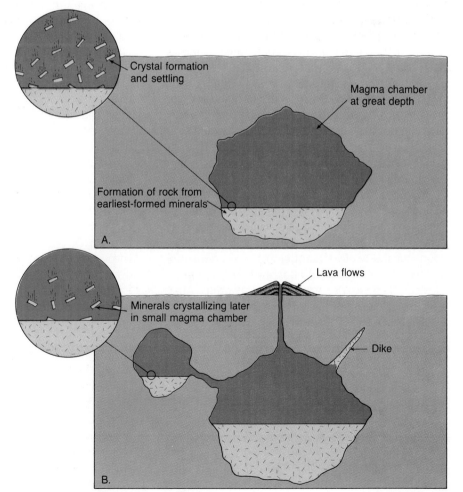

FIGURE 2.7 Separation of minerals by fractional crystallization. **A.** Illustration of how the earliest-formed minerals can be separated from a magma by settling. **B.** The remaining melt could migrate to a number of locations and, upon further crystallization, generate rocks having a composition much different from the parent magma.

Crystal formation and settling

Magma chamber at great depth

Formation of rock from earliest-formed minerals

A.

Lava flows

Minerals crystallizing later in small magma chamber

Dike

B.

sity of igneous rocks? Apparently, at one or more stages in the crystallization process, the solid and liquid components of a magma frequently separate. This can happen, for example, if the earlier-formed minerals are heavier than the liquid portion and settle to the bottom of the magma chamber as shown in Figure 2.7A (page 39). This settling is thought to occur frequently with the dark silicates, such as olivine. When the remaining melt crystallizes, either in place or in a new location if it migrates out of the chamber, it will form a rock with a chemical composition much different from the original magma (Figure 2.7B). In many instances the melt which has migrated from the initial magma chamber will undergo further segregation. As crystallization progresses in the "new" magma, the solid particles may accumulate into rocklike masses surrounded by pockets of the still molten material. It is very likely that some of this melt will be squeezed from the mixture into the cracks which develop in the surrounding rock. This process will generate an igneous rock of yet another composition.

The process involving the segregation of minerals by differential crystallization and separation is called **fractional crystallization**. At any stage in the crystallization process the melt might be separated from the solid portion of the magma. Consequently, fractional crystallization can produce igneous rocks having a wide range of compositions.

Bowen successfully demonstrated that through fractional crystallization one magma can generate several different igneous rocks. However, more recent work has indicated that this process cannot account for the relative quantities of the various rock types known to exist. Although more than one rock type can be generated from a single magma, apparently other mechanisms also exist to generate magmas of quite varied chemical compositions. We will examine some of these mechanisms at the end of the next chapter.

NAMING IGNEOUS ROCKS

As was stated previously, igneous rocks are most often classified, or grouped, on the basis of their texture and mineral composition. The various igneous textures result from different cooling histories, whereas the mineral composition of an igneous rock is the consequence of the chemical makeup of the parent magma and the environment of crystallization. As we might expect from the results of Bowen's work, minerals that crystallize under similar conditions are most often found together composing the same igneous rock. Hence, the classification of igneous rocks closely corresponds to Bowen's reaction series (see Figure 2.6).

The first minerals to crystallize—calcium feldspar, pyroxene, and olivine—are high in iron, magnesium, and calcium, and low in silicon. Basalt is a common extrusive rock of this composition; thus, the term *basaltic* is used to denote rocks of this type. Due to their iron content, basaltic rocks are typically darker in color and slightly heavier than other igneous rocks commonly found at the earth's surface.

Among the last minerals to crystallize are potassium feldspar and quartz. Igneous rocks in which these two minerals predominate are referred to as having a *granitic* composition. Intermediate igneous rocks are made up of minerals found near the middle of Bowen's reaction series. Amphibole along with the intermediate plagioclase feldspars are the main constituents of this rock group. We will refer to those rocks that have a composition between that of granite and basalt as being *andesitic*.

Although each basic rock group is composed mainly of minerals located in a specific region of Bowen's reaction series, other constituents are usually present in lesser amounts. For example, granitic rocks are composed mainly of quartz and potassium feldspar (K feldspar), but may also contain muscovite, biotite, amphibole, and sodium feldspar (Na feldspar). See Table 2.1.

This discussion has concentrated on only three mineral compositions, yet it is important to note that gradations among these types also exist (Figure 2.8). For example, an abundant intrusive igneous rock called *granodiorite* has a mineral composition between that of rocks with a granitic composition and those with an andesitic composition. Another important igneous rock, called *peridotite*, contains mostly olivine and thus falls near the very beginning of Bowen's reaction series. Peridotite is believed to be a major constituent of the upper mantle.

An important aspect of the mineral composition of igneous rocks is silica (SiO_2) content. Recall that

TABLE 2.1 Common igneous rocks.

		Granitic	Andesitic	Basaltic
Intrusive		Granite	Diorite	Gabbro
Extrusive		Rhyolite	Andesite	Basalt
Mineral Composition		Quartz Potassium feldspar Sodium feldspar	Amphibole Intermediate plagioclase feldspar Biotite	Calcium feldspar Pyroxene
Minor Mineral Constituents		Muscovite Biotite Amphibole	Pyroxene	Olivine Amphibole

most of the minerals in igneous rocks contain some silica. Typically, the silica content of crustal rocks ranges from a low of 50 percent in basaltic rocks to a high of about 70 percent in granitic rocks. The percentage of silica in igneous rocks actually varies in a systematic manner which parallels the abundance of the other elements. For example, rocks low in silica contain large amounts of calcium, iron, and magnesium. Consequently, the chemical makeup of an igneous rock can be inferred directly from its silica content. Further, the amount of silica present in magma strongly influences its behavior. Granitic magma, which has a high silica content, is quite viscous and exists as a fluid at temperatures as low as 800°C. On the other hand, basaltic magmas are low in silica and generally quite fluid. Basaltic magmas are also extruded at much higher temperatures, often 1200°C or higher.

It is not at all uncommon for two rocks to have the same mineral constituents and yet have different names. This resulted, in part, from the fact that many igneous rocks have ancient names which were given to them on the basis of their overall appearance rather than on their mineral composition. For example, the coarse-grained intrusive rock granite has a fine-grained volcanic equivalent called rhyolite. Although these rocks are mineralogically the same, they have different textures and do not look at all

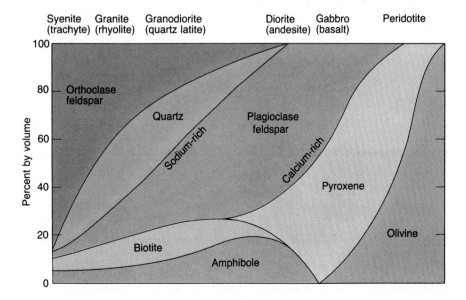

FIGURE 2.8 Mineralogy of the common igneous rocks. Parentheses indicate the name of equivalent extrusive rock. (After Turekian).

A.

B.

FIGURE 2.9 A. Granite, one of the more common coarse-grained igneous rocks. **B.** Rhyolite, the fine-grained equivalent of granite, is far less abundant. (Photos by E. J. Tarbuck)

alike. Therefore the rocks were given different names (Figure 2.9).

Granitic Rocks

Granite is perhaps the best known of all igneous rocks (Figure 2.9A). This is partly because of its natural beauty, which is enhanced when it is polished, and partly because of its abundance. Slabs of pol-

ished granite are commonly used for tombstones, monuments, and as building stones.

Granite is a phaneritic rock composed of up to 25 percent quartz and over 50 percent potassium feldspar and sodium-rich feldspar. The quartz crystals, which are roughly spherical in shape, are most often clear to light gray in color. In contrast to quartz, the feldspar crystals in granite are not as glassy, but rectangular in shape and generally salmon pink to white in color. Other common constituents of granite are muscovite and the dark silicates, particularly biotite and amphibole. Although the dark components of granite make up less than 20 percent of most samples, dark minerals appear to be more prominent than their percentage would indicate. In some granites, K feldspar is dominant and very dark pink in color, so that the rock appears almost reddish. This variety is very popular as a building stone. However, most often the feldspar grains are white, so that when viewed at a distance granite appears light gray in color. Granite may also have a porphyritic texture, in which feldspar crystals a centimeter or more in length are scattered among a coarse-grained groundmass of quartz and amphibole.

Granite is often produced by the processes which generate mountains. Because granite is a by-product of mountain building and is very resistant to weathering and erosion, it frequently forms the core of eroded mountains. For example, Pikes Peak in the Rockies, Mount Rushmore in the Black Hills, the White Mountains of New Hampshire, Stone Mountain in Georgia, and Yosemite National Park in the Sierra Nevada are all areas where large quantities of granite are exposed at the surface (Figure 2.10). As we can see from these examples, granite is a very abundant rock. However, it has become common practice among geologists to apply the term *granite* to any coarse-grained intrusive rock composed predominantly of light silicate minerals. We will follow this practice for the sake of simplicity. The student should keep in mind that this use of the term *granite* covers rock having a range of mineral compositions.

Magma having a granitic composition contains up to 5 percent water. Because water will not crystallize in the magma chamber, it can make up a much higher percentage of the melt during the final phase of solidification. Crystallization in a water-rich environment, where ion migration is enhanced, is be-

FIGURE 2.10 Yosemite National Park, located in the Sierra Nevada, is one of many areas where vast amounts of granite are exposed at the surface. (Photograph used by permission of Dennis Tasa)

lieved to result in the formation of crystals several centimeters, or even a few meters, in length. The resulting rocks, called **pegmatites**, are composed of unusually large crystals.

Some of the largest crystals ever uncovered have been found in pegmatites. Feldspar masses the size of houses have been quarried from a pegmatite located in North Carolina. Gigantic hexagonal crystals of muscovite measuring a few meters across have been found in Ontario, Canada. In the Black Hills, crystals as large as telephone poles of the lithium-bearing mineral spodumene have been mined. The largest of these was more than 12 meters long. Not all pegmatites contain such large crystals, but these examples emphasize the special conditions that must exist during the formation of pegmatites.

Although most pegmatites are granitic in composition and consist of unusually large crystals of quartz, feldspar, and muscovite, pegmatites of other compositions also exist. Some granitic pegmatites are commercially valuable. The feldspar is used in the production of ceramics and the muscovite is used for isinglass, electrical insulation, and glitter. Further, because pegmatites form at the end of the crystallization process, they often contain some of the least abundant elements. Thus, some rare minerals may also be found in pegmatites. In addition to the common silicates, some pegmatites contain semiprecious gems such as beryl, topaz, and tourmaline. Also, minerals containing the elements lithium, cesium, uranium, and the rare earths* are occasionally found. Most pegmatites are located within large igneous masses or as veins which cut into the rock that surrounds the magma chamber. In the latter case, hydrothermal (hot water) solutions are thought to have deposited the minerals in cracks that penetrated into the country rock.

---

*The rare earths are a group of fifteen elements (atomic numbers 57 through 71) that possess similar properties. They are useful catalysts in petroleum refining and are used to improve color retention in television picture tubes.

Rhyolite is the volcanic equivalent of granite. Like granite, rhyolite is composed primarily of the light-colored silicates (Figure 2.9B). This fact accounts for its color, which is usually buff to pink or occasionally very light gray. Rhyolite is usually aphanitic and frequently contains glassy fragments and voids indicating rapid cooling in a surface environment. In those instances when rhyolite contains phenocrysts, they are usually small and composed of either quartz or potassium feldspar. In contrast to granite, rhyolite is rather uncommon. Yellowstone Park is one well-known exception. Here rhyolitic lava flows and ash deposits of similar composition are widespread.

Obsidian is a dark-colored, glassy rock which forms when lava is quenched very quickly (Figure 2.11A). In contrast to the orderly arrangement of ions that is characteristic of minerals, the ions in glass are unordered. Consequently, glassy rocks such as obsidian are not composed of minerals in the same sense as most other rocks.

Although usually black or reddish-brown in color, obsidian has a high silica content. Thus, its composition is more akin to the light igneous rocks such as granite than to the dark rocks of basaltic composition. By itself, silica is clear like window glass; the dark color results from the presence of metallic ions. If you examine a thin edge of a piece of obsidian, it will be nearly transparent. Because of its excellent conchoidal fracture, obsidian was a prized material from which American Indians made arrowheads and cutting tools.

Pumice is a volcanic rock which, like obsidian, has a glassy texture. Usually found with obsidian, pumice forms when large amounts of gas escape through lava to generate a gray, frothy mass (Figure 2.11B). This material is similar to the foam which flows from a newly opened bottle of champagne. In some samples, the voids are quite noticeable, while in others the pumice resembles fine shards of intertwined glass. Due to a high percentage of voids, many samples of pumice will float in water. Oftentimes flow lines are visible in pumice, indicating some movement before solidification was complete. Moreover, pumice and obsidian often form in the same rock mass, where they exist in alternating layers.

Andesitic Rocks

Andesite is a medium gray, fine-grained rock of volcanic origin. The name derives from the Andes Mountains, where numerous volcanoes are composed of this rock type. In addition to the volcanoes of the Andes, many volcanic structures encircling the Pacific Ocean are of andesitic composition. Andesite quite commonly exhibits a porphyritic texture (Figure 2.12). In these cases, the phenocrysts are often

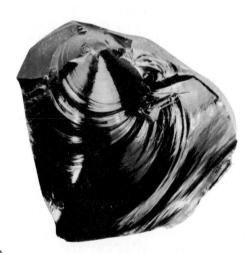

A.   B.

FIGURE 2.11 A. Obsidian, a glassy volcanic rock. (Courtesy of Ward's Natural Science Establishment, Inc., Rochester, N.Y.). **B.** Pumice, a glassy rock containing numerous tiny voids. (Photo by E. J. Tarbuck)

FIGURE 2.12 Andesite porphyry, a common volcanic rock. (Photo by E. J. Tarbuck)

FIGURE 2.13 Basalt flows of the Columbia Plateau. (Photo by E. T. Jones, U.S. Geological Survey)

light, rectangular crystals of plagioclase feldspar or black, elongated hornblende crystals.

Diorite is a coarse-grained intrusive rock that resembles gray granite. However, it can be distinguished from granite by the absence of visible quartz crystals. The mineral makeup of diorite is primarily sodium-rich plagioclase and amphibole, with lesser amounts of biotite. Because the white feldspar grains and dark amphibole crystals are roughly equal in abundance, diorite has a "salt and pepper" appearance.

Basaltic Rocks

Basalt is a very dark green to black, fine-grained volcanic rock composed primarily of pyroxene and calcium-rich feldspar, with lesser amounts of olivine and amphibole present (Figure 2.13). When porphyritic, basalt commonly contains small, light-colored calcium feldspar phenocrysts or glassy-appearing olivine phenocrysts embedded in a dark groundmass.

Basalt is the most common extrusive igneous rock. Many volcanic islands, such as the Hawaiian Islands and Iceland, are composed mainly of basalt. Further, the upper layers of the oceanic crust consist of basalt. In the United States, large portions of central Oregon and Washington were the sites of extensive basaltic outpourings (Figure 2.13). At some locations these once fluid basaltic flows have accumulated to thicknesses approaching 2 kilometers.

Gabbro is the intrusive equivalent of basalt. Like basalt, it is very dark green to black in color and composed primarily of pyroxene and calcium-rich plagioclase. Although gabbro is not a common constituent of the continental crust, it undoubtedly makes up a significant percentage of the oceanic crust. Here large portions of the magma found in underground reservoirs that once fed basalt flows eventually solidified at depth to form gabbro.

Pyroclastic Rocks

Pyroclastic rocks are those which form from fragments ejected during a volcanic eruption. One very common pyroclastic rock, called *tuff*, is composed of tiny ash-sized fragments that were later cemented together. In situations where the ash particles remained hot enough to fuse, the rock is generally called *welded tuff.* Since welded tuffs consist of glass shards, their appearance may closely resemble pumice. Deposits of partially welded tuffs are easily quarried and used as a durable building material.

Several villages in Cappadocia in central Turkey, which date as far back as the fourth century, have been carved into vertical cliffs composed of this material.

Pyroclastic rocks composed of particles larger than ash are called *volcanic breccia*. The particles in volcanic breccia can consist of streamlined fragments that solidified in air, blocks broken from the walls of the vent, crystals, and glass fragments. Unlike the other igneous rock names, the terms *tuff* and *volcanic breccia* do not denote mineral composition.

REVIEW QUESTIONS

1 How does lava differ from magma?

2 How does the rate of cooling influence crystallization?

3 In addition to the rate of cooling, what other factors influence crystallization?

4 The classification of igneous rocks is based largely upon two criteria. Name these criteria.

5 The statements that follow relate to terms describing igneous rock textures. For each statement, identify the appropriate term.
a Openings produced by escaping gases.
b Obsidian exhibits this texture.
c A matrix of fine crystals surrounding phenocrysts.
d Crystals are too small to be seen with the unaided eye.
e A texture characterized by two distinctively different crystal sizes.
f Coarse grained, with crystals of roughly equal size.

6 What does a porphyritic texture indicate about an igneous rock?

7 What is fractional crystallization? How might fractional crystallization lead to the formation of several different igneous rocks from a single magma?

8 Relate the classification of igneous rocks to Bowen's reaction series.

9 How are granite and rhyolite different? In what way are they similar?

10 Why are the crystals in pegmatites so large?

11 Compare and contrast each of the following pairs of rocks:
a Granite and diorite.
b Basalt and gabbro.
c Andesite and rhyolite.

12 How do tuff and volcanic breccia differ from other igneous rocks such as granite and basalt?

CHAPTER THREE

IGNEOUS ACTIVITY

At 8:32 A.M. on Sunday, May 18, 1980, the largest volcanic eruption to occur in North America in recent times transformed a picturesque volcano into a decapitated remnant (Figure 3.1). On this date in southwestern Washington state, Mount St. Helens erupted with a force hundreds of times greater than that of the atomic bombs dropped on Japan during World War II. The blast blew out the entire north flank of the volcano, leaving a gaping hole. A once prominent volcano that had grown to more than 2900 meters (9500 feet) had in one brief moment been lowered by about 410 meters (1350 feet).

The early morning blast totally devastated a wide swath of timber-rich land on the north side of the mountain (Figure 3.2). Trees within a 400-square-kilometer area lay intertwined and flattened, stripped of their branches and appearing from the air like toothpicks strewn about. The immense force caused trees as far away as 25 kilometers to topple. The gases and ash unleashed from the volcano had temperatures that probably exceeded 800°C (1470°F)! As many as 59 persons were killed by the eruption. Some died from the intense heat and the suffocating cloud of ash and gases. Others perished as they were hurled from the mountain by the force of the blast. Still others were trapped by debris-laden mudflows.

The blast and accompanying mudflows carried ash, trees, and water-saturated rock debris 29 kilometers down the Toutle River. The river quickly became a mud-filled torrent and reached depths of 60 meters in some places. Further, a debris dam was deposited at the outlet of Spirit Lake, causing its level to rise by more than 30 meters (100 feet). For several days the threat of pent-up waters breaching the dam posed another potential hazard.

The eruption of May 18th ejected an estimated three to four cubic kilometers of ash and rock debris. By comparison, this is roughly equal to the quantity of ash that buried the city of Pompeii during the historic eruption of Mount Vesuvius in 79 A.D.

Following the devastating explosion, Mount St. Helens continued to emit great quantities of hot gases and ash. Only minutes after the eruption began, a dark plume rose from the volcano. The force of the blast was so strong that some ash was propelled high into the stratosphere, more than 18,000 meters (60,000 feet) above the ground. During the next hours and days, this very fine grained material was carried around the earth by strong upper air winds. Measurable deposits were reported as far away as Oklahoma and Minnesota. Meanwhile, ash fallout in the immediate vicinity accumulated to depths exceeding 2 meters, and the air over Yakima, Washington, 130 kilometers to the east, was so filled with ash that residents experienced midnight-like darkness at noon. Crop damage from the volcanic fallout was reported as far away as central Montana.

The events leading to the May 18th eruption began about two months earlier, on March 20th, as a series of minor earth tremors centered beneath the awakening mountain. The first volcanic activity took place on March 27th, when a small amount of ash and steam rose from the summit. Over the next several weeks, sporadic eruptions of varied intensity occurred.

Prior to the main eruption, the primary concern had been the potential hazard of mudflows. These moving lobes of saturated debris were created when ice and snow were melted by heat from the magma within the volcano. The only sign of a potentially hazardous eruption was a bulge on the volcano's

Mount Shasta, California, one of the most picturesque volcanoes of the Cascade Range. (Photo by John S. Shelton)

FIGURE 3.1 Before and after photographs show the transformation of Mount St. Helens caused by the May 18, .1980 eruption. ("Before" photo by Roland V. Emetaz, courtesy USDA Forest Service; "after" photo by Jim Hughes)

north flank. Careful monitoring of this dome-shaped structure indicated a very slow but steady growth rate of a few meters per day. Geologists monitoring the activity suggested that if the growth rate of the bulge changed appreciably, an eruption might quickly follow. Unfortunately, no such variation was detected prior to the explosion. In fact, the seismic activity decreased during the two days preceding the huge blast.

"Vancouver, Vancouver, this is it!" was the only warning to precede the unleashing of tremendous quantities of pent-up gases. The trigger was an earth tremor with a magnitude of 5.1 on the Richter scale. The vibrations sent the north slope of the cone plummeting into the Toutle River, effectively remov-

ing the overburden which had trapped the magma below. With the pressure reduced, the water-rich magma is thought to have ruptured like an over-heated steam boiler. Since the eruption originated in the vicinity of the bulge, which was several hundred meters below the summit, the main impact of the eruption was directed laterally rather than vertically. Had the full force of the eruption been upward, far less destruction would have occurred.

Mount St. Helens is only one of the 15 large volcanoes and innumerable smaller ones extending from British Columbia to northern California. Eight of the larger cones have been active in the past few hundred years, whereas the last eruptive phase of Mount St. Helens came to an end in 1857. Of the

FIGURE 3.2 Forest lands and battered van northwest of Mount St. Helens destroyed by lateral blast of May 18, 1980. (Photo courtesy of USDA Forest Service)

remaining seven "active" volcanoes, Mount Baker, Mount Shasta, Lassen Peak, and Mount Rainier are believed most likely to erupt again. It is hoped that the eruptions of Mount St. Helens will provide geologists with enough data to more effectively evaluate the potential hazards of future volcanic eruptions.

THE NATURE OF VOLCANIC ACTIVITY

Volcanic activity is commonly perceived as a process that produces a picturesque, cone-shaped structure which periodically erupts in a violent manner. However, although some eruptions may be cataclysmic, many are relatively quiescent. The primary factors which determine the nature of volcanic eruptions include the magma's composition, its temperature, and the amount of dissolved gases it contains. These factors affect the magma's mobility, or **viscosity**. The more viscous the material, the greater its resistance to flow. For example, molasses is more viscous than water. The effect of temperature on viscosity is easily seen. Just as heating molasses makes it more fluid, the mobility of lava is also influenced by temperature changes. As a lava flow cools and begins to congeal, its mobility decreases and eventually the flowing halts.

The chemical composition of magmas was discussed in Chapter Two with the classification of igneous rocks. One major difference between various igneous rocks and therefore between the magmas from which they originate is their silica (SiO_2) content (Table 3.1). Magmas that produce basaltic rocks contain about 50 percent silica, whereas rocks of granitic composition (granite and its extrusive equiv-

alent, rhyolite) contain over 70 percent silica. The intermediate rock types, andesite and diorite, contain around 60 percent silica. It is important to note that a magma's viscosity is directly related to its silica content. In general, the higher the percentage of silica in magma, the greater its viscosity. It is believed that the flow of magma is impeded because the silica molecules link into long chains even before crystallization begins. Consequently, because of their lower silica content, basaltic lavas tend to be quite fluid, whereas lavas of granitic composition are very viscous and incapable of flow over appreciable distances even at relatively high temperatures.

The gas content of a magma also affects its mobility. Dissolved gases tend to increase the fluidity of magma. Of far greater consequence is the fact that escaping gases provide enough force to propel molten rock from a volcanic vent. As magma moves into a near-surface environment, such as within a volcano, the confining pressure in the uppermost portion of the magma body is greatly reduced. This reduced confining pressure allows the gases, which had been dissolved when they were at greater depths, to be released suddenly. At temperatures of 1000°C (1830°F) and low, near-surface pressures, these gases will expand to occupy hundreds of times their original volume. Very fluid basaltic magmas allow the expanding gases to migrate upward and escape from the vent with relative ease. As they escape, the gases will often carry incandescent lava hundreds of meters into the air, producing lava fountains. Although spectacular, such fountains are not generally associated with major explosive events of the type which cause great loss of life and property. Rather, eruptions of fluid basaltic lavas, such as those that occur in Hawaii, are relatively quiescent.

TABLE 3.1 Variations in properties among magmas of differing compositions.

Property	Basaltic	Andesitic	Granitic
Silica content	Least (about 50%)	Intermediate (about 60%)	Most (about 70%)
Viscosity	Least	Intermediate	Highest
Tendency to form lavas	Highest	Intermediate	Least
Tendency to form pyroclastics	Least	Intermediate	Highest
Melting point	Highest	Intermediate	Lowest

At the other extreme, highly viscous magmas impede the upward migration of gases. As a consequence, gases collect as bubbles and pockets that increase in size and pressure until they explosively eject the semimolten rock from the volcano.

To summarize, we have seen that the quantity of dissolved gases, as well as the ease with which the gases can escape, largely determine the nature of a volcanic eruption. We can now understand why the volcanic eruptions on Hawaii are relatively quiet, whereas the volcanoes bordering the Pacific are explosive and pose the greatest threat to people, because these latter volcanoes generally contain great quantities of gas and emit viscous lavas.

MATERIALS EXTRUDED DURING AN ERUPTION

Many people believe that lava is the primary material extruded from a volcano. This is not always the case. Explosive eruptions which eject huge quantities of broken rock, lava bombs, and fine ash and dust, occur just as frequently. Moreover, all volcanic eruptions emit large amounts of gas. In this section we will examine each of these materials associated with a volcanic eruption.

Lava Flows

Due to their low silica content, basaltic lavas are usually very fluid and flow in thin, broad sheets or tongues. On the island of Hawaii such lavas have been clocked at speeds of 30 kilometers (20 miles) per hour on steep slopes. These velocities are rare, however, and flow rates of 10 to 300 meters per hour are more common. In contrast, the movement of silica-rich lava is often too slow to be perceptible.

When fluid basaltic lavas of the Hawaiian type congeal, they often form a relatively smooth skin that sometimes wrinkles as the still-molten subsurface lava continues to advance (Figure 3.3A). These are known as **pahoehoe flows** and resemble the twisted braids in ropes. Another common type of basaltic lava has a surface of rough, jagged blocks with dangerously sharp edges and spiny projections (Figure 3.3B). The name **aa** (pronounced "ah ah") is given to these flows. Active aa flows are relatively cool and thick and, depending upon the slope, advance at rates from 5 to 50 meters per hour. Further, escaping gases fragment the cool surface and produce numerous voids and sharp spines in the congealing lava. As the molten interior advances, the outer crust is broken further, giving the flow the appearance of an advancing mass of lava rubble.

FIGURE 3.3 **A.** Active pahoehoe (ropy) flow extruded during the 1983 eruption of Kilauea, Hawaii. **B.** Advancing aa flow in the Royal Gardens subdivision, Hawaii. (Photos by J. D. Griggs, U.S. Geological Survey)

A.

B.

Gases

Magmas contain varied amounts of dissolved gases held in the molten rock by confining pressure, just as carbon dioxide is held in soft drinks. As with soft drinks, as soon as the pressure is reduced, the gases begin to escape. Obtaining samples from an erupting volcano is very difficult and dangerous, so geologists usually only estimate the amount of gas originally contained within the magma.

The gaseous portion of most magmas is believed to compose from 1 to 5 percent of the total weight, and most of this is in the form of water vapor. Although the percentage may be small, the actual quantity of emitted gas can exceed thousands of tons per day. The composition of the gases is also of interest to scientists, since much evidence points to these as the source of the oceans and the earth's atmosphere. Analysis of samples taken during Hawaiian eruptions indicated that the gases emitted there consist of about 70 percent water vapor, 15 percent carbon dioxide, 5 percent each of nitrogen and sulfur compounds, and lesser amounts of chlorine, hydrogen, and argon. Sulfur compounds are easily recognized by their pungent odor and because they readily form sulfuric acid, which when inhaled produces a burning sensation.

Pyroclastic Materials

When basaltic lava is extruded, the dissolved gases escape quite freely and continually. As stated earlier, these gases often carry incandescent blobs of lava to great heights, thereby producing spectacular lava fountains (see Figure 3.7). Some ejected material may land near the vent and produce a cone structure, whereas smaller particles will be carried great distances by the wind. The gases in highly viscous magmas, on the other hand, are less able to escape and may build up an internal pressure capable of producing a violent eruption. Upon release, these superheated gases expand a thousandfold as they blow pulverized rock and lava from the vent. The particles produced by these processes are called **pyroclastics**. These ejected lava fragments range in size from very fine dust and sand-sized volcanic ash, to large volcanic bombs and blocks.

The fine *ash* particles are produced when the extruded lava contains so many gas bubbles that it re-

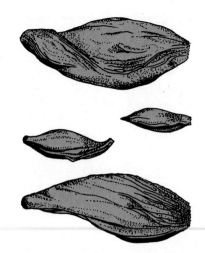

FIGURE 3.4 Volcanic bombs. Ejected lava fragments take on a streamlined shape as they sail through the air. (From Foster, *Physical Geology*, 4th ed., Columbus, Ohio: Charles E. Merrill, 1983)

sembles the froth flowing from a newly opened bottle of champagne. As the hot gases expand explosively, the lava is disseminated into very fine fragments. When the hot ash falls, the glassy shards often fuse to form *welded tuff*. Sheets of this material, as well as ash deposits that consolidate later, cover vast portions of the western United States. In some instances the froth-like lava is ejected in larger pieces called *pumice*. This material has so many voids that it is often light enough to float in water.

Pyroclastics the size of walnuts, called *lapilli* ("little stones"), and pea-sized particles called *cinders* are also very common. Cinders contain numerous voids and form when ejected lava blobs are pulverized by escaping gases. Particles larger than lapilli are called *blocks* when they are made of hardened lava and *bombs* when they are ejected as incandescent lava. Since volcanic bombs are semimolten upon ejection, they often take on a streamlined shape as shown in Figure 3.4.

VOLCANOES AND VOLCANIC ERUPTIONS

Successive eruptions from a central vent result in a mountainous accumulation of material known as a **volcano**. Located at the summit of many volcanoes is a steep-walled depression called a **crater**, which is connected to a magma chamber via a pipelike conduit, or vent. Some volcanoes have unusually large summit depressions that exceed one kilometer in diameter and are known as **calderas**. When fluid lava leaves a conduit, it is often stored in the crater or caldera until it overflows. On the other hand, lava that is very viscous forms a plug in the pipe which rises slowly or is blown out, often enlarging the crater. However, lava does not always issue from a central crater. Sometimes it is easier for the magma or escaping gases to push through fissures on the volcano's flanks. Continued activity from a flank eruption may build a so-called *parasitic cone*. Mount Etna in Italy, for example, has more than 200 secondary vents. Some of these secondary vents extrude only gases and are appropriately called *fumaroles*.

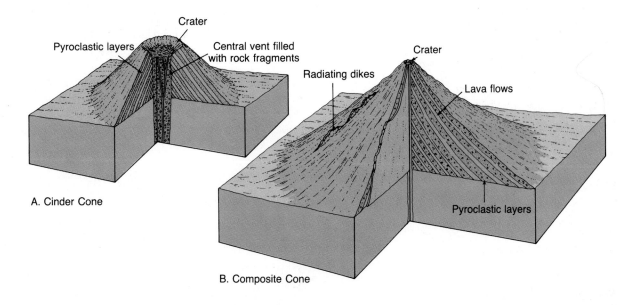

A. Cinder Cone

B. Composite Cone

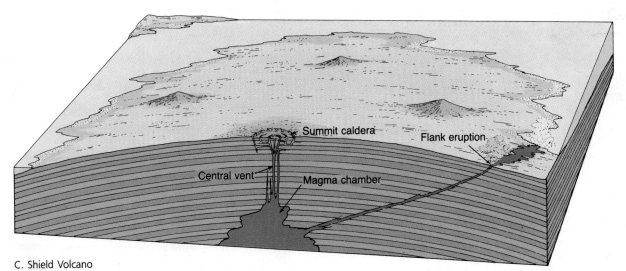

C. Shield Volcano

FIGURE 3.5 The three basic types of volcanic structures. **A.** Cinder cone. **B.** Composite cone. **C.** Shield volcano.

The eruptive history of each volcano is unique; consequently, all volcanoes are somewhat different in form and size. Nevertheless, volcanologists have recognized that volcanoes exhibiting somewhat similar eruptive styles can be grouped. Based on their "typical" eruptive patterns and characteristic form, three groups of volcanoes are generally recognized: shield volcanoes, cinder cones, and composite cones (Figure 3.5, page 55).

Shield Volcanoes

When fluid lava is extruded, the volcano takes the shape of a broad, slightly domed structure called a **shield volcano** (Figure 3.5C). Shield volcanoes are built primarily of basaltic lava flows and contain only a small percentage of pyroclastic material. Typically they have a slope of a few degrees at their flanks and generally do not exceed 15 degrees near their summit, as exemplified by the volcanoes of the Hawaiian Islands. Mauna Loa, probably the largest volcano on earth, is one of the five shield volcanoes that together make up the island of Hawaii (Figure 3.6). Its base rests on the ocean floor 5000 meters below sea level, while its summit reaches a height of

4170 meters (13,680 feet) above the water. Nearly one million years and numerous eruptive cycles were required to build this truly gigantic pile of volcanic rock. Many other volcanic structures, including Midway Island and the Galapagos Islands, have been built in a similar manner from the ocean's depths.

Perhaps the most active and intensively studied shield volcano is Kilauea, located on the island of Hawaii on the southeastern flank of the larger volcano Mauna Loa. Kilauea has erupted more than 50 times in recorded history and is still active today. A testimony to the relatively quiescent nature of these eruptions is the fact that the U.S. Geological Survey's Hawaiian Volcano Observatory is situated on the very rim of the summit caldera. Several months before each eruptive phase, the summit inflates as magma rises from its source 60 to 100 kilometers below the surface. This molten rock gradually works its way upward and accumulates in smaller reservoirs 3 to 5 kilometers below the summit. For up to 24 hours in advance of each eruption, swarms of earth tremors warn of the impending activity. The 1983 eruption occurred along a 6.5-kilometer (4-mile) fissure located east of the summit caldera (Figure 3.7). Here fountains of fluid lava with heights approach-

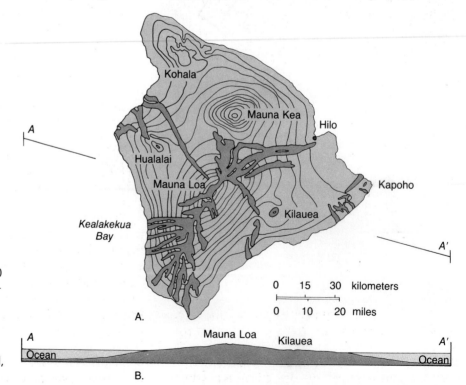

FIGURE 3.6 Map of the island of Hawaii. **A.** Five volcanoes collectively make up the island. Contour interval is 300 meters (1000 feet). **B.** Illustration of the very gentle slope which is characteristic of a shield volcano (no vertical exaggeration). (After H. T. Stearns and G. A. MacDonald, U.S. Geological Survey)

FIGURE 3.7 Lava fountain along the east rift zone of Kilauea, 1983. (Photo by J. D. Griggs, U.S. Geological Survey)

ing 100 meters (330 feet) fed a lava flow that extended for 6 kilometers and eventually inundated a few dwellings in the sparsely populated settlement of Royal Gardens. This eruptive phase, like most others, was accompanied by a gradual subsidence of the summit area, nearly equal in volume to the extruded lava.

Cinder Cones

As the name suggests, **cinder cones** are built from ejected lava fragments. Because unconsolidated pyroclastic material maintains a high angle of repose (between 30 and 40 degrees), volcanoes of this type have very steep slopes. Cinder cones are rather

small, usually less than 300 meters (1000 feet) high, and often form as parasitic cones on or near larger volcanoes. In addition, they frequently occur in groups, where the cones apparently represent the last phase of activity in a region of older basaltic flows. This may result because the contributing magma has cooled and become more viscous.

One of the very few volcanoes whose formation has been observed by geologists from beginning to end is a cinder cone called Parícutin. This volcano's history serves to illustrate the formation and structure of a relatively large cinder cone.

In 1943, about 200 miles west of Mexico City, the volcano Parícutin was born (see Figure 2.1). The eruption site was a cornfield owned by Dionisio Pulido, who with his wife, Paula, witnessed the event as they were preparing the field for planting. For two weeks prior to the first eruption, numerous earth tremors caused apprehension in the village of Parícutin about 3.5 kilometers away. Then around 4:00 P.M. on February 20th, smoke with a sulfurous odor began billowing from a small hole that had been in the cornfield for as long as Dionisio could remember. During the night hot, glowing rock fragments thrown into the air from the hole produced a spectacular fireworks display. By the next day the cone had grown to a height of 40 meters and by the fifth day it was over 100 meters high. At this time explosive eruptions were throwing hot fragments 1000 meters (3300 feet) above the crater rim. The larger fragments fell near the crater, some remaining incandescent as they rolled down the slope. These fragments built an aesthetically pleasing cone, while finer ash fell over a much larger area, burning and eventually covering the village of Parícutin. Within two years the cone had grown to 400 meters (1310 feet) high and would rise only a few tens of meters more.

The first lava flow came from a fissure that had opened just north of the cone; but after a few months of activity, flows began to emerge from the base of the cone itself. In June of 1944, a clinkery flow 10 meters thick moved over the village of San Juan Parangaricutiro, leaving only the church steeple exposed (Figure 3.8). After nine years the activity ceased almost as quickly as it began. Now Parícutin is just another one of the numerous cinder cones dotting the landscape in this region of Mexico. Like the others, it will probably not erupt again.

FIGURE 3.8 The village of San Juan Parangaricutiro engulfed by lava from Parícutin, shown in the background. Only the church towers remain. (Photo by Tad Nichols)

Composite Cones

The earth's most picturesque volcanoes are **composite cones**, or **stratovolcanoes**. Just as shield volcanoes owe their shape to the fluid nature of the extruded lavas, so too do composite cones reflect the nature of the erupted material. Composite cones are produced when relatively viscous lavas of andesitic composition are extruded. A composite cone may extrude viscous lava for long periods. Then suddenly the eruptive style changes and the volcano violently ejects pyroclastic material. Most of the ejected pyroclastic material falls near the summit, building a steep-sided mound of cinders. In time this debris will be covered by lava. Occasionally both activities occur simultaneously, and the resulting structure consists of alternating layers of lava and pyroclastics. Two of the most perfect cones, Mount Mayon in the Philippines and Fujiyama in Japan, exhibit the classic form of the composite cone with its steep summit area and rather gently sloping flanks.

Although composite cones are the most picturesque, they also represent the most violent type of volcanic activity. Their eruption can be unexpected and devastating as was the 79 A.D. eruption of the Italian volcano we now call Vesuvius. Prior to this eruption, Vesuvius was dormant for centuries. Although minor earthquakes probably warned of the events to follow, Vesuvius was covered with a heavy coat of vegetation and hardly looked threatening. On August 24th, however, the tranquility ended, and in the next three days the city of Pompeii (near Naples) and more than 2000 of its 20,000 residents were buried. They remained so for nearly seventeen centuries, until the city was rediscovered and excavated.

The destruction of Pompeii was truly catastrophic, yet eruptions of a more devastating nature occur when hot gases infused with incandescent ash are ejected, producing a fiery cloud called **nuée ardente**. Also referred to as *glowing avalanches,* these flows, which are black in daylight and glow red at night, move down steep volcanic slopes at speeds exceeding 150 kilometers per hour (Figure 3.9). These glowing avalanches are very dense, but their

FIGURE 3.9 Nuée ardente races down the slope of Mount Ngauruhoe, New Zealand, 1974. This volcano has had over 50 explosive eruptions since 1839. (Photo courtesy of NOAA)

weight is supported by the expanding gases emitted from the hot lava particles. In this way, the material, which can be composed of rather large lava fragments, flows downslope in an almost frictionless environment cushioned by the expanding gases.

In 1902, a nuée ardente from Mount Pelée, a small volcano on the Caribbean island of Martinique, destroyed the port town of St. Pierre. The destruction was instantaneous and so devastating that almost all of St. Pierre's 28,000 inhabitants were killed. Reportedly only a prisoner protected in a dungeon, a shoemaker, and a few people on ships in the harbor were spared (Figure 3.10).

When we compare the destruction of St. Pierre with that of Pompeii, several differences are noticed. Pompeii was totally buried by an event lasting three days, whereas St. Pierre was destroyed in a brief instant and its remains were only mantled by a thin layer of volcanic debris. Also, the structures of Pompeii remained intact except for roofs that collapsed under the weight of the ash. In St. Pierre masonry walls nearly one meter thick were knocked over like dominoes; large trees were uprooted and cannons were torn from their mounts.

As with all land areas, volcanoes are continually lowered by the forces of weathering and erosion.

Cinder cones are easily eroded because they are composed of unconsolidated materials. However, all volcanic structures will eventually be worn away. As erosion progresses, the rock occupying the vent is often more resistant and may remain standing above the terrain long after most of the cone has vanished. Shiprock, New Mexico, is thought to be such a feature, called a **volcanic neck** (see Figure 3.15). This structure, higher than many skyscrapers, is but one of many that protrude conspicuously from the red desert landscape of the southwestern United States.

Formation of Calderas

Earlier it was pointed out that some volcanoes have unusually large craters known as calderas. Some calderas are thought to form when the summit of a volcano collapses into the partially emptied magma chamber below (Figure 3.11). Crater Lake in Oregon occupies a depression 8–10 kilometers wide and up to 1300 meters deep (Figure 3.12). The creation of Crater Lake began about 7000 years ago when the volcano, later to be named Mount Mazama, put forth a violent ash eruption much like that of Vesuvius. However, this ancient eruption was on a much larger scale, extruding an estimated 50–70 cubic

FIGURE 3.10 St. Pierre as it appeared shortly after the eruption of Mount Pelée, 1902. (Reproduced from the collection of the Library of Congress)

kilometers of volcanic material. With the loss of support, 1500 meters of this once prominent 3600-meter cone collapsed. After the collapse, rainwater filled the caldera. Later activity built a small cinder cone called Wizard Island, which today provides a mute reminder of past activity.

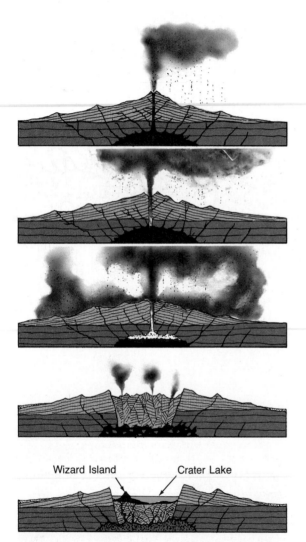

Wizard Island   Crater Lake

FIGURE 3.11 Sequence of events that formed Crater Lake, Oregon. About 7000 years ago, the summit of former Mount Mazama collapsed following a violent eruption which partly emptied the magma chamber. Subsequent eruptions produced the cinder cone called Wizard Island. Rainfall and groundwater contributed to form the lake. (After H. Williams, *The Ancient Volcanoes of Oregon,* p. 47. Courtesy of the University of Oregon)

Calderas of varying sizes are known to exist, the largest more than 20 kilometers (12 miles) across. Some, such as the calderas of Mauna Loa and Kilauea, clearly result from subsidence that occurred as supporting magma was diverted to a flank eruption. Others were formed by an explosive event that blasted the upper portion of the volcano away, as was the case at Mount St. Helens in 1980.

FISSURE ERUPTIONS

Although volcanic eruptions from a central vent are the most familiar, by far the largest amounts of volcanic material are extruded from cracks in the crust called **fissures**. Rather than building a cone, these long, narrow cracks distribute volcanic materials over a wide area. An extensive region in the northwestern United States known as the Columbia Plateau was formed in this manner. Here numerous fissure eruptions extruded very fluid basaltic lava. Successive flows, some 50 meters thick, buried the old landscape as they built a lava plain, which in some places is nearly a mile thick (Figure 3.13). The fluidity is evident, since some lava remained molten long enough to flow 150 kilometers from its source. The term **flood basalts** appropriately describes these waterlike flows.

When silica-rich magma is extruded from fissures, **pyroclastic flows** consisting largely of ash and pumice fragments usually result. When these pyroclastic materials are ejected, they move away from the vent at high speeds and may blanket extensive areas before coming to rest. Once deposited, the pyroclastic materials closely resemble lava flows.

Extensive pyroclastic flow deposits are found in many parts of the world and are most often associated with large calderas. Perhaps the best-known region of pyroclastic flows is the Yellowstone Plateau in northwestern Wyoming. Here a large magma body, rich in silica, still exists a few kilometers below the surface. Several times over the past two million years, fracturing of the rocks overlying the magma chamber has resulted in huge eruptions accompanied by the formation of calderas. In the northwestern portion of Yellowstone National Park, 27 fossil forests have been discovered, one resting upon another. During periods of inactivity, a forest developed upon the newly-formed volcanic surface, only

FIGURE 3.12 Crater Lake occupies a caldera about 10 kilometers (6 miles) in diameter. (Courtesy of the National Park Service)

to be covered by ash from the next eruptive phase. Fortunately, no eruption of this type has occurred in modern times.

VOLCANOES AND CLIMATE

The idea that explosive volcanic eruptions change the earth's climate was first proposed many years ago and is still regarded as an explanation for some aspects of climatic variability. Explosive eruptions emit huge quantities of gases and fine-grained debris into the atmosphere. The greatest eruptions are sufficiently powerful to inject material high into the stratosphere, where it spreads around the globe and remains for many months or even years. The basic premise is that this suspended volcanic material

(most importantly, droplets of sulfuric acid) will filter out a portion of the incoming solar radiation which, in turn, will lower air temperatures.

Perhaps the most notable cool period linked to a volcanic event is the "year without a summer" that followed the 1815 eruption of Mount Tambora in Indonesia. In many Northern Hemisphere locations, including New England, the abnormally cold spring and summer of 1816 were believed to be caused by the cloud of volcanic debris ejected from Tambora.

When Mount St. Helens erupted on May 18, 1980, there was almost immediate speculation about the possible effects of this event on our climate. Can an eruption such as this change our climate? Although spectacular, a single explosive volcanic eruption of the magnitude of Mount St. Helens occurs somewhere in the world every 2 to 3 years.

FIGURE 3.13 Basalt flows of the Columbia Plateau. (Photo by E. T. Jones, U.S. Geological Survey)

Studies of these events indicate that a very slight cooling of the lower atmosphere does occur. However, it is believed that the cooling is so slight, less than one-tenth of one degree Celsius, as to be inconsequential.

On April 4, 1982, El Chichón, a little-known volcano on Mexico's Yucatan peninsula, erupted. The cloud of debris and sulfur gases lofted into the atmosphere was huge, probably 20 times greater than the cloud from Mount St. Helens. Following the blast, scientists predicted a gradual lowering of temperatures in the Northern Hemisphere, perhaps as great as 0.3–0.5°C. Such a change is large enough to be distinguishable from normal temperature fluctuations but is probably too small to affect our lifestyles. Nevertheless, many scientists agree that such a hemispheric cooling could alter the general pattern of atmospheric circulation for a limited period. Such a change, in turn, could have an effect on the weather in some regions. However, the prediction of specific regional effects still presents a considerable challenge to atmospheric scientists.

INTRUSIVE IGNEOUS ACTIVITY

Volcanic eruptions can be among the most violent and spectacular events in nature and are therefore worthy of detailed study. Yet most magma is believed to be emplaced at depth. An understanding of intrusive igneous activity is therefore as important to geologists as the study of volcanic events. Figure 3.14 shows several types of intrusive igneous bodies that form when magma crystallizes within the earth's crust. Notice that some of these structures have a tabular shape, while others are quite massive. Also observe that some of these bodies cut across existing structures, such as layers of sedimentary rocks, while others form when magma is injected between sedimentary layers. Due to these differences, intrusive igneous bodies are generally classified according to their shape as either *tabular* or *massive,* and by their orientation with respect to the country (host) rock. Intrusive igneous bodies are said to be *discordant* if they cut across existing sedimentary beds and *concordant* if they form parallel to the existing sedimentary beds.

Intrusive igneous bodies occur in a great variety of sizes and shapes. **Dikes** are discordant masses produced when magma is injected into fractures that cut across rock layers. Once crystallized, these tabular structures have thicknesses ranging from less than one centimeter to more than one kilometer. The largest have lengths of one hundred kilometers or more. Dikes are often oriented vertically and represent pathways followed by molten rock which fed ancient lava flows. Frequently dikes are more resistant to weathering than the surrounding rock. When exposed, these dikes have the appearance of a wall as shown in Figure 3.15.

Sills are tabular bodies formed when magma is injected along sedimentary bedding surfaces (Figure 3.16). Horizontal sills are the most common although all orientations, even vertical, are known to exist where the strata have been tilted. Due to their relatively uniform thickness and large extent, sills are believed to form from very fluid magma. As we may expect, sills are most often composed of basaltic magma, which is typically quite fluid. The emplacement of a sill requires that the overlying sedimentary rock be lifted to a height equal to the thickness of the sill. Consequently, sills form only at rather shallow depths where the pressure exerted by the weight of overlying strata is relatively low.

INTRUSIVE IGNEOUS ACTIVITY

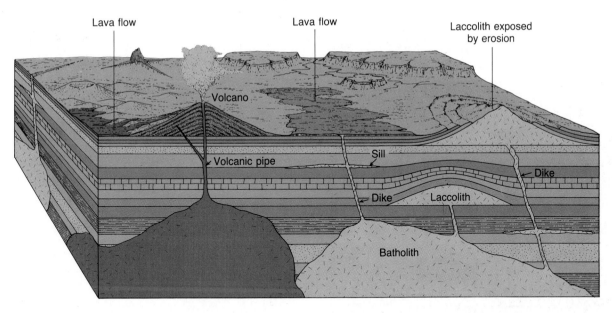

FIGURE 3.14 Block diagram illustrating intrusive and extrusive igneous structures.

FIGURE 3.15 The tabular structure in the foreground is a dike. In the background is Shiprock, New Mexico, the remnant of a pipe which once fed a volcano that has long since eroded away. (Courtesy of Ward's Natural Science Establishment, Inc., Rochester, N.Y.)

FIGURE 3.16 Taylor Glacier region, Antarctica. The dark horizontal band is a sill intruded into flat-lying sandstone. (Photo by W. B. Hamilton, U.S. Geological Survey)

Laccoliths are similar to sills because they form when magma is intruded between sedimentary layers in a near-surface environment. However, unlike sills, the magma that generates laccoliths is believed to be quite viscous. This thick, nonfluid magma collects as a lens-shaped mass that arches the overlying strata upward. Consequently, a laccolith can be detected because of the dome it creates at the surface even before the overlying rock is stripped away by erosional forces.

By far the largest intrusive igneous bodies are **batholiths**. The Idaho batholith, for example, encompasses an area of more than 40,000 square kilometers. These massive, discordant bodies are usually composed of rock types having mineral compositions near the granitic end of the igneous rock spectrum. Small batholiths can be relatively simple structures composed almost entirely of one rock type. However, studies of large batholiths have shown that they resulted from several distinct events that occurred over millions of years (see Figure 2.10). Batholiths frequently compose the cores of mountain systems. Here uplift and erosion have removed the surrounding rock to expose the resistant igneous body. Some of the highest mountain peaks, such as Mount Whitney in the Sierra Nevada, are carved from such a granitic mass. Large expanses of granitic rock are also exposed in the stable interiors of the continents, such as the Canadian Shield of North America. These relatively flat outcrops are believed to be the remnants of ancient mountains that erosion has long since leveled.

IGNEOUS ACTIVITY AND PLATE TECTONICS

The origin of magma has been a controversial topic in geology almost from the very beginning of the science. How do magmas of different compositions form? Why do volcanoes located in the deep-ocean basins primarily extrude basaltic lava, whereas those adjacent to oceanic trenches extrude mainly andesitic lava? Why does an area of igneous activity, commonly called the Ring of Fire, surround the Pacific Ocean? New insights gained from the theory of plate tectonics are providing answers to these questions. We will first examine the origin of magma and then look at the global distribution of volcanic activity as viewed from the model provided by plate tectonics.

Origin of Magma

We know that magma can be produced when rock is heated to its melting point. In a surface environment, rocks of granitic composition begin to melt at temperatures near 750°C (1400°F), whereas basaltic rocks must reach temperatures above 1000°C (1850°F) before melting will begin. One important difference exists between the melting of a substance that consists of a single compound, such as ice, and the melting of igneous rocks, which are mixtures of several minerals. Whereas ice melts at 0°C, most igneous rocks melt over a temperature range of a few hundred degrees. As a rock is heated, the first liquid to form will contain a higher percentage of the low-melting-point minerals than the original rock. If melting continues, the composition of the melt will steadily approach the overall composition of the rock from which it is derived. Most often, however, melting is not complete. This process, known as **partial melting**, produces most, if not all, magma.

A significant result of partial melting is the production of a melt with a higher silica content than the parent rock. Recall that basaltic rocks have a relatively low silica content and that granitic rocks have a much higher silica content. Consequently, magmas generated by partial melting are nearer the granitic end of the compositional spectrum than the parent material from which they formed. As we shall see, this idea will help us to understand the global distribution of the various types of volcanic activity.

What is the heat source to melt rock? One source is the heat liberated during the decay of radioactive elements that are thought to be concentrated in the upper mantle and crust. Workers in underground mines have long recognized that temperatures increase with depth.

If temperature were the sole factor to determine whether or not a rock melts, the earth would be a molten ball covered with only a thin, solid outer shell. However, pressure also increases with depth. Since rock expands when heated, extra heat is needed to melt buried rocks in order to overcome the effect of confining pressure. In general, an increase in the confining pressure causes an increase in the rock's melting point.

In nature, deeply buried rocks melt for one of two reasons. First, rocks melt when they are heated to their melting points. Second, without increasing temperature, a reduction in the confining pressure can lower the melting temperature sufficiently to trigger melting. Both processes are thought to play significant roles in magma formation.

Distribution of Igneous Activity

Most of the more than 600 active volcanoes that have been identified are located in the vicinity of convergent plate margins. Further, extensive volcanic activity occurs out of view along spreading centers of the oceanic ridge system. In this section we will examine three zones of volcanic activity and relate them to global tectonic activity. These active areas are found along the oceanic ridges, adjacent

to ocean trenches, and within the plates themselves (Figure 3.17).

Spreading Center Volcanism. The greatest volume of volcanic rock is produced along the oceanic ridge system, where sea-floor spreading is active (Figure 3.17). As the rigid lithosphere pulls apart, the pressure on the underlying rocks is lessened. This reduced pressure, in turn, lowers the melting point of the mantle rocks. Partial melting of these rocks (primarily peridotite) generates large quantities of basaltic magma that moves upward to fill the newly formed cracks.

Some of the molten basalt reaches the ocean floor, where it produces extensive lava flows or occasionally grows into a volcanic pile. Sometimes this activity produces a volcanic cone that rises above sea level as the island of Surtsey did in 1963 (Figure 3.18). Numerous submerged volcanic cones also dot the flanks of the ridge system and the adjacent deep-ocean floor. Many of these formed along

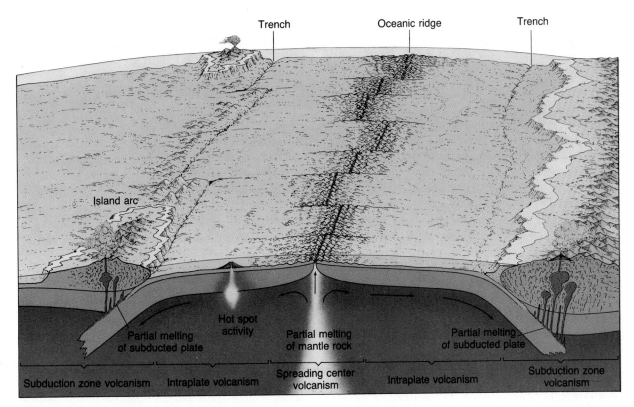

FIGURE 3.17 Three zones of volcanism. Two of these zones are plate boundaries, and the third includes areas within the plates themselves.

FIGURE 3.18 Surtsey emerged from the ocean just south of Iceland in 1963. (Courtesy of Sólarfilma)

the ridge crests and were moved away as new oceanic crust was created by the process of sea-floor spreading.

Subduction Zone Volcanism. Rocks with andesitic to granitic composition are confined to the continents and to volcanic island chains, such as the Aleutians, which lie along oceanic margins. Only very small amounts are found as part of the volcanoes in the deep-ocean basins. Further, most active volcanoes that extrude andesitic magma are found on continental areas or island arcs located adjacent to deep-ocean trenches. Recall that ocean trenches are sites where slabs of oceanic crust are bent and move downward into the upper mantle.

When cold oceanic lithosphere reaches depths of about 125 kilometers (80 miles), melting is believed to take place. The partial melting of these wet, sediment-laden basalts yields a magma of andesitic composition. After a sufficient quantity has melted, this magma buoys upward, because it is less dense than the surrounding rock. The volcanoes of the Andes Mountains, from which andesite obtains its name, are examples of this mechanism at work.

The *Ring of Fire* is associated with subduction and melting of the Pacific plate. The volcanoes in this very active zone primarily extrude magma having an intermediate silica content. The volcanoes of the Cascade Range in the northwestern United States, including Mounts St. Helens, Rainier, and Shasta, are all of this type.

Intraplate Volcanism. The processes that actually trigger volcanic activity within a rigid plate are difficult to establish. Activity such as in the Yellowstone region and other nearby areas produced rhyolitic lava, pumice, and ash flows, while extensive basaltic flows cover vast portions of our Northwest. Yet these rocks of greatly varying compositions actually overlie one another in several locations.

Because basaltic extrusions occur on the continents as well as within the ocean basins, the partial melting of upper mantle rocks is the most probable source for this activity. One proposal is that a small percentage of the rocks of the asthenosphere exists in the molten state. From these molten pockets, called *hot spots*, plumes of magma are thought to migrate upward where they often penetrate to the surface. Hot spots are believed to be located beneath Hawaii and Iceland, and may have formerly existed beneath the Columbia Plateau.

Generally lavas and ash of granitic composition are extruded from vents located landward of the continental margins. This suggests that remelting of the continental crust may be one mechanism for the formation of these silica-rich magmas. But what mechanism causes large quantities of continental material to melt? One proposal is that a thick segment of continental crust occasionally becomes situated over a plume of rising magma, that is, a hot spot. Rather than producing vast outpourings of basaltic lava as occurs at oceanic sites such as Hawaii, the magma from the rising plume is emplaced at depth. Here the incorporation and melting of sur-

rounding country rock results in the formation of a secondary, silica-rich magma which slowly migrates upward. Continued hot spot activity supplies heat to the rising mass, thereby aiding its ascent. The activity in the Yellowstone region may have resulted from just this type of activity.

Although the theory of plate tectonics has answered many of the questions which have plagued volcanologists for decades, many new questions have arisen; for example, Why does sea-floor spreading occur in some areas and not others? How do hot spots originate? These are just two of the many unanswered questions.

REVIEW QUESTIONS

1 What is the difference between magma and lava?

2 What three factors determine the nature of a volcanic eruption? What role does each play?

3 Why is a volcano fed by highly viscous magma likely to be a greater threat than a volcano supplied with very fluid magma?

4 Describe pahoehoe and aa lava.

5 List the main gases released during a volcanic eruption.

6 Describe each type of pyroclastic material.

7 Compare and contrast the main types of volcanoes (size, shape, eruptive style, and so forth).

8 Name one example of each of the three types of volcanoes.

9 Compare the formation of Hawaii with that of Parícutin.

10 How is a caldera different from a crater?

11 Describe the formation of Crater Lake.

12 What is Shiprock, New Mexico, and how did it form?

13 Describe each of the four intrusive features discussed in the text.

14 Why might a laccolith be detected at the earth's surface before being exposed by erosion?

15 What is the largest of all intrusive igneous bodies? Is it tabular or massive? Concordant or discordant (see Figure 3.14)?

16 Explain how most magma is thought to originate.

CHAPTER FOUR

WEATHERING AND SEDIMENTARY ROCKS

To the casual observer the face of the earth may appear to be without change, unaffected by time. For that matter, less than 200 years ago most people believed that mountains, lakes, and deserts were permanent features of an earth that was thought to be no more than a few thousand years old. Today, however, we know that mountains eventually succumb to weathering and erosion and are washed into the sea, lakes fill with sediment and vegetation or are drained by streams, and deserts come and go as relatively minor climatic changes occur.

The earth is indeed a dynamic body. Volcanic and tectonic activities* are elevating parts of the earth's surface, while opposing processes are continually removing materials from higher elevations and moving them to lower elevations. The latter processes include:

1 **Weathering**—disintegration and decomposition of rock at or near the surface of the earth.
2 **Erosion**—incorporation and transportation of material by a mobile agent, usually water, wind, or ice.
3 **Mass wasting**—transfer of rock material downslope under the influence of gravity.

We will first turn our attention to the process of weathering and the products generated by this activity. However, weathering cannot be easily separated from the other two processes because as weathering breaks rocks apart, it facilitates the movement of rock debris by erosion and mass wast-

---

*Tectonic activities** are activities that result in the deformation of the earth's crust.

Bryce Canyon National Park. Weathering accentuates differences in rocks to produce some of our most spectacular scenery. (Photo by Stephen Trimble)

ing. On the other hand, the transport of material by erosion and mass wasting furthers the disintegration and decomposition of rock.

WEATHERING

All materials are susceptible to weathering. Consider, for example, the fabricated product concrete, which closely resembles the sedimentary rock called conglomerate (see Figure 4.10C). A newly-poured concrete sidewalk has a smooth, fresh, unweathered look. However, not many years later the same sidewalk will appear chipped, cracked, and rough, with pebbles exposed at the surface. If a tree is nearby, its roots may heave and buckle the concrete as well. The same natural processes that eventually destroy a concrete sidewalk also act to disintegrate rock.

Why does rock weather? Simply, weathering is the response of earth materials to a changing environment. For instance, after millions of years of uplift and erosion, the rocks overlying a large intrusive igneous body may be removed, exposing it at the surface. This mass of crystalline rock, which formed in a high-temperature, high-pressure environment perhaps several kilometers below ground, is now subjected to a very different and comparatively hostile surface environment. In response, this rock mass will gradually change until it is once again in equilibrium, or balance, with its new environment. This transformation of rock is what we call weathering.

In the following sections we will discuss the various types of mechanical and chemical weathering. Although we will consider these two processes separately, keep in mind that they usually work simultaneously in nature.

Mechanical Weathering

When a rock undergoes **mechanical weathering** it is broken into smaller and smaller pieces, each

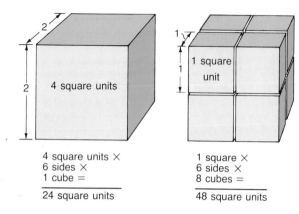

4 square units ×
6 sides ×
1 cube =
———————
24 square units

1 square ×
6 sides ×
8 cubes =
———————
48 square units

FIGURE 4.1 Mechanical weathering increases the surface area available for chemical attack.

retaining the characteristics of the original material. The end result is many small pieces from a single large one. Figure 4.1 shows that breaking a rock into smaller pieces increases the surface area available for chemical attack. An analogous situation occurs when sugar is added to a liquid. In this situation, a cube of sugar will dissolve much slower than an equal volume of granules because of the vast difference in surface area. Hence, by breaking rocks into smaller pieces, mechanical weathering increases the amount of surface area available for chemical weathering.

In nature four important physical processes lead to the fragmentation of rock: frost wedging, expansion resulting from unloading, thermal expansion, and organic activity.

Frost Wedging. Alternate freezing and thawing of water is one of the most important processes of mechanical weathering. Water has the unique property of expanding about 9 percent as it freezes. This increase in volume occurs because as water solidifies, the water molecules arrange themselves into a very open crystalline structure. As a result, when water freezes it expands and exerts a tremendous outward force. This can be verified by filling a container with water and freezing it. If sufficient volume does not exist in the container, it will shatter.

In nature, water works its way into cracks or voids in rock and, upon freezing, expands and wedges the rock apart. This process is appropriately called **frost**

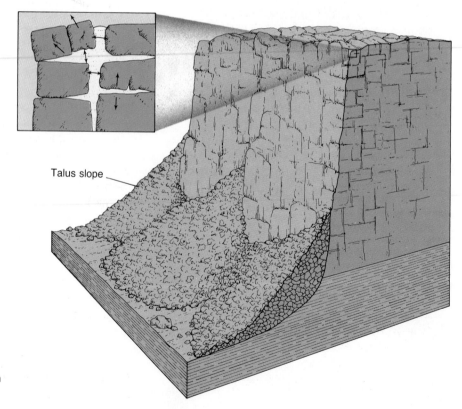

Talus slope

FIGURE 4.2 Frost wedging. As water freezes it expands, exerting a force great enough to break rock.

wedging (Figure 4.2). Frost wedging is very pronounced in mountainous regions in the middle latitudes where a daily freeze-thaw cycle often exists. Here sections of rock are wedged loose and may tumble into large piles called **talus slopes** that often form at the base of steep rock outcrops (Figure 4.2).

Unloading. When large igneous bodies, particularly those composed of granite, are exposed by erosion, concentric slabs begin to break loose. The process generating these onionlike layers is called **sheeting** and is thought to occur, at least in part, because of the great reduction in pressure when the overlying rock is stripped away. Accompanying the unloading, the outer layers expand more than the rock below, and thus separate from the rock body. The fractures typically develop parallel to the surface topography and give the exhumed igneous body a domed shape. Continued weathering eventually causes the slabs produced by sheeting to separate and spall off these large structures known as **exfoliation domes**. Excellent examples of exfoliation domes are Stone Mountain, Georgia, and Half Dome and Liberty Cap in Yosemite National Park (Figure 4.3).

Mine shafts provide us with a view of how rocks behave once the confining pressure is removed. Large rock slabs have been known to explode off the walls of newly cut mine shafts because of the reduced pressure. Evidence of this type, plus the fact that fracturing occurs parallel to the floor of a rock quarry when large blocks are removed, strongly supports the process of unloading as the cause of sheeting.

Although many fractures are created by expansion, others are produced by contraction during the cystallization of magma and still others by tectonic forces during mountain building. Fractures produced by these activities generally form a definite pattern and are called *joints*. Joints are important rock structures which allow water to penetrate to depth and start the process of weathering long before the rock reaches the surface.

Thermal Expansion. The daily cycle of temperature change is thought to weaken rocks, particularly in hot, dry regions where daily variations often exceed 30°C (54°F). Heating a rock causes it to expand, and cooling causes it to contract. Although this process was once thought to be of major importance in the disintegration of rock, laboratory experiments have not substantiated this idea. In one test rocks were heated to temperatures much higher than those normally experienced on the earth's surface and then were cooled. This procedure was repeated many times to simulate hundreds of years of weathering, but the rocks showed little apparent change. Additional work is needed before the impact of daily temperature variations on weathering can be determined.

Organic Activity. Weathering is also accomplished by the activities of organisms, including plants, burrowing animals, and humans. Plant roots in search of minerals and water grow into fractures, and as the roots grow they wedge the rock apart. Burrowing animals further break down rock by moving fresh material to the surface, where physical and chemical processes can more effectively attack it. Further, decayed organisms produce acids, which contribute to chemical weathering. Where rock has been blasted in search of minerals or for road construction, the impact is quite noticeable, but on a worldwide scale human activity probably ranks behind burrowing animals in earth-moving accomplishments.

Chemical Weathering

Chemical weathering involves the complex processes that alter the internal structures of minerals

FIGURE 4.3 Summit of Half Dome, an exfoliation dome in Yosemite National Park. (Photo by Stephen Trimble)

by removing and/or adding elements. Water is by far the most important agent of chemical weathering. Although pure water is nonreactive, a small amount of dissolved materials is generally all that is needed to activate it. Oxygen dissolved in water will *oxidize* some materials. For example, when an iron nail is found in the soil it will have a coating of rust (iron oxide), and if the time of exposure has been long, the nail will be so weak that it can be broken as easily as a toothpick. When rocks containing iron-rich minerals oxidize, a yellow to reddish brown rust will appear on the surface.

Carbon dioxide (CO_2) dissolved in water (H_2O) forms carbonic acid (H_2CO_3), the same weak acid produced when soft drinks are carbonated. Rain dissolves some carbon dioxide as it falls through the atmosphere, and additional amounts released by decaying organic matter are acquired as the water percolates through the soil. Carbonic acid ionizes to form the very reactive hydrogen ion (H^+) and the bicarbonate ion (HCO_3^-).

To illustrate how rock chemically weathers when attacked by carbonic acid, we will consider the weathering of granite, the most abundant continental rock. Recall that granite consists mainly of quartz and potassium feldspar. The weathering of the potassium feldspar component of granite takes place as follows:

$$2KAlSi_3O_8 + 2(H^+ + HCO_3^-) + H_2O \longrightarrow$$
$$\underset{\text{potassium feldspar}}{} \quad \underset{\text{carbonic acid}}{} \quad \underset{\text{water}}{}$$

$$Al_2Si_2O_5(OH)_4 + 2KHCO_3 + 4SiO_2$$
$$\underset{\text{clay mineral}}{} \quad \underset{\text{potassium bicarbonate}}{} \quad \underset{\text{silica}}{}$$

In this reaction, the hydrogen ions (H^+) attack and replace potassium ions (K^+) in the feldspar structure, thereby disrupting the crystalline network. Once removed, the potassium is available as a nutrient for plants or becomes the soluble salt potassium bicarbonate ($KHCO_3$), which may be incorporated into other minerals or carried to the ocean in dissolved form by streams. The most abundant by-products of the chemical breakdown of feldspar are residual clay minerals. Because clay minerals are the end product of weathering, they are very stable under surface conditions. Consequently, clay minerals make up a high percentage of the inorganic material in soil. The most abundant sedimentary rock, shale, is also composed of clay minerals. Some of the

silica removed from the feldspar structure will go into solution with the groundwater (water beneath the earth's surface). This silica will eventually precipitate to produce nodules of chert or flint, fill in the pore spaces between sediment grains, or be carried to the ocean, where microscopic animals such as diatoms will remove it to build silica shells.

To summarize, the weathering of potassium feldspar generates a residual clay mineral, a soluble salt (potassium bicarbonate), and some silica which enters into solution.

Quartz, the other main component of granite, is very resistant to chemical weathering; hence, it remains substantially unaltered when attacked by weakly acidic solutions. As a result, when granite weathers, the feldspar crystals dull and slowly turn to clay, releasing the once-interlocked quartz grains, which still retain their fresh, glassy appearance. Although some quartz remains in the soil, much is transported to the sea, where it becomes the main constituent of sandy beaches and in time may be lithified to form the sedimentary rock sandstone.

Table 4.1 lists the weathered products of some of the more common silicate minerals. Remember that silicate minerals make up most of the earth's crust and that these minerals are essentially composed of only eight elements. When chemically weathered, the silicate minerals yield sodium, calcium, potassium, and magnesium ions that form soluble products which may be removed by groundwater. The element iron combines with oxygen, producing rela-

TABLE 4.1 Products of weathering.

Mineral	Residual Products	Material in Solution
Quartz	Quartz grains	Silica
Feldspars	Clay minerals	Silica K^+, Na^+, Ca^{2+}
Hornblende	Clay minerals Limonite Hematite	Silica Ca^{2+}, Mg^{2+}
Olivine	Limonite Hematite	Silica Mg^{2+}

tively insoluble iron oxides, most notably hematite and limonite, which give soil a reddish-brown or yellowish color. Under most conditions the three remaining elements, aluminum, silicon, and oxygen, join with water to produce residual clay minerals. However, even the highly insoluble clay minerals are very slowly removed by subsurface water.

As chemical weathering alters the internal structure of minerals, physical changes are also occurring.

FIGURE 4.4 Successive shells are loosened as the weathering process continues to penetrate ever deeper into the rock. (Photo by Kenneth Hasson)

For example, when angular rock fragments produced by regular joint systems are attacked by chemical weathering, the fragments take on a spherical shape. Any process that tends to give the weathered rock a spherical shape is called **spheroidal weathering**. One type of spheroidal weathering resembles sheeting, except that much smaller rocks are involved. As the minerals in the rock weather to clay, they increase in size because of the addition of water into their structure. This increase in bulk exerts an outward force that is thought to cause concentric layers of rock to break loose and fall off (Figure 4.4). Hence chemical weathering can produce forces great enough to cause mechanical weathering.

Rates of Weathering

The rate at which rock weathers depends on many factors. We have already seen how the particle size influences the rate of weathering. The mineral make-up of a rock is also a very important factor, which can be demonstrated by comparing headstones carved from different rock types. Headstones made of granite are relatively resistant to chemical weathering as we can see by examining the inscription on the headstone in Figure 4.5A. This is not true of the marble headstone (Figure 4.5B), which shows signs

FIGURE 4.5 An examination of headstones reveals the rate of chemical weathering on diverse rock types. The granite headstone (left) was erected a few years before the marble headstone (right), whose inscription date of 1892 is nearly illegible. (Photos by E. J. Tarbuck)

A.

B.

of extensive chemical alteration over a relatively short period. Marble is composed of calcite, which readily dissolves even in a weakly acidic solution.

The silicates, the most abundant mineral group, weather in essentially the same order as their order of crystallization. The minerals that crystallize first form at much higher temperatures than those that crystallize last. Consequently, these early-forming minerals are not as stable at the earth's surface, where the temperatures and pressures are radically different from those in the environment where they form. By examining Bowen's reaction series (see Figure 2.6), we can see that olivine crystallizes first and is therefore the least resistant to weathering, whereas quartz, which crystallizes last, is the most resistant.

Climatic factors, particularly temperature and moisture, are of primary significance to the rate of rock weathering. The optimum environment for chemical weathering is a combination of warm temperatures and abundant moisture. In polar regions and at high altitudes, chemical weathering is ineffective because frigid temperatures keep the available moisture locked up as ice, while in arid regions there is insufficient moisture to foster much chemical weathering. A classic example of how climate affects the rate of weathering was provided when Cleopatra's Needle, a granite obelisk, was moved from Egypt to New York City. After withstanding approximately 3500 years of exposure in the dry climate of Egypt, the hieroglyphics were almost completely removed from the windward side in less than 75 years in the wet, chemical-laden air of New York City (Figure 4.6).

The sum of these factors determines the type and rate of rock weathering for a given region. However, there is generally enough variation, even within a relatively small area, for the rocks to exhibit some differential weathering. Differential weathering and subsequent erosion are responsible for many unusual and often spectacular landforms. Included are features such as natural bridges like those found in Arches National Park and sculptured rock pinnacles such as those found in Bryce Canyon National Park (see chapter-opening photo).

SOIL

Soil has accurately been called "the bridge between life and the inanimate world." All life owes its exist-

FIGURE 4.6 Chemical weathering of Cleopatra's Needle, a granite obelisk. **A.** Before it was removed from Egypt. (Courtesy of The Metropolitan Museum of Art). **B.** After a span of 75 years in New York City's Central Park. After having survived intact for about 35 centuries in Egypt, the windward side has been almost completely defaced in less than a century. (Courtesy of New York City Parks)

A.

B.

ence to a dozen or so elements that must ultimately come from the earth's crust. Once weathering and other processes create soil, plants carry out the intermediary role of assimilating the necessary elements and making them available to animals and people.

With few exceptions, the earth's land surface is covered by **regolith**, the layer of rock and mineral fragments produced by weathering. Some would call this material soil, but soil is more than an accumulation of weathered debris. **Soil** is a combination of mineral and organic matter, water, and air—that portion of the regolith that supports the growth of plants. Although the proportions vary, the major components do not (Figure 4.7). About one-half of the total volume of a good quality surface soil is a mixture of disintegrated and decomposed rock (mineral matter) and **humus**, the decayed remains of animal and plant life (organic matter). The remaining half consists of pore spaces, where air and water circulate.

Although the mineral portion of the soil is usually much greater than the organic portion, humus is a very significant component. In addition to being an important source of plant nutrients, humus enhances the soil's ability to retain water. Since plants require air and water to live and grow, the portion of the soil consisting of pore spaces that allow for the circulation of these fluids is as vital as the solid soil

constituents. Soil water is not "pure" water; instead, it is a complex solution containing many soluble nutrients. Soil water not only provides the necessary moisture for the chemical reactions that sustain life, it also supplies plants with nutrients in a form they can use. The pore spaces not filled with water contain air. This air is the source of necessary oxygen and carbon dioxide for most microorganisms and plants that live in the soil.

Controls of Soil Formation

Soil is the product of the complex interplay of several factors, including parent material, time, climate, plants and animals, and slope. Although all of these factors are interdependent, it will be helpful to examine their roles separately.

Parent Material. The **parent material** from which a soil has evolved may either be underlying bedrock or a layer of unconsolidated deposits. Soils formed on bedrock are termed **residual soils**, whereas those developed on unconsolidated sediment are called **transported soils** (Figure 4.8).

The nature of the parent material influences soils in two ways. First, the type of parent material to some degree will affect the rate of weathering and thus the rate of soil formation. For example, the mineral composition of the parent material will influence the rate of chemical weathering. Also, since unconsolidated deposits are already partly weathered, soil development on such material will likely progress more rapidly than when bedrock is the parent material. Second, the chemical makeup of the parent material will affect the fertility of the soil. If it lacks the necessary elements for plant growth, its usefulness is obviously diminished.

At one time it was thought that the parent material was the primary factor causing differences among soils. Today soil scientists realize that other factors, especially climate, are more important. In fact, it has been found that similar soils are often produced from different parent materials and that dissimilar soils have developed from the same parent material. Such discoveries reinforce the importance of the other soil-forming factors.

Time. If weathering has been going on for a comparatively short time, the character of the parent

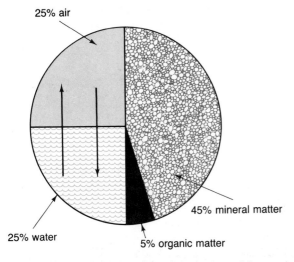

FIGURE 4.7 Composition (by volume) of a soil in good condition for plant growth. Although the percentages vary, each soil is composed of mineral and organic matter, water, and air.

25% air

45% mineral matter

25% water

5% organic matter

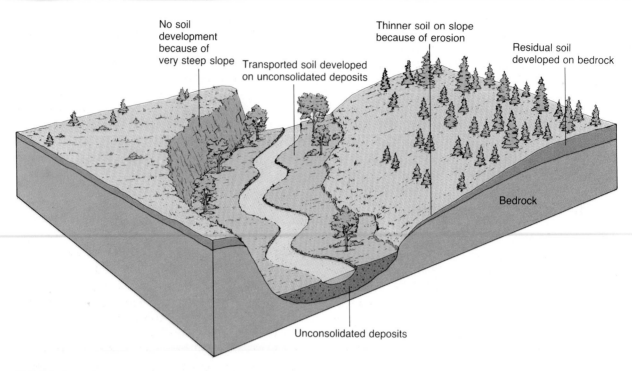

No soil development because of very steep slope

Transported soil developed on unconsolidated deposits

Thinner soil on slope because of erosion

Residual soil developed on bedrock

Bedrock

Unconsolidated deposits

FIGURE 4.8 The parent material for residual soils is the underlying bedrock, whereas transported soils form on unconsolidated deposits. Also note that soils are thinner, or nonexistent, on the slopes.

material determines to a large extent the characteristics of the soil. As the weathering process continues, the influence of parent material on soil is overshadowed by the other soil-forming factors. Therefore time is an important control of soil formation. It is not possible to list the length of time required for various soils to evolve, because the soil-forming processes act at varying rates under different circumstances. However, it is safe to say that the longer a soil has been forming, the thicker it becomes and the less it resembles the parent material from which it formed.

Climate. Climate is considered to be the most important control of soil formation. It determines whether chemical or mechanical weathering will predominate and also greatly influences the rate and depth of weathering. For instance, a hot and wet climate may produce a thick layer of chemically weathered soil in the same amount of time that a cold and less humid climate produces a thin mantle

of mechanically weathered debris. Furthermore, the amount of precipitation influences the degree to which various materials are removed from the soil, thereby affecting soil fertility. Finally, climatic conditions are an important control on the type of plant and animal life present.

Plants and Animals. The chief function of plants and animals is to supply organic matter to the soil. Certain bog soils are composed almost entirely of organic matter, while desert soils may contain only a very small percentage. Although the quantity of organic matter present varies substantially among soils, no soil completely lacks it.

The primary source of organic matter is plants, although animals and the uncountable numbers of microorganisms also contribute. When organic matter is decomposed, it supplies important nutrients for plants, as well as food for the animals and microorganisms living in the soil. Consequently, soil fertility is in part related to the amount of organic matter present. Furthermore, the decay of plant and animal remains causes various organic acids to form. These complex acids hasten the weathering process. Organic matter has a high water-holding ability and thus aids water retention in a soil.

Microorganisms, including fungi, bacteria, and the single-celled protozoa, play the active role in the decay of plant and animal remains. The end product is humus, a material that no longer resembles the plants and animals from which it formed. In addition, certain microorganisms aid soil fertility because they have the ability to *fix* (change) atmospheric nitrogen into soil nitrogen.

Earthworms and other burrowing animals act to mix the mineral and organic portions of a soil. Earthworms, for example, feed on the organic matter in the soil and thoroughly mix soils in which they live, often moving and enriching many tons per acre each year. Burrows and holes also aid the passage of water and air through the soil.

Slope. Slope significantly affects the amount of erosion and the water content of soil. On steep slopes soils are often poorly developed. In such situations the quantity of water soaking in is slight, and as a result, the moisture content of the soil may not be sufficient for vigorous plant growth. Further, because of accelerated erosion on steep slopes, the soils there are thin, or in some cases, nonexistent (Figure 4.8). On the other hand, poorly drained and waterlogged soils found in bottomlands have a much different character. Such soils are usually very thick and very dark, the dark color resulting from the large quantity of organic matter that accumulates because saturated conditions retard the decay of vegetation. The optimum slope for soil development is a flat-to-undulating, upland surface. Here we find good drainage, minimum erosion, and sufficient infiltration of water into the soil.

Slope orientation, the direction the slope faces, is another aspect worthy of mention. In the mid-latitudes of the Northern Hemisphere a south-facing slope will receive a great deal more sunlight than a north-facing slope. In fact, a steep north-facing slope may receive no direct sunlight at all. The difference in the amount of solar radiation received will cause differences in soil temperature and moisture, which in turn may influence the nature of the vegetation and the character of the soil.

Although this section has dealt separately with each of the soil-forming factors, remember that all of them work together to form soil. No single factor is responsible for a soil being as it is, but rather the combined influence of parent material, time, climate, plants and animals, and slope.

The Soil Profile

Since soil-forming processes operate from the surface downward, variations in composition, texture, structure, and color gradually evolve at varying depths. These vertical differences, which usually become more pronounced as time passes, divide the soil into zones or layers known as **horizons**. If you were to dig a trench in soil, you would see that its walls are layered. Such a vertical section through all of the soil horizons constitutes the **soil profile** (Figure 4.9).

Four basic horizons are identified and from top to bottom are designated as *O*, *A*, *B*, and *C*, respectively. The three upper layers may be further divided.

Unlike the layers beneath it, which consist mainly of mineral matter, the *O* horizon consists largely of organic material. The upper portion of this horizon is primarily plant litter such as loose leaves and other organic debris that are still recognizable. By contrast, the lower portion of the *O* horizon is made up of partly decomposed organic matter (humus) in which plant structures can no longer be identified.

Underlying the organic-rich *O* horizon is the *A* horizon. This zone is largely mineral matter, yet biological activity is high and humus is generally present— up to 30 percent in some instances. As water percolates downward from the surface through the *A* horizon, finer particles are carried away with it. This washing out of fine soil components is termed **eluviation**. As a consequence of eluviation, the texture of the *A* horizon gradually becomes coarser, because a portion of the fine particles is removed. Water percolating downward also dissolves soluble inorganic soil components and carries them to deeper zones. This depletion of soluble materials from the upper soil is termed **leaching**.

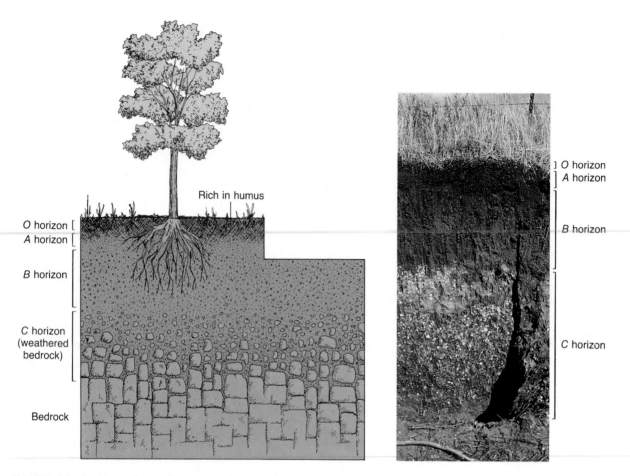

Rich in humus

O horizon [
A horizon [

B horizon

C horizon
(weathered
bedrock)

Bedrock

O horizon
A horizon

B horizon

C horizon

FIGURE 4.9 Mature soils are characterized by a series of horizontal layers called horizons, which comprise the soil profile.

Immediately below the A horizon is the B horizon, or *subsoil*. Much of the material removed from the A horizon by eluviation is deposited in the B horizon, which is often referred to as the *zone of accumulation*. The accumulation of fine clay particles derived from the A horizon enhances water retention in the subsoil. However, in extreme cases, clay accumulation can form an extremely dense and impermeable layer called *hardpan*. Since the B horizon has an intermediate position in the soil profile, it may be considered, at least in part, a transitional zone. For example, living organisms and organic matter are more abundant in the B than in the C horizon, but considerably less so than in the A horizon. The O, A, and B horizons together constitute the **solum**, or "true soil." It is in the solum that the soil-forming pro-

cesses are active and that living roots and other plant and animal life are largely confined.

Below the solum is the C horizon, a layer characterized by partially altered parent material and little if any organic matter. While the parent material may be so dramatically altered in the solum that its original character is not recognizable, it is easily identifiable in the C horizon.

The boundaries between soil horizons may be very sharp, or the horizons may blend gradually from one to another. Furthermore, some soils lack horizons altogether. Such soils are called *immature* because soil building has been going on for only a short time. Immature soils are also characteristic of steep slopes where erosion continually strips away the soil, preventing full development.

Soil Types

In the discussion which follows we shall briefly examine some common soil types. As you read, notice that the characteristics of each of the soil types are primarily manifestations of the prevailing climatic conditions. A summary of the characteristics of the soils discussed in this section is provided in Table 4.2.

The term **pedalfer** gives a clue to the basic characteristic of this soil type. The word is derived from the Greek **ped**on, meaning "soil," and the chemical symbols **Al** (aluminum) and **Fe** (iron). Pedalfers are characterized by an accumulation of iron oxides and aluminum-rich clays in the B horizon. In mid-latitude areas where the annual rainfall exceeds 63 centimeters (25 inches) most of the soluble materials, such as calcium carbonate, are leached from the soil and carried away by underground water. The less soluble iron oxides and clays are carried from the A horizon and deposited in the B horizon, giving it a brown to red-brown color. These soils are best developed under forest vegetation where large quantities of decomposing organic matter provide the acid conditions necessary for leaching. In the United States pedalfers are found east of a line extending from northwestern Minnesota to south-central Texas.

Pedocal is derived from the Greek **ped**on, meaning "soil," and the first three letters of **cal**cite (calcium carbonate). As the name implies, pedocals are characterized by an accumulation of calcium carbonate. This soil type is found in the drier western United States in association with grassland and brush vegetation. Here rainwater percolating through the soil often evaporates before it can remove soluble materials, chiefly calcium carbonate. The result is a whitish accumulation called *caliche*. In addition, since chemical weathering is less intense in drier areas, pedocals generally contain a smaller percentage of clay minerals than pedalfers.

In the hot wet climates of the tropics, soils called **laterites** may develop. Since chemical weathering is intense under such climatic conditions, these soils are usually deeper than soils developing over a similar period of time in the mid-latitudes. Not only does

TABLE 4.2 Summary of soil types.

Climate	Temperate humid (>63 cm rainfall)	Temperate dry (>63 cm rainfall)	Tropical (heavy rainfall)		Extreme arctic or desert
Vegetation	Forest	Grass and brush	Grass and trees		Almost none, so no humus develops
Typical Area	Eastern U.S.	Western U.S.			
Soil Type	Pedalfer	Pedocal	Laterite		
Topsoil	Sandy; light colored; acid	Commonly enriched in calcite, whitish color	Zones not developed	Enriched in iron (and aluminum); brick red color	No real soil forms because there is no organic material. Chemical weathering is very slow
Subsoil	Enriched in aluminum, iron, and clay; brown color	Enriched in calcite; whitish color		All other elements removed by leaching	
Remarks	Extreme development in conifer forests, because abundant humus makes groundwater very acid. Produces light gray soil because of removal of iron	Caliche is name applied to the accumulation of calcite	Apparently bacteria destroy humus, so no acid is available to remove iron		

leaching remove the soluble materials such as calcite, but the great quantities of percolating water also remove much of the silica, with the result that oxides of iron and aluminum become concentrated in the soil. Iron gives soil a distinctive red color. When dried, laterites are very hard. In fact, some people use this soil for making bricks. If the parent rock contained little iron, the product of weathering is an aluminum-rich accumulation called *bauxite.* Bauxite is the primary ore of aluminum.

Since bacterial activity is very high in the tropics, laterites contain practically no humus. This fact, coupled with the highly leached and bricklike nature of these soils, make laterites poor for growing crops. The infertility of these soils has been borne out repeatedly in tropical countries where cultivation has been expanded into such areas.

In cold or dry climates soils are generally very thin and poorly developed. The reasons for this are fairly obvious. Chemical weathering progresses very slowly in such climates, and the scanty plant life yields very little organic matter.

SEDIMENTARY ROCKS

The products of mechanical and chemical weathering constitute the raw materials for sedimentary rocks. The word *sedimentary* indicates the nature of these rocks, for it is derived from the Latin *sedimentum,* which means "settling," a reference to solid material settling out of a fluid. Most but not all sediment is deposited in this fashion. Weathered debris is constantly being swept from bedrock, carried away, and eventually deposited in lakes, river valleys, seas, and countless other places. The particles in a desert sand dune, the mud on the floor of a swamp, the gravels in a stream bed, and even household dust are examples of this never-ending process. Since the weathering of bedrock and the transport and deposition of the weathering products are continuous, sediment is found almost everywhere. As piles of sediment accumulate, the materials near the bottom are compacted. Over long periods, these sediments are cemented together by mineral matter deposited in the spaces between particles to form solid rock.

Geologists estimate that sedimentary rocks account for only about 5 percent (by volume) of the earth's outer 16 kilometers (10 miles). However, the importance of this group of rocks is far greater than this percentage would imply. If we were to sample the rocks exposed at the earth's surface, we would find that the great majority are sedimentary. Indeed, about 75 percent of all rock outcrops on the continents are sedimentary. Therefore, we may think of sedimentary rocks as comprising a relatively thin and somewhat discontinuous layer in the uppermost portion of the crust. This fact is readily understood when we consider that sediment accumulates at the surface of the earth.

Since sediments accumulate at the earth's surface, the rock layers that they eventually form contain evidence of past events at the surface. By their very nature, sedimentary rocks contain within them indications of past environments in which their particles were deposited and in some cases, clues to the mechanisms involved in their transport. Furthermore, it is sedimentary rocks that contain fossils, which are vital tools in the study of the geologic past. Thus, it is largely from this group of rocks that geologists must reconstruct the details of earth history.

Finally, it should be mentioned that many sedimentary rocks are very important economically. Coal, for example, is classified as a sedimentary rock, whereas our other major energy resources, petroleum and natural gas, are found in association with sedimentary rocks. Still others represent major sources of iron, aluminum, manganese, and fertilizer as well as numerous materials essential to the construction industry.

TYPES OF SEDIMENTARY ROCKS

Materials accumulating as sediment have two principle sources. First, sediments may be accumulations of materials that originate and are transported as solid particles derived from both mechanical and chemical weathering. Deposits of this type are termed *detrital* and the sedimentary rocks that they form are called **detrital sedimentary rocks**. The second major source of sediment is soluble material produced largely by chemical weathering. When these dissolved substances are precipitated by either inorganic or organic processes, the material is known as chemical sediment and the rocks formed from it are called **chemical sedimentary rocks**.

Detrital Sedimentary Rocks

Though a wide variety of minerals and rock fragments may be found in detrital rocks, clay minerals and quartz are the chief constituents of most sedimentary rocks in this category. Recall that clay minerals are the most abundant product of the chemical weathering of silicate minerals, especially the feldspars. Clays are fine-grained minerals with sheetlike crystalline structures similar to the micas. The other common mineral, quartz, is abundant because it is very resistant to chemical weathering. Thus, when igneous rocks such as granite are attacked by weathering processes, individual quartz grains are freed.

Other common minerals in detrital rocks are the feldspars and micas. Since chemical weathering rapidly transforms these minerals into new substances, their presence in a sedimentary rock indicates that mechanical rather than chemical weathering was responsible for creating the sediment.

Particle size is the primary basis for distinguishing among various detrital sedimentary rocks. Table 4.3 presents the size categories for particles making up detrital rocks. Note that in this context the term *clay* refers only to a particular size and not to the minerals of the same name. Although most clay minerals are of clay size, not all clay-sized sediment consists of clay minerals.

The size of the particles in a detrital rock can often be related to the energy of the transporting medium. Currents of water or air sort the particles by size; the stronger the current, the larger the particle size carried. Gravels, for example, are moved by swiftly flowing rivers as well as by landslides and glaciers. Less energy is required to transport sand, thus it is common to such features as windblown dunes, as well as some river deposits and beaches. Since silts and clays settle very slowly, accumulations of these materials are generally associated with the quiet waters of a lake, lagoon, swamp, or marine environment.

Shale. *Shale* is a sedimentary rock consisting of silt- and clay-sized particles (Figure 4.10A). These fine-grained detrital rocks account for an estimated 70 percent of all sedimentary rocks. The particles in these rocks are so small that they cannot be readily identified without great magnification. Since the sediments are not only microscopic but tend to be flat or tabular as well, they usually become tightly packed. As a result there is very little space through which solutions containing cementing material can circulate. Therefore, shales are commonly not well cemented and readily crumble. Although shale is very abundant, it is still the least well known common sedimentary rock. Shale does not form prominent outcrops as do sandstone and other common sedimentary rocks. In addition, shale weathers easily and usually forms a cover of soil that hides the unweathered rock below.

The word *shale* is often applied to all fine-grained detrital rocks, yet many geologists have a more restricted use of the term. In this more restricted usage, shale must exhibit the ability to split into thin layers along well-developed, closely spaced planes. If the rock breaks into chunks or blocks, the name *mudstone* is applied.

Sandstone. *Sandstone* is the name given rocks when sand-sized grains predominate (Figure 4.10B). After shale, sandstone is the most abundant sedimentary rock, accounting for approximately 20 percent of the total group. In most sandstones, quartz is the predominant mineral. When this is the case, the rock may simply be called *quartz sandstone*. If the sandstone contains appreciable quantities of feldspar, the rock is called *arkose*. The presence of abundant feldspar is an indication that the sediment was subjected to little chemical weathering. Finally, a third type of sandstone is known as *graywacke*. In addition to containing quartz and feldspar, this dark-

TABLE 4.3 Particle size classification for detrital rocks.

Size Range (millimeters)	Particle Name	Common Sediment Name	Detrital Rock
>256 64–256 4–64 2–4	Boulder Cobble Pebble Granule	Gravel	Conglomerate or breccia
1/16–2	Sand	Sand	Sandstone
1/256–1/16 <1/256	Silt Clay	Mud	Shale or mudstone

FIGURE 4.10 Common detrital sedimentary rocks. **A.** Shale. **B.** Sandstone.
C. Conglomerate. **D.** Breccia. (Photos by E. J. Tarbuck)

colored rock contains abundant angular rock fragments and clay. Because its particles are characteristically poorly sorted, graywacke is often referred to
as "dirty" sandstone.

Conglomerate. *Conglomerate* consists largely of
gravels (Figure 4.10C). As Table 4.3 indicates, these
particles may range in size from large boulders to
particles as small as garden peas. The large particles
in a conglomerate are commonly rock fragments.
Usually the openings between the gravel particles
are filled with mud and sand, then the entire mass is
cemented into a hard rock. If the large particles are

angular rather than rounded, the rock is called *breccia* (Figure 4.10D).

Chemical Sedimentary Rocks

In contrast to detrital rocks, which form from the
solid products of weathering, chemical sediments
derive from material that is carried in solution to
lakes and seas. This material does not remain dissolved in the water indefinitely, however. Rather,
some of it precipitates to form chemical sediments.
This precipitation of material may occur directly as
the result of inorganic processes or indirectly as the

result of the life processes of water-dwelling organisms. Sediment formed in this second way is said to have a *biochemical* origin.

An example of a deposit resulting from inorganic chemical processes is the salt left behind as a body of salt water evaporates. In contrast, many water-dwelling animals and plants extract dissolved mineral matter to form shells and other hard parts. After the organisms die, their skeletons collect on the floor of a lake or ocean.

Limestone. Representing about 10 percent of the total volume of all sedimentary rocks, *limestone* is the most abundant chemical sedimentary rock. It is composed chiefly of the mineral calcite ($CaCO_3$) and forms by either inorganic means or as the result of biochemical processes. Limestones having a biochemical origin are by far the most common. As much as 90 percent of the world's limestone may have originated as accumulations of biochemical sediment.

Although most limestone is the product of biological processes, this origin is not always evident because shells and skeletons may undergo considerable change before being converted to rock. However, one easily identified biochemical limestone is *coquina,* a coarse rock composed of poorly cemented shells and shell fragments. Another less obvious, but nevertheless familiar example is *chalk,* a

FIGURE 4.11 The White Chalk Cliffs of Dover. (Courtesy of British Tourist Authority)

rock made up almost entirely of the hard parts of foraminifera, which are microscopic organisms no larger than the head of a pin (Figure 4.11).

Limestones having an inorganic origin form when evaporation and/or high water temperatures increase the concentration of calcium carbonate to the point that it precipitates. *Travertine,* the type of limestone commonly seen in caves, is one example, as is *oolitic limestone.* Travertine is deposited when groundwater containing calcium carbonate evaporates. The formation of oolitic limestone, a rock composed of small spherical grains called *oolites,* is somewhat more complex. Oolites form in shallow marine waters as tiny "seed" particles are moved back and forth by currents and become coated with layer upon layer of calcium carbonate as they roll on the ocean floor.

Dolomite. Closely related to limestone is *dolomite,* a rock composed of the calcium-magnesium carbonate mineral of the same name. Dolomite is one of the rare cases where the same term is used to designate both a mineral and a rock. To avoid confusion, some geologists refer to the rock as *dolostone.* Although dolomite can form by direct precipitation from seawater, it is thought that most originates when magnesium in seawater replaces some of the calcium in limestone. The latter theory is reinforced by the fact that there are practically no young dolomite rocks. Rather, most dolomites are ancient rocks in which there was ample time for magnesium to replace calcium.

Chert. *Chert* is a name used for a variety of very dense, hard rocks made of microcrystalline silica (SiO_2). One well-known form is *flint,* whose dark color results from the organic matter it contains. *Jasper,* a red variety, gets its bright color from the iron oxide it contains.

Chert deposits are commonly found in one of two situations: as irregularly shaped nodules in limestone and as layers of rock. The silica composing most chert nodules is believed to have been deposited directly from water. Thus, these nodules have an inorganic origin. However, it is unlikely that a very high percentage of chert layers was precipitated directly from seawater, because seawater is not generally saturated with silica. Hence, beds of chert are thought to have originated largely as biochemical sediment. Although most water-dwelling organisms that produce hard parts secrete shells made of calcium carbonate, some, such as diatoms and radiolarians, produce glasslike silica skeletons. These tiny organisms are able to extract silica from very unsaturated solutions. It is from these remains that beds of chert are made. Note that when a chert specimen is being examined, there are few reliable criteria by which the mode of origin (inorganic versus biochemical) can be determined.

Rock Salt and Rock Gypsum. Very often evaporation is the mechanism triggering deposition of chemical precipitates. Minerals commonly precipitated in this fashion include halite (sodium chloride), the chief component of *rock salt,* and gypsum (hydrous calcium sulfate), the main ingredient of *rock gypsum.*

In the geologic past, many areas that are now dry land were covered by shallow arms of the sea that had only narrow connections to the open ocean. Under these conditions, water continually moved into the bay to replace water lost by evaporation. Eventually the waters of the bay became saturated and salt deposition began. Such deposits are called **evaporites**. Today such deposits serve as an important source of many chemicals. Similar deposits may be seen in such places as Death Valley, California.

Here, following rains or periods of snowmelt in the mountains, streams flow from surrounding mountains into an enclosed basin. As the water evaporates, *salt flats* form from dissolved materials left behind as a white crust on the ground (Figure 4.12).

Coal. *Coal* is quite different from any other sedimentary rock. Coal is often grouped with biochemical sedimentary rocks, but unlike other rocks in this category, which are composed of the calcite-rich or silica-rich hard parts secreted by water-dwelling plants and animals, coal is made of organic matter. A close examination of a piece of coal under a microscope or magnifying glass often reveals the presence of various plant structures such as leaves, bark, and wood that have been chemically altered but are nevertheless still identifiable. This supports the conclusion that coal is the end product of the burial of large amounts of plant material for extended periods of time. Under the right circumstances a great volume of plants may accumulate rather than totally decompose, which is normally the case. One such environment is an oxygen-poor swamp. At various times during earth history such environments have been relatively common. With each successive stage in coal formation, higher temperatures and pressures drive off impurities and volatiles as follows:

FIGURE 4.12 These salt flats (composed of rock gypsum and rock salt) in Death Valley, California, are an example of evaporite deposits. (Photo by James E. Patterson)

PEAT $\xrightarrow[\text{pressure}]{\text{burial}}$ LIGNITE $\xrightarrow[\text{and pressure}]{\text{greater burial}}$

(partially altered   (soft, brown coal)
plant material)

BITUMINOUS $\xrightarrow{\text{metamorphism}}$ ANTHRACITE

(soft, black coal)   (hard, black coal)

Bituminous coal is by far the most important coal resource. *Anthracite* forms when bituminous coal undergoes metamorphism, and although it is cleaner burning, anthracite is not as widespread and is more expensive to mine.

Organic matter is present in small quantities in many other sedimentary rocks. When present, it usually colors the rock black. In such cases the term **carbonaceous** is often applied to describe the rock. Thus a black, organic-rich shale would be called carbonaceous shale.

TURNING SEDIMENT INTO SEDIMENTARY ROCK

Lithification refers to the processes by which unconsolidated sediments are transformed into solid sedimentary rocks. One very common process affecting sediments is **compaction**. As sediments accumulate through time, the weight of overlying material compresses the deeper sediments. As the grains are pressed closer and closer, there is a considerable reduction in pore space. For example, when clays are buried beneath several thousand meters of material the volume of the clay may be reduced by as much as 40 percent. Since sands and other coarse sediments are only slightly compressible, compaction is most significant as a lithification process in fine-grained sedimentary rocks such as shale.

Cementation is another important means by which sediments are converted to sedimentary rocks. The cementing materials are carried in solution by water percolating through the open spaces between particles. Through time, the cement precipitates onto the sediment grains, fills the open spaces, and joins the particles. Calcite, silica, and iron oxide are the most common cements. The identification of the cementing material is a relatively simple matter. Calcite cement will effervesce with dilute hydrochlo-

ric acid. Silica is the hardest cement and thus produces the hardest sedimentary rocks. When a sedimentary rock has an orange or dark red color, this usually means that iron oxide is present. In some instances, cements can even be economically significant. For example, the iron oxide cement in the Clinton Formation in the Appalachians was rich enough to make this sandstone into an important iron ore.

Although most sedimentary rocks are lithified by compaction, cementation, or a combination of both, some are made of interlocking crystals. This type of lithification is confined largely to certain chemical sedimentary rocks.

CLASSIFICATION OF SEDIMENTARY ROCKS

The classification scheme in Table 4.4 divides sedimentary rocks into two major groups: detrital and chemical. Further, we can see that the main criterion for dividing detrital rocks is particle size, whereas the primary basis for distinguishing among different rocks in the chemical group is their mineral composition.

As is the case with many (perhaps most) classifications of natural phenomena, the categories presented in Table 4.4 (page 86) are more rigid than the actual state of nature. In reality, many of the sedimentary rocks classified into the chemical group also contain at least small quantities of detrital sediment. Many limestones, for example, contain varying amounts of mud or sand, giving them a ''sandy'' or ''shaly'' quality. On the other hand, since practically all detrital rocks are cemented with material that was originally dissolved in water, they too are far from being ''pure.''

As was the case with the igneous rocks examined in Chapter Two, texture is an important aspect of sedimentary rock classification. There are two major textures used in the classification of sedimentary rocks: clastic and nonclastic. The term **clastic** is taken from a Greek word meaning ''broken.'' Thus clastic rocks are made of broken fragments. An examination of Table 4.4 reveals that all detrital rocks have a clastic texture. The table also shows that some chemical sedimentary rocks may also exhibit this texture. For example, coquina, the limestone

TABLE 4.4 Classification of sedimentary rocks.

DETRITAL ROCKS

Texture	Sediment Name and Particle Size	Comments	Rock Name
Clastic	Gravel >2 mm	Rounded rock fragments	Conglomerate
		Angular rock fragments	Breccia
	Sand 1/16–2 mm	Quartz predominates	Quartz sandstone
		Quartz with considerable feldspar	Arkose
		Dark color; quartz with considerable feldspar, clay, and rock fragments	Graywacke
	Mud <1/16 mm	Splits into thin layers	Shale
		Breaks into clumps or blocks	Mudstone

CHEMICAL ROCKS

Group	Texture	Composition	Rock Name
Inorganic	Clastic or nonclastic	Calcite, $CaCO_3$	Limestone
	Clastic or nonclastic	Dolomite, $CaMg(CO_3)_2$	Dolomite (dolostone)
	Nonclastic	Microcrystalline quartz, SiO_2	Chert
	Nonclastic	Halite, NaCl	Rock salt
	Nonclastic	Gypsum, $CaSO_4 \cdot 2H_2O$	Rock gypsum
Biochemical	Clastic or nonclastic	Calcite, $CaCO_3$	Limestone
	Nonclastic	Microcrystalline quartz, SiO_2	Chert
	Nonclastic	Altered plant remains	Coal

composed of shells and shell fragments, is obviously as clastic as a conglomerate or sandstone. The same applies for some varieties of oolitic limestone.

Some chemical sedimentary rocks have a **nonclastic** texture in which the minerals form a pattern of interlocking crystals. Because of this, some non-clastic sedimentary rocks may look more like igneous rocks, which are also composed of intergrown crystals. However, the two are usually easy to distinguish between because the minerals that compose nonclastic sedimentary rocks are quite different from the minerals found in igneous rocks.

FEATURES OF SEDIMENTARY ROCKS

As stated earlier, sedimentary rocks are particularly important in interpreting earth history. These rocks form at the earth's surface, and as layer upon layer of sediment accumulates, each records the nature of the environment at the time the sediment was deposited. These layers, called **strata**, or **beds**, are probably the single most characteristic feature of sedimentary rocks (Figure 4.13).

The thickness of beds ranges from microscopically thin to tens of meters thick. Separating the strata are **bedding planes**, flat surfaces along which rocks

FIGURE 4.13 This outcrop of sedimentary strata illustrates the characteristic layering of this group of rocks. (Photo by G. K. Gilbert, U.S. Geological Survey)

tend to separate or break. Changes in the grain size or in the composition of the sediment being deposited can create bedding planes. Pauses in deposition can also lead to layering because chances are slight that newly deposited material will be exactly the same as previously deposited sediment. Generally each bedding plane marks the end of one deposit and the beginning of another.

As geologists examine sedimentary rocks, much can be deduced. A conglomerate, for example, may indicate a high-energy environment, such as a rushing stream, where only the coarse materials can settle out (Figure 4.14). If the rock is arkose, it may signify a dry climate, where little chemical alteration of feldspar is possible. Carbonaceous shale is a sign of a low-energy, organic-rich environment such as a swamp or lagoon. Other features found in some sedimentary rocks also give clues to past environments. An example of one such feature is **ripple marks**. A corrugated surface, such as the one pictured in Figure 4.15A, may be produced by stream or tidal currents flowing across a sandy bottom or by wind blowing over a sand dune. Some ripple marks can be used to determine the direction of movement of ancient currents and winds.

Mud cracks (Figure 4.15B) indicate that the sediment in which they formed was alternately wet and dry. When exposed to air, wet mud dries out and shrinks, producing cracks. Their presence is associated with such environments as shallow lakes, tidal flats, and desert basins.

Sometimes when a bed of sedimentary rock is examined, we can see layers inclined at a steep angle to the horizontal. Such layering is termed **cross-**

FIGURE 4.14 Cross section of an ancient stream channel filled with conglomerate. In a high-energy environment, such as a rushing stream, only the coarse materials can settle out. (Photo by W. R. Hansen, U.S. Geological Survey)

A.

B.

FIGURE 4.15 A. Ripple marks may indicate a beach or stream channel environment. (Photo by Stephen Trimble) **B.** Mud cracks form when wet mud or clay dries and shrinks, perhaps signifying a tidal flat or desert basin. (Photo by Garrett Deckert) **C.** The cross-bedding in this sandstone indicates that it was once a sand dune. (Photograph used by permission of Dennis Tasa)

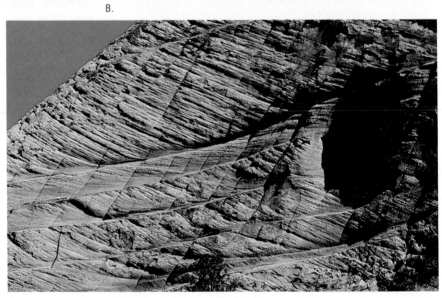

C.

bedding and is most characteristic of river deltas and sand dunes (Figure 4.15C).

Fossils, the evidence or remains of prehistoric life, are perhaps the most important inclusions found in sedimentary rock. Fossils are important tools used to interpret the geologic past. Knowing the nature of the life forms that existed at a particu-lar time may help to answer questions about the environment. Further, fossils are important time indi-cators and play a key roll in correlating rocks of simi-lar ages but from different places. Fossils will be ex-amined in more detail in Chapters Fifteen and Sixteen.

REVIEW QUESTIONS

1 If two identical rocks were weathered, one mechani-cally and the other chemically, how would the prod-ucts of weathering for the two rocks differ?

2 How does mechanical weathering add to the effec-tiveness of chemical weathering?

3 Granite and basalt are exposed at the surface in a hot, wet region.
   a Which type of weathering will predominate?
   b Which of the rocks will weather most rapidly? Why?

4 Heat speeds up a chemical reaction. Why then does chemical weathering proceed slowly in a hot desert?

5 How is carbonic acid (H_2CO_3) formed in nature? What results when this acid reacts with potassium feldspar?

6 What is the difference between soil and regolith?

7 What factors might cause different soils to develop from the same parent material, or similar soils to form from different parent materials?

8 Which of the controls of soil formation is most impor-tant? Explain.

9 List the characteristics associated with each of the horizons in a well-developed soil profile. Which hori-zons constitute the solum? Under what circum-stances do soils lack horizons?

10 How can slope affect soil development? What is meant by the term *slope orientation?*

11 Distinguish between pedalfers and pedocals.

12 Soils formed in the humid tropics and the Arctic both contain little organic matter. Do both lack humus for the same reasons?

13 How does the volume of sedimentary rocks in the earth's crust compare with the volume of igneous rocks in the crust? Are sedimentary rocks evenly dis-tributed throughout the crust?

14 What minerals are most common in detrital sedimen-tary rocks? Why are these minerals so abundant?

15 What is the primary basis for distinguishing among various detrital sedimentary rocks?

16 The term *clay* can be used in two different ways. De-scribe the two meanings of this term.

17 Why does shale usually crumble quite easily?

18 Distinguish between conglomerate and breccia (see Figure 4.10).

19 Distinguish between the two categories of chemical sedimentary rocks.

20 What are evaporite deposits? Name a rock that is an evaporite.

21 Compaction is an important lithification process with which sediment size?

22 List three common cements for sedimentary rocks. How might each be identified?

23 What is the primary basis for distinguishing among different chemical sedimentary rocks?

24 Distinguish between clastic and nonclastic textures. What type of texture is common to all detrital sedi-mentary rocks?

25 What is probably the single most characteristic fea-ture of sedimentary rocks?

CHAPTER FIVE

METAMORPHIC ROCKS

METAMORPHISM

The process of metamorphism involves the transformation of pre-existing rock. Metamorphic rocks can form from igneous, sedimentary, or even from other metamorphic rocks. The term for this process is very appropriate because it literally means to "change form." The agents of change include heat, pressure, and chemically active fluids, whereas the changes that occur are textural as well as mineralogical.

In some instances metamorphic rocks are only slightly changed, becoming more compact. In other cases the transformation is so complete that the identity of the original rock cannot be determined. In high-grade metamorphism, such features as bedding planes, fossils, and vesicles that may have existed in the parent rock are completely destroyed. Further, when subjected to intense heat and directional pressure, these rocks behave plastically and bend into intricate folds (Figure 5.1). In the most extreme metamorphic environments, the temperatures approach those at which rocks melt. However, during metamorphism the deformed material must remain solid, for once melting occurs, we have entered the realm of igneous activity.

The process of metamorphism takes place when rock is subjected to conditions unlike those in which it formed; the rock becomes unstable and gradually changes until a state of equilibrium with the new environment is reached. The changes occur at the temperatures and pressures existing in the region extending from a few kilometers below the earth's surface to the crust-mantle boundary. Since the formation of metamorphic rocks is completely hidden from view (which is not the case for many sedimentary and some igneous rocks), metamorphism is undoubtedly a very difficult process for geologists to study.

Metamorphism most often occurs in one of three settings. First, during mountain building great quantities of rock are subjected to the intense stresses and temperatures associated with large-scale deformation. The end result may be extensive areas of metamorphic rocks that are said to have undergone **regional metamorphism**. The greatest volume of metamorphic rock is produced in this fashion. Second, when rock is in contact or close proximity to a mass of magma, **contact metamorphism** takes place. In this circumstance the changes are caused primarily by the high temperatures of the molten material, which in effect "bake" the surrounding rock. The third and least common type of metamorphism occurs along fault zones. Here rock is broken and distorted as crustal blocks on opposite sides of a fault grind past one another.

AGENTS OF METAMORPHISM

As stated earlier, the agents of metamorphism include heat, pressure, and chemically active fluids.

FIGURE 5.1 Intricately folded rock of Cabbage Island, Maine. (Photo by W. B. Hamilton, U.S. Geological Survey)

Ancient Precambrian rocks exposed at the bottom of the Grand Canyon. (Photo by E. J. Tarbuck)

During metamorphism, rocks are often subjected to all three metamorphic agents simultaneously. However, the degree of metamorphism and the contribution of each agent varies greatly from one environment to another. In low-grade metamorphism, rocks are subjected to temperatures and pressures only slightly greater than those associated with the lithification of sediments. High-grade metamorphism, on the other hand, involves extreme conditions closer to those at which rocks melt.

Heat as a Metamorphic Agent

Perhaps the most important agent of metamorphism is heat. Rocks formed near the earth's surface may be subjected to intense heat when they are intruded by molten material rising from below. The effects of contact metamorphism are most apparent when it occurs at or near the surface, where the temperature contrast between the molten, intrusive rock and the country rock is most pronounced. Here the adjacent country rock is "baked" by the emplaced magma. In this high-temperature, high-pressure environment, the boundary that forms between the intrusive igneous body and the altered rocks is usually quite distinct.

Rocks that originate in a surface environment may also be subjected to extreme temperatures if they are subsequently buried deep within the earth. Recall that temperatures increase with depth at a rate known as the thermal gradient. In the upper crust, this increase in temperature averages about 30°C per kilometer. As we discussed earlier, earth materials are continually being transported to great depths at convergent plate boundaries. When buried to a depth of only a few kilometers, certain minerals, such as clay, become unstable and begin to recrystallize into minerals that are stable in this environment. Other minerals, particularly those found in crystalline igneous rocks, are stable at relatively high temperatures and pressures and therefore require burial to 20 kilometers or more before metamorphism will occur.

Pressure as a Metamorphic Agent

Pressure, like temperature, also increases with depth. Buried rocks are subjected to the force exerted by the load above. This confining pressure is

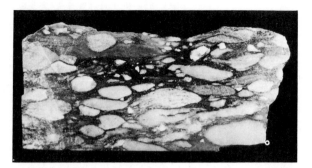

FIGURE 5.2 Metaconglomerate. These once-rounded rock fragments have been elongated as if squeezed in a gigantic vise. (Photo by E. J. Tarbuck)

analogous to air pressure where the force is applied equally in all directions.

In addition to the pressure exerted by the load of material above, rocks are subjected to **stress** during the process of mountain building. Here, the applied force is directional, squeezing the material as if it had been placed in a vise. Rock located at great depth is quite warm and behaves plastically during deformation (Figure 5.2). By contrast cooler, near-surface rocks will usually **shear** during deformation. Shearing results when relatively brittle rock is broken into thin slabs that are able to slide past each other. This phenomenon can be demonstrated using a deck of playing cards. Shearing is similar to the slippages that occur between individual cards when the deck is held between your hands and the top of the deck is moved relative to the bottom.

Chemical Activity and Metamorphism

Chemically active fluids, most commonly water containing ions in solution, also enhance the metamorphic process. Some water is contained in the pore spaces of virtually every rock. In addition, many minerals are hydrated, and thus contain water within their crystalline structures. When deep burial occurs, water is forced out of the mineral structures and is then available to aid in chemical reactions. Water that surrounds the crystals acts as a catalyst by aiding ion migration. In some instances the minerals recrystallize to form more stable configurations. In other cases, ion exchange among minerals results in the formation of completely new minerals. Complete alteration of rock by hot, mineral-rich water has been observed in the near-surface environment of Yellowstone National Park. Further, along oceanic ridge systems seawater circulating through the still-hot rock helps transform dark basaltic minerals into metamorphic minerals such as serpentine and talc.

TEXTURAL AND MINERALOGICAL CHANGES

The degree of metamorphism is reflected in the texture and mineralogy of metamorphic rocks. When rocks are subjected to very low-grade metamorphism, they become more compact and thus more dense. A common example is the metamorphic rock slate, formed from the further compaction of shale.

Under more extreme conditions, pressure causes certain minerals to recrystallize. As described earlier, water is believed to play a very important role in the recrystallization process by aiding the migration of ions. In general, recrystallization encourages the growth of larger crystals. Consequently, many metamorphic rocks consist of visible crystals, much like phaneritic igneous rocks. The crystals of some minerals, such as micas, which have a sheet structure, and hornblende, which has an elongated structure, will recrystallize with a preferred orientation. The new orientations will be essentially perpendicular to the direction of the compressional force. The resulting mineral alignment usually gives the rock a layered or banded appearance termed **foliation** (Figure 5.3).

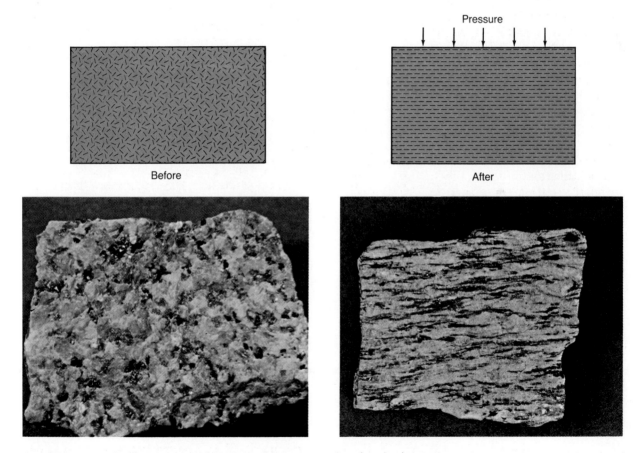

FIGURE 5.3 Under the pressures of metamorphism, some mineral grains become reoriented and aligned at right angles to the pressure. The resulting linear orientation of mineral grains gives the rock a foliated texture. If the coarse-grained igneous rock (granite) on the left underwent intense metamorphism, it could end up closely resembling the sample on the right (gneiss).

Various types of foliation exist, depending to a large extent upon the degree of metamorphism. For example, during the transformation of shale to slate, clay minerals, which are stable at the surface, recrystallize into minute mica flakes, which are stable at much higher temperatures and pressures. Further, during recrystallization these fine-grained mica crystals become aligned so that their flat surfaces are nearly parallel. Consequently, slate can be split easily along these layers of mica grains into rather flat slabs. This property is called **rock cleavage**, to differentiate it from the type of cleavage exhibited by minerals (Figure 5.4). Since the mica flakes composing slate are tiny, slate is usually not visibly foliated. However, because it exhibits excellent rock cleavage, which is evidence that its minerals are aligned, slate is considered to be foliated.

Under more extreme temperature-pressure regimes, the very fine mica grains of slate will grow many times larger. These mica crystals, which are about a centimeter in diameter, will give the rock a platy or scaly appearance. This type of foliation is called **schistosity**, and a rock having this texture is called *schist*. Many types of schist exist and are

FIGURE 5.4 Rock cleavage in schist, northeastern California. (Photo by E. J. Tarbuck)

named according to their mineral constituents. By far the most abundant are the mica schists, which are usually composed of either muscovite or biotite, or both (Figure 5.5).

During high-grade metamorphism, ion migrations can be extreme enough to cause minerals to segregate. An example of a metamorphic rock that exhibits mineral segregation is shown in Figure 5.3. Notice that the dark and light silicate minerals have separated, giving the rock a banded appearance. Metamorphic rocks with this texture are called *gneiss* (pronounced "nice") and are quite common. Gneiss usually forms from the metamorphism of granite or diorite, but can form from gabbro or even by the high-grade metamorphism of shale. Although banded, gneiss will not usually split parallel to the crystals as easily as slate.

Not all metamorphic rocks have a foliated texture. Such rocks are said to be **nonfoliated**. Metamorphic rocks composed of only one mineral which forms equidimensional crystals are as a rule not visibly foliated. For example, when a fine-grained limestone is metamorphosed, the small calcite crystals combine to form relatively large interlocking crystals. The resulting rock has an appearance similar to a coarse-grained igneous rock. This metamorphic equivalent of limestone is *marble*. Although it is considered to be nonfoliated, microscopic investigation of marble may reveal some flattening and parallelism of the grains. Further, some limestones contain thin layers of clay minerals which may become distorted during metamorphism. These "impurities" will often appear as curved bands of dark material flowing through the marble.

In the metamorphism of shale to slate we saw that clay minerals recrystallized to form mica crystals. In most instances, including this example, the chemical composition of the rock does not change during recrystallization. Rather, the existing minerals and available ions in the water will recombine to form minerals that are stable in the new environment. A common example is the formation of the metamorphic mineral *wollastonite*. Wollastonite is generated when limestone ($CaCO_3$), which contains abundant sandy material in the form of quartz (SiO_2), is subjected to high temperatures during contact metamorphism. The calcite and quartz crystals chemically react to form wollastonite ($CaSiO_3$), while carbon dioxide is liberated.

In some environments, however, new materials are actually introduced during the metamorphic process. For example, country rock adjacent to a large magma body would be altered by ion-rich **hydrothermal** (hot water) **solutions** released during the latter stages of crystallization. Many metallic ore deposits were formed by the deposition of minerals from hydrothermal solutions. Further, the seawater percolating through newly formed oceanic

FIGURE 5.5 Mica schist, a common metamorphic rock composed of shiny mica flakes. (Photo by E. J. Tarbuck)

crust contains numerous active ions that chemically react with existing rocks. Some of the earth's richest copper ores have been formed in this manner.

In summary, the metamorphic process causes many changes in rocks, including increased density, growth of larger crystals, reorientation of the mineral grains into a layered or banded appearance known as foliation, and the transformation of low-temperature minerals into high-temperature minerals. Further, the introduction of ions generates new minerals, some of which are economically important.

COMMON METAMORPHIC ROCKS

Foliated Rocks

Slate is a very fine-grained foliated rock composed of minute mica flakes. The most noteworthy characteristic of slate is its excellent rock cleavage. This prop-

erty has made slate a most useful rock for roof and floor tile, blackboards, and billiard tables. Slate is most often generated by the low-grade metamorphism of shale, although less frequently it forms from the metamorphism of volcanic ash. Slate can be almost any color depending on its mineral constituents. Black (carbonaceous) slate contains organic material, red slate gets it color from iron oxide, and green slate is usually composed of chlorite, a mica-like mineral formed by the metamorphism of iron-rich silicates. Because slate forms during low-grade metamorphism, evidence of the original bedding planes is often preserved. However, the orientation of slate's rock cleavage generally trends across the original sedimentary layering (Figure 5.6). Thus, unlike shale, which splits along bedding planes, slate splits across them.

Phyllite represents a gradation in metamorphism between slate and schist. Its constituent platy miner-

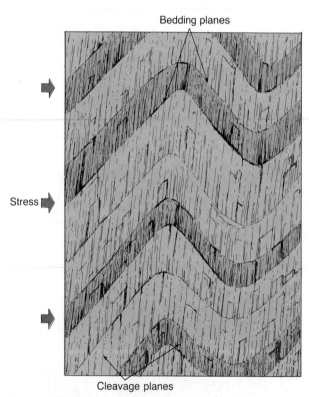

A.

B.

FIGURE 5.6 Illustration and photograph showing the relationship between slate cleavage and original bedding planes. (Photo by G. K. Gilbert, U.S. Geological Survey)

FIGURE 5.7 Phyllite (left) can be distinguished from slate by its glossy sheen. (Photo by E. J. Tarbuck)

als are larger than those in slate, but not yet large enough to be clearly identifiable. Although phyllite appears similar to slate, it can be easily distinguished from slate by its glossy sheen (Figure 5.7). Phyllite usually exhibits rock cleavage and is composed mainly of very fine crystals of either muscovite or chlorite.

Schists are distinctive metamorphic rocks and are almost as common as gneisses. By definition, schists contain more than 50 percent platy minerals, most commonly muscovite and biotite. Like slate, the parent material from which most schists originate is shale, but in the case of schist, the metamorphism is more intense. If the parent rock contained abundant silica, schist will often contain thin layers of quartz and possibly feldspar as well.

Schists are named according to their mineral composition. Those composed primarily of muscovite and biotite with lesser amounts of quartz and feldspar are called *mica schist.* Depending upon the degree of metamorphism, mica schists often contain accessory minerals quite unique to metamorphic rocks. Some common accessory minerals include garnet, staurolite, and sillimanite (Figure 5.8). Some schists contain graphite, which is recovered for use as pencil lead, graphite fibers, and lubricant. In addition, schists may be composed largely of the minerals chlorite or talc, in which case they are called *chlo-*

rite schist and *talc schist,* respectively. Both chlorite and talc schists form when rocks with a basaltic composition undergo metamorphism.

Gneiss is the term applied to foliated metamorphic rocks that contain mostly granular, as opposed to platy, minerals. The most common minerals found in gneisses are quartz, potassium feldspar, and sodium feldspar. In addition, lesser amounts of muscovite, biotite, and hornblende are common. The segregation of light and dark silicates gives gneisses a

FIGURE 5.8 Garnet-mica schist. (Courtesy of the American Museum of Natural History)

FIGURE 5.9 A contorted gneiss found on the north rim of the Black Canyon of the Gunnison River in Colorado. (Photo by W. R. Hansen, U.S. Geological Survey)

characteristic foliated texture. Thus, most gneisses consist of alternating bands of white or reddish feldspar-rich zones and layers of dark ferromagnesian minerals. These banded gneisses are often deformed while in a plastic state into rather intricate folds (Figure 5.9). Some gneisses will split readily along the layers of platy minerals, but most break in an irregular fashion, like other crystalline rocks.

Gneisses generally have a composition similar to granite and are probably derived from granite or its aphanitic equivalent. However, they may also form from the high-grade metamorphism of shale. In this instance, gneiss represents the last rock in the sequence of shale, slate, phyllite, schist, and gneiss. Like schists, gneisses may also include large crystals of accessory minerals such as garnet and staurolite. When banded rocks are made up predominately of dark minerals, such as those common in basalt, they are called *amphibolites,* after the mineral amphibole.

Nonfoliated Rocks

Marble is a coarse, crystalline rock whose parent rock was limestone or dolomite (Figure 5.10). When a hand sample is examined, marble closely resembles crystalline limestone. Pure marble is white and composed essentially of the mineral calcite. Due to its color and relative softness (hardness of 3), marble is a popular building stone. White marble is particularly prized as a stone from which to carve monuments and statues, such as the famous statue of David by Michelangelo.

Often the limestone from which marble forms contains impurities that color the marble. Thus, marble can be pink, gray, green, or even black. Also, when impure limestone is metamorphosed, it may yield a variety of accessory minerals including chlorite, mica, garnet, and commonly wollastonite. When marble forms from limestone interbedded with shales, it will appear banded. Occasionally, marble will split along these bands and reveal the mica minerals that crystallized from clay minerals. Under extreme deformation, the bands in marble may become highly contorted and give the rock a rather artistic design.

Quartzite is a very hard metamorphic rock most often formed from quartz sandstone. Under moderate-to-high-grade metamorphism, the quartz grains in sandstone fuse. The recrystallization is so com-

FIGURE 5.10 Marble, a crystalline rock formed by the metamorphism of limestone. (Photo by E. J. Tarbuck)

plete that when broken, quartzite will split across the original quartz grains, rather than between them. In some instances, such sedimentary structures as cross-bedding are preserved and give the rock a banded appearance.

Although pure quartzite is white, it commonly contains some iron oxide stain and thus appears pink to red. Occasionally quartzite may contain a small percentage of dark minerals which give it a gray color.

Quartzite, like marble, is made up of only one mineral, which develops equidimensional crystals. Consequently, it does not develop any parallelism of mineral grains and therefore lacks foliation.

OCCURRENCES OF METAMORPHIC ROCKS

Recall that metamorphic rocks commonly form in one of three environments: along fault zones, in contact with igneous bodies, or during dynamic episodes associated with mountain building.

Metamorphism along Fault Zones

When faulting occurs near the surface, the stress and frictional heat produced along the fault zone generate a loosely coherent rock composed of broken or distorted rock fragments. When these rocks are composed of angular granules, they are called *fault breccia*. Metamorphic rocks produced along fault zones which are located at depth often have elongated grains and more closely resemble the rocks formed by the other metamorphic processes. Consequently, their origin may not be discernible when hand samples are examined.

The quantity of metamorphic rock generated solely by faulting is relatively insignificant when compared to the amount generated by the other two processes. Nevertheless, in some areas these granulated rocks are quite abundant. For example, movements along California's San Andreas fault have created a zone of fault breccia and related rock types nearly 1000 kilometers long and up to 3 kilometers wide.

Contact Metamorphism

Contact metamorphism occurs when molten rock comes into contact with cooler rock. Contact metamorphism is clearly distinguishable only when it occurs at the surface or in a near-surface environment where the temperature contrast between the magma and host rock is great (Figure 5.11). Undoubtedly, contact metamorphism is also an active

FIGURE 5.11 Close-up view of a contact zone where light-colored igneous rock has invaded and metamorphosed the dark-colored host rock. (Photo by John S. Shelton)

process at great depth. However, its effect is blurred due to the general alteration of regional metamorphism.

During contact metamorphism, a zone of alteration called an **aureole** (or halo) forms around the emplaced magma. Small intrusive bodies such as dikes and sills have aureoles only a few centimeters thick. On the other hand, large igneous bodies such as batholiths and laccoliths may form zones of metamorphic rocks a few kilometers or more in thickness (Figure 5.12). These large aureoles often consist of definite zones of metamorphism. Near the magma body, high-temperature minerals such as garnet may form, while farther away such low-grade minerals as chlorite are produced. In addition to the size of the intrusive magma body, the mineral composition of the country rock and the availability of water greatly

affect the size of the aureole produced. In chemically active rock such as limestone, the zone of alteration can extend to distances of 10 kilometers or more from the igneous body. Here the occurrence of minerals such as garnet and wollastonite marks the extent of metamorphism.

Most contact metamorphic rocks are fine-grained, dense, tough rocks of various chemical compositions. For example, during contact metamorphism, clay minerals are baked, as if placed in a kiln, and can generate a very hard, fine-grained rock not unlike porcelain. Since directional pressure is not a major factor in the formation of these rocks, they generally are not foliated. The name applied to a wide variety of rather hard, nonfoliated metamorphic rocks is *hornfels.*

When larger igneous masses are involved in contact metamorphism, hydrothermal solutions which originate within the magma can migrate great distances. As these solutions percolate through the host rock, chemical reactions with this rock greatly enhance the metamorphic process. Further, ores of numerous metals are thought to result from the emplacement of metallic ions whose source is hydrothermal solutions. These deposits include ores of copper, zinc, lead, iron, and gold.

Regional Metamorphism

By far the greatest quantity of metamorphic rock is produced during regional metamorphism. As stated earlier, regional metamorphism takes place at considerable depths over an extensive area and is associated with the process of mountain building. During mountain building, a large segment of the earth's crust is intensely squeezed into a highly deformed mass. As the rocks are folded and faulted, the crust is shortened and thickened. This general thickening of the crust results in mountain terrains that stand high above sea level. Although material is obviously elevated to great heights during mountain building, an equally large quantity of rock is forced downward, where it is exposed to high temperatures and pressures. Here in the "roots" of mountains, the most intense metamorphic activity occurs. Some of the deformed rock is thought to be heated enough to melt and thereby generate magma. This magma, being less dense than the surrounding rock, will mi-

FIGURE 5.12 The dark layer, called a roof pendant, consists of metamorphosed country rock adjacent to the upper part of an igneous pluton. The term *roof pendant* implies that the rock was once the roof of a magma chamber. This roof pendant is in the Sierra Nevada of eastern California, near Split Mountain. (Photo by John S. Shelton)

grate upward. Magmas emplaced in a near-surface environment will cause contact metamorphism within the zone of regional metamorphism. Consequently, the cores of many mountain ranges consist of intrusive igneous bodies surrounded by high-grade metamorphic rocks. As these deformed rock masses are uplifted, erosion removes the overlying material to expose the igneous and metamorphic rocks composing the central core of the mountain range.

Since metamorphic rocks that form during regional metamorphism are deformed by directional stress, they are usually foliated. Further, in regional metamorphism, there usually exists a gradation in the intensity of metamorphism. As we progress from areas of low-grade metamorphism to areas of high-grade metamorphism, changes in mineralogy and rock texture can be observed.

A somewhat oversimplified example of progressive metamorphism can be made using the sedimentary rock shale, which under low-grade metamorphism yields the metamorphic rock slate. In high-

temperature, high-pressure environments, slate will turn into mica schist. Under more extreme conditions, the micas in schist will recrystallize into minerals such as feldspar and hornblende, and eventually generate a gneiss. We can see certain aspects of this transition as we approach the Appalachian Mountains from the west. Beds of shale, which once extended over large areas of the eastern United States, can still be found as nearly flat-lying strata in Ohio. However, in the broadly folded Appalachians of central Pennsylvania, these beds are inclined and composed of low-grade slate. As we progress farther eastward to the intensely deformed crystalline Appalachians, we find large outcrops of schist and gneiss, some of which are perhaps remnants of once flat-lying shale beds. The most intense zones of metamorphism are found in regions such as Vermont and New Hampshire, often in close association with igneous intrusions.

In addition to the textural changes already considered, changes in mineralogy will be encountered as we progress from areas of low-grade metamorphism to areas of high-grade metamorphism. The typical transition in mineralogy that would result from the regional metamorphism of shale is shown in Figure 5.13. The first new mineral to be produced in the formation of slate is chlorite, which as we move toward the region of high-grade metamorphism would be replaced by ever greater amounts of muscovite and biotite. Mica schists are formed under more extreme conditions and may contain garnet and staurolite crystals as well. At temperatures and

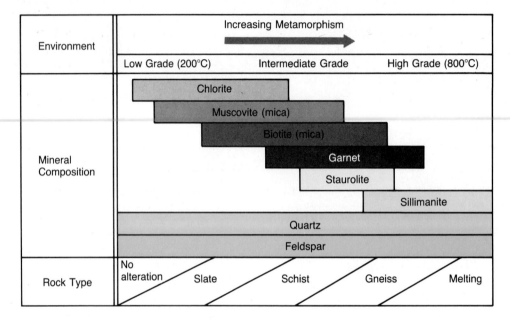

FIGURE 5.13 The typical transition in mineralogy that results from progressive metamorphism of shale.

FIGURE 5.14 Migmatite. The lightest colored layers are igneous rock composed of quartz and feldspar, while the darker layers have a metamorphic origin. (Photo by J. B. Hadley, U.S. Geological Survey)

pressures approaching the melting point of rock, sillimanite forms. Sillimanite is a high-temperature metamorphic mineral used to make refractory porcelains such as those used in the casings of spark plugs.

In the most extreme environments even the highest-grade metamorphic rocks are subjected to change. In a low-pressure environment where temperatures exceed 800°C, a schist or gneiss having a chemical composition similar to granite will begin to melt. However, recall from our discussion of igneous rocks that not all minerals melt at the same temperature. The light-colored silicates, usually quartz and potassium feldspar, will melt first, whereas the dark silicates, such as amphibole and biotite, will remain solid. If this partially melted rock cools, the light bands will be made of crystalline igneous rock while the dark bands will consist of unmelted metamorphic material. Rocks of this type fall into a transitional zone somewhere between "true" igneous rocks and "true" metamorphic rocks and are called **migmatites** (Figure 5.14).

REVIEW QUESTIONS

1 What is metamorphism? What are the agents of change?

2 What is foliation? Distinguish between *rock cleavage* and *schistosity.*

3 List some changes that might occur to a rock in response to metamorphic processes.

4 Slate and phyllite resemble each other. How might you distinguish one from the other?

5 Each of the following statements describes one or more characteristics of a particular metamorphic rock. For each statement, name the metamorphic rock that is being described.
 a Calcite-rich and nonfoliated.
 b Foliated and composed mainly of granular materials.
 c Represents a grade of metamorphism between slate and schist.
 d Very fine-grained and foliated; excellent rock cleavage.
 e Foliated and composed of more than 50 percent platy minerals.
 f Often composed of alternating bands of light and dark silicate minerals.
 g Hard, nonfoliated rock resulting from contact metamorphism.

6 Distinguish between contact metamorphism and regional metamorphism. Which creates the greatest quantity of metamorphic rock?

7 What feature would make schist and gneiss easily distinguishable from quartzite and marble?

8 Briefly describe the textural and mineralogical differences among slate, mica schist, and gneiss. Which one of these rocks represents the highest degree of metamorphism?

9 Are migmatites associated with high-grade or low-grade metamorphism?

CHAPTER SIX

MASS WASTING AND THE WORK OF RUNNING WATER

MASS WASTING

Landslides are spectacular examples of a common geologic process called mass wasting. **Mass wasting** refers to the downslope movement of rock, regolith, and soil under the direct influence of gravity. Once weathering breaks up rock, gravity pulls the material to lower elevations, where streams usually carry it away, eventually depositing it in the ocean. In this manner the earth's landscape is slowly being shaped.

Although gravity is the controlling force of mass wasting, other factors play an important part in bringing about the downslope movement of material. Water is one of these factors. When the pores in sediment become filled with water, the cohesion between the particles is destroyed, allowing them to slide past one another with relative ease. For example, when sand is slightly moist, it may stick together quite well. However, if more water is added, filling the openings between the grains, the sand will ooze out in all directions. Thus, saturation reduces the internal resistance of materials, which are then easily set in motion by the force of gravity. When clay is wetted, it becomes very slick—another example of the "lubricating" effect of water. Water also adds considerable weight to a mass of material. The added weight in itself may be enough to cause the material to slide or flow downslope.

Oversteepening of slopes is another cause for many mass movements. Loose, undisturbed particles assume a stable slope called the **angle of repose**, the steepest angle at which material remains stable. Depending upon the size and shape of the particles, the angle varies from 25 to 40 degrees. The larger, more angular particles maintain the steepest slopes. If the angle is increased, the rock debris will adjust by moving downslope. There are many situations in nature where this takes place. A stream undercutting a valley wall and waves pounding against the base of a cliff are but two familiar examples. Furthermore, through their activities, people often create oversteepened and unstable slopes that become prime sites for mass wasting.

Classification of Mass Wasting Processes

There is a broad complex of activities that geologists call mass wasting. Generally the different types are divided and described on the basis of the type of material involved, the kind of motion that is displayed, and by the velocity of the movement.

The classification of mass wasting processes based on the material involved in the movement depends upon whether the descending mass began as unconsolidated material or as bedrock. If soil and regolith dominate, terms such as "debris," "mud," or "earth" are used in the description. On the other hand, when a mass of bedrock breaks loose and moves downslope, the term "rock" may be part of the description.

In addition to the type of material involved in a mass wasting event, the way in which material moves may also be important. Generally the kind of motion is described as either a fall, a slide, or a flow.

When the movement involves the free-fall of detached individual pieces of any size, it is termed a *fall*. Fall is a common form of movement on slopes that are so steep that loose material cannot remain on the surface. The rock may fall directly to the base of the slope or move in a series of leaps and bounds over other rocks along the way. Many falls result when freeze and thaw cycles or the action of plant roots loosen rock to the point that gravity takes over.

The Lower Falls of the Grand Canyon of the Yellowstone. An excellent example of a valley in the youthful stage of development. (Photograph used by permission of Dennis Tasa)

Many mass wasting processes are described as *slides*. Slides occur whenever material remains fairly coherent and moves along a well-defined surface. Sometimes the surface is a joint, fault, or bedding plane that is approximately parallel to the slope. However, in the case of the movement called slump, the descending mass moves along a curved surface of rupture. A note of clarification is appropriate at this point. Sometimes the word "slide" is used as a synonym for the word "landslide." It should be pointed out that although many people, including geologists, use the term, the word "landslide" has no specific definition in geology. Rather it should be considered as a popular nontechnical term to describe all perceptible forms of mass wasting, including those in which sliding does not occur.

The third type of movement common to mass wasting is termed *flow*. Flow occurs when material moves downslope as a viscous fluid. Most flows are saturated with water and typically move as lobes or tongues.

When mass wasting events make the news, a large quantity of material has in all likelihood moved rapidly downslope and had a disastrous effect upon people and property. Indeed, during events called *rock avalanches*, rock and debris can move downslope at speeds well in excess of 200 kilometers (125 miles) per hour. Many researchers believe that rock avalanches, such as the one that produced the scene in Figure 6.1, must literally "float on air" as they move downslope. That is, high velocities result when air becomes trapped and compressed beneath the

FIGURE 6.1 Debris deposited atop Sherman Glacier in Alaska by a rock avalanche. The event was triggered by a tremendous earthquake in March, 1964. (Photo by Austin Post, U.S. Geological Survey)

falling mass of debris, allowing it to move as a buoyant, flexible sheet across the surface.

Most mass movements, however, do not move with the speed of a rock avalanche. In fact, a great deal of mass wasting is imperceptibly slow. One process that we will examine later, termed creep, results in particle movements that are usually measured in millimeters or centimeters per year. Thus, as you can see, rates of movement can be spectacularly sudden or exceptionally gradual. Although various types of mass wasting are often classified as either rapid or slow, such a distinction is highly subjective because there is a wide range of rates between the two extremes. Even the velocity of a single process at a particular site can vary considerably from one time to another.

Slump

Slump refers to the downward slipping of a mass of rock or unconsolidated material moving as a unit along a curved surface (Figure 6.2). Usually the slumped material does not travel spectacularly fast nor very far. This is a very common form of mass wasting, especially in thick accumulations of cohesive materials such as clay. The surface of rupture beneath the slump block is characteristically spoon-shaped and concave upward or outward. As the movement occurs, a crescent-shaped scarp (cliff) is

created at the head and the block's upper surface is tilted backwards.

Slump commonly occurs because a slope has been oversteepened. The material on the upper portion of a slope is held in place by the material at the bottom of the slope. As this anchoring material at the base is removed, the material above is made unstable and reacts to the pull of gravity. Slumping may also occur when a slope is overloaded, causing internal stress on the material below. This type of slump often occurs where weak, clay-rich material underlies layers of stronger, more resistant rock such as sandstone. The seepage of water through the upper layers reduces the strength of the clay below and slope failure results.

Rockslide

Rockslides occur when blocks of bedrock break loose and slide downslope. Such events are among the fastest and most destructive mass movements. Usually rockslides take place in a geologic setting where the rock strata are inclined, or joints and fractures exist, parallel to the slope. If the rock is undercut at the base of the slope, it loses support and the rock eventually gives way. Sometimes the rockslide is triggered when rain or melting snow lubricates the underlying surface to the point where friction is no

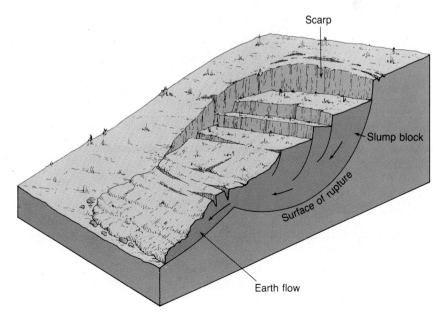

FIGURE 6.2 Slump occurs when material slips downslope en masse along a curved surface of rupture.

longer sufficient to hold the rock unit in place. As a result, rockslides tend to be most prevalent during the spring, when heavy rains and melting snow are greatest. Earthquakes are another mechanism that often triggers rockslides and other mass movements. The 1811 earthquake at New Madrid, Missouri, for example, caused landslides in an area of more than 13,000 square kilometers (5000 square miles) along the Mississippi River valley. In terms of property damage, the most devastating effects of the famous 1964 Alaska earthquake were the landslides in Anchorage, some 130 kilometers (80 miles) from the center of the quake.

The Gros Ventre River flows west from the Wind River Range in northwestern Wyoming, through Grand Teton National Park, eventually emptying into the Snake River. On June 23, 1925, a classic rockslide took place in its valley, just east of the small town of Kelly. In the span of only a few minutes a great mass of sandstone, shale, and soil crashed down the south side of the valley, carrying with it a dense pine forest. The volume of debris, estimated at 38 million cubic meters (50 million cubic yards) created a 67–75-meter-(220–250-foot-) high dam on the Gros Ventre River (Figure 6.3). The river was completely blocked, creating a lake, which filled so quickly that a house that had been 18 meters (60 feet) above the river was floated off its foundation 18 hours after the slide. In 1927 the lake overflowed the dam, partially draining the lake and resulting in a devastating flood downstream.

Why did the Gros Ventre rockslide take place? Figure 6.4 is a diagrammatic cross-sectional view of the geology of the valley. Notice the following points: (1) The sedimentary strata in this area dip (tilt) 15–21 degrees; (2) Underlying the bed of sandstone is a relatively thin layer of clay; and (3) At the bottom of the valley the river had cut through much of the sandstone layer. During the spring of 1925 water from the heavy rains and melting snow seeped through the sandstone, saturating the clay below. Since much of the sandstone layer had been cut through by the Gros Ventre River, the layer had virtually no support at the bottom of the slope. Eventually the sandstone could no longer hold its position on the wetted clay, and gravity pulled the mass down the side of the valley. The circumstances at this location were such that a rockslide was inevitable.

Mudflow

Mudflow is a relatively rapid type of mass wasting that involves a flowage of debris containing a large amount of water. Mudflows are most characteristic of canyons and gullies in semiarid mountainous regions (Figure 6.5). When such an area experiences a

FIGURE 6.3 The scar, 2.4 kilometers (1.5 miles) long and 0.8 kilometer (0.5 mile) wide, left on the side of Sheep Mountain by the Gros Ventre rockslide. (Courtesy of Ward's Natural Science Establishment, Inc., Rochester, N.Y.)

FIGURE 6.4 Cross-sectional view of the Gros Ventre rock-slide. The slide occurred when the tilted and undercut sandstone bed could no longer maintain its position atop the saturated bed of clay. [After W. C. Alden, "Landslide and Flood at Gros Ventre, Wyoming." *Transactions* (AIME) 76 (1928) : 348]

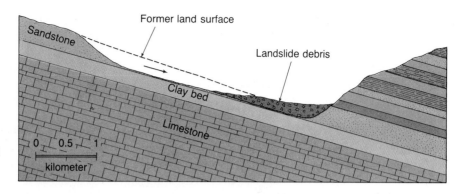

heavy rain, large quantities of sediment are washed into the channel from the valley walls, which usually have little or no vegetation to anchor the loose material. The end product is a rapidly flowing tongue of well-mixed mud, soil, rock, and water. Its consistency may range from that of wet concrete to a soupy mixture not much thicker than muddy water. Because of its high density a mudflow can often carry or push large boulders, trees, and even houses with relative ease. Upon reaching the end of a steep, narrow canyon, the mudflow spreads out, covering the area beyond the mouth of the canyon with a mixture of wet debris.

Mudflows are also common in some volcanic regions. Here the layers of fine-grained volcanic ash which mantle the steep volcanic slopes may form large mudflows following sudden heavy rains or periods of snowmelt. Flows such as this were triggered by the period of activity at Mount St. Helens in 1980. The increased heat flow from the volcano caused rapid melting of great quantities of snow. Some of the ash-rich mudflows that resulted were quite extensive.

Earthflow

Unlike mudflows, which are usually confined to channels in semiarid regions, **earthflows** most often occur on hillsides in humid areas as the result of excessive rainfall. When water saturates clay-rich regolith on a hillslope, the material may break away and flow a short distance downslope, leaving a scar

FIGURE 6.5 View of the great Slumgullion mudflow from its source to Lake San Cristobal, a vertical distance of about 780 meters (2600 feet). Since it occurred in prehistoric times, there is no direct information about its rate of movement. (Photo by W. Cross, U.S. Geological Survey)

FIGURE 6.6 Earthflow.
(Photo by E. J. Tarbuck)

on the hillside (Figure 6.6). Depending upon the steepness of the slope and the material's consistency, the speed of an earthflow may vary from a few meters per hour to several meters per minute. However, since earthflows are quite viscous, they generally move more slowly than the more fluid mudflows described earlier. In addition to occurring as isolated hillside phenomena, earthflows commonly take place in association with large slumps. In this situation, they may be seen as tonguelike flows at the base of the slump block (see Figure 6.2).

Creep

Movements such as rockslides and rock avalanches are certainly the most spectacular and catastrophic forms of mass wasting. Since these events have been known to kill thousands, they deserve intensive study so that, through more effective prediction, timely warnings and better controls can save lives. However, because of their large size and spectacular nature they give us a false impression of their importance as a mass wasting process. Indeed, sudden movements are responsible for moving less material than the slow and far more subtle action of creep. Whereas rapid types of mass wasting are characteristic of mountains and steep hillsides, creep can take place on gentle slopes and is thus much more widespread.

Creep is a type of mass wasting that involves the gradual downhill movement of soil and regolith. One of the primary causes of creep is the alternate expansion and contraction of surface material caused by freezing and thawing or wetting and drying. As shown in Figure 6.7, freezing or wetting lifts the soil at right angles to the slope (solid lines), and thawing or drying allows the particles to fall back to a slightly lower level (dashed lines). Each cycle therefore moves the material a short distance downhill. Creep may also be initiated if the ground becomes saturated with water. Following a heavy rain or snowmelt, a waterlogged soil may lose its internal cohesion, allowing gravity to pull the material down-

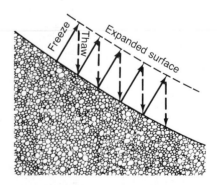

FIGURE 6.7 The repeated expansion and contraction of the surface material causes a net downward migration of rock particles—a process called creep.

slope. Although the movement is imperceptibly slow, its effects are recognizable. Creep causes fences and telephone poles to tilt, and tree trunks will often be bent as a consequence of this movement.

Solifluction

In the frigid zones found at high latitudes **solifluction** is of major importance. In such regions only the ice in the upper few meters of regolith melts during the spring and summer, while the ground below remains permanently frozen. Since the water in the upper zone has nowhere to go, this layer remains saturated and slowly flows down even the gentlest slopes (Figure 6.8). In this manner the overburden is removed and the unweathered rock below is exposed. When the newly exposed rock is eventually weathered, it too will be removed by solifluction.

THE HYDROLOGIC CYCLE

The amount of water on earth is immense, an estimated 1.36 billion cubic kilometers (326 million cubic miles). Of this total, the vast bulk—97.2 percent—is part of the world ocean. Icecaps and glaciers account for another 2.15 percent, leaving only 0.65 percent to be divided among lakes, streams, subsurface water, and the atmosphere. Although the percentage of the earth's total water found in each of the latter sources is but a small fraction of the total inventory, the absolute quantities are great.

An adequate supply of water is vital to life on earth. With increasing demands on this finite resource, scientists have given a great deal of attention to the continuous exchanges of water among the oceans, the atmosphere, and the continents. This unending circulation of the earth's water supply has come to be called the **hydrologic cycle**. It is a gigantic system powered by energy from the sun in which the atmosphere provides the vital link between the oceans and continents. Water from the oceans, and to a much lesser extent from the continents, is constantly evaporating into the atmosphere. Winds transport the moisture-laden air, often great distances, until the complex processes of cloud formation are set in motion that eventually result in precipitation. The precipitation that falls into the ocean has ended its cycle and is ready to begin an-

FIGURE 6.8 Solifluction lobes northeast of Fairbanks, Alaska. (Photo by James E. Patterson)

other. The water that falls on the continents, however, must still make its way back to the ocean.

What happens to precipitation once it has fallen on land? A portion of the water soaks into the ground, some of it moving downward, then laterally, finally seeping into lakes, streams, or directly into the ocean. When the rate of rainfall is greater than the earth's ability to absorb it, the additional water flows over the surface into lakes and streams. Much of the water which soaks in **(infiltration)** or runs off **(runoff)** eventually finds its way back to the atmosphere because of evaporation from the soil, lakes, and streams. Also, some of the water that infiltrates the ground surface is absorbed by plants, which then release it into the atmosphere. This process is called **transpiration**. Each year a field of crops may transpire an amount of water equivalent to a layer 60 centimeters (2 feet) deep over the entire field, whereas a forest may pump twice this amount into the atmosphere.

When precipitation falls at high elevations or high latitudes, the water may not immediately soak in, run off, or evaporate. Instead it may become part of a snowfield or glacier. Glaciers store large quantities of water on land. If present-day glaciers were to melt and release their storage of water, sea level would rise by several tens of meters and submerge many heavily populated coastal areas. As we shall see in Chapter Eight, over the past two million years, huge continental glaciers have formed and melted on several occasions.

A diagram of the earth's water balance, a quantitative view of the hydrologic cycle, is shown in Figure 6.9. Although the amount of water vapor in the air at any one time is but a minute fraction of the earth's total water supply, the absolute quantities

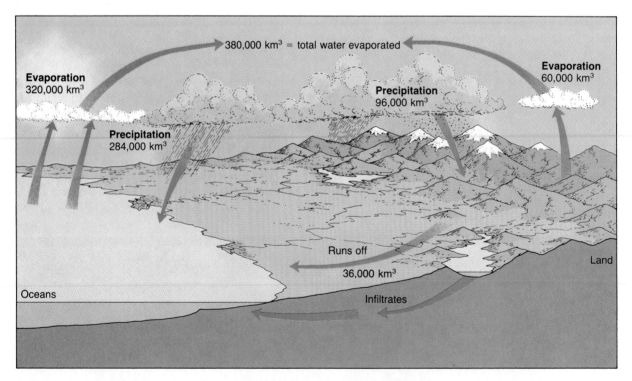

FIGURE 6.9 The earth's water balance. About 320,000 cubic kilometers of water are evaporated each year from the oceans, while evaporation from the land (including lakes and streams) contributes 60,000 cubic kilometers of water. Of this total of 380,000 cubic kilometers of water, about 284,000 cubic kilometers fall back to the ocean, and the remaining 96,000 cubic kilometers fall onto the earth's land surface. Since 60,000 cubic kilometers of water evaporate from the land, 36,000 cubic kilometers of water remain to erode the land during the journey back to the oceans.

that are cycled through the atmosphere over a one-year period are immense—some 380,000 cubic kilometers—enough to cover the earth's surface to a depth of about one meter (39 inches). Since the total amount of water vapor in the atmosphere remains about the same, average annual precipitation over the earth must be equal to the quantity of water evaporated. However, for all of the continents taken together, precipitation exceeds evaporation. Conversely, over the oceans, evaporation exceeds precipitation. Since the level of the world ocean is not dropping, runoff from land areas must balance the deficit of precipitation over the oceans.

The hydrologic cycle represents the continuous movement of water from the oceans to the atmosphere, from the atmosphere to the land, and from the land back to the sea. The wearing down of the earth's land surface is primarily attributable to the last of the steps in this cycle.

RUNNING WATER

Among geologic processes, running water is of great importance to people. We depend upon rivers for energy, travel, and irrigation; their fertile floodplains have fostered human progress since the dawn of civilization. As the dominant agent of landscape alteration, streams have sculptured much of our physical environment.

Although people have always depended to a great extent on running water, its source eluded them for centuries. It was not until the sixteenth century that they first realized streams were supplied by runoff and underground water, which ultimately had their sources as rain and snow. Runoff initially flows in broad sheets that enter small rills, which carry it to a stream. The word *stream* is used to denote channelized flow of any size, from the smallest brook to the mighty Amazon. Although the terms *river* and *stream* are used synonymously, the term *river* is often preferred when describing a main stream into which several tributaries flow.

Streamflow

Flowing water makes its way to the sea under the influence of gravity. The time required for the journey depends upon the velocity of the stream, which is measured in terms of the distance the water travels in a given unit of time. Some sluggish streams travel at less than 0.8 kilometer (0.5 mile) per hour, while a few rapid ones reach speeds as high as 32 kilometers (20 miles) per hour. Velocities are determined at gauging stations where measurements are taken at several locations across the channel and then averaged. Along straight stretches the highest velocities are near the center of the channel just below the surface, where friction is lowest. But when a stream curves, its zone of maximum speed shifts toward its outer bank.

The ability of a stream to erode and transport materials is directly related to its velocity; thus, it is a very important characteristic. Even slight variations in velocity can lead to significant changes in the load of sediment transported by the water. Several factors determine the velocity of a stream and therefore control the amount of erosional work a stream may accomplish. These factors include the following: (1) Gradient, (2) Shape, size, and roughness of the channel, and (3) Discharge.

Certainly one of the more obvious factors controlling stream velocity is the **gradient**, or slope, of a stream channel. Gradient is typically expressed as the vertical drop of a stream over a fixed distance. Gradients may vary considerably from one stream to another as well as along the course of a given stream. Portions of the lower Mississippi River, for example, have gradients of 10 centimeters per kilometer and less. By way of contrast, some mountain stream channels decrease in elevation at a rate of more than 40 meters per kilometer, 400 times more abruptly than the lower Mississippi. The higher the gradient, the more energy available for streamflow. If two streams were identical in every respect except gradient, the stream with the higher gradient would obviously have the greater velocity.

The cross-sectional shape of a channel determines the amount of water in contact with the channel and hence affects the frictional drag. The most efficient channel is one with the least perimeter for its cross-sectional area. Figure 6.10 compares three types of channels. Although the cross-sectional area of all three is identical, the semicircular shape has less water in contact with the channel than the others and therefore less frictional drag. As a result, if all other factors are equal, the water will flow more rapidly in the semicircular channel.

FIGURE 6.10 Influence of channel shape on velocity. Although the cross-sectional area of all three channels is the same, the semicircular channel has less water in contact with the channel, and hence less frictional drag. As a result, the water will flow more rapidly in this channel, all other factors being equal.

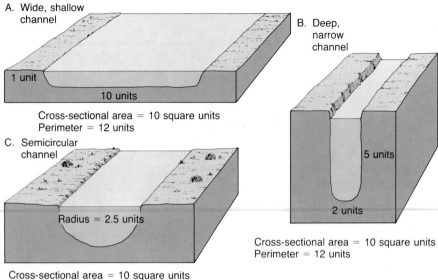

A. Wide, shallow channel

1 unit

10 units

Cross-sectional area = 10 square units
Perimeter = 12 units

B. Deep, narrow channel

5 units

2 units

Cross-sectional area = 10 square units
Perimeter = 12 units

C. Semicircular channel

Radius = 2.5 units

Cross-sectional area = 10 square units
Perimeter = 7.9 units

The size and roughness of the channel also affect the amount of friction. An increase in the size of a channel reduces the ratio of perimeter to cross-sectional area and therefore increases the efficiency of flow. The effect of roughness is obvious. A smooth channel promotes a more uniform flow, while an irregular channel filled with boulders creates enough turbulence to significantly retard the stream's forward motion.

The **discharge** of a stream is the amount of water flowing past a certain point in a given unit of time. This is usually measured in cubic meters per second or cubic feet per second. Discharge is found by multiplying a stream's cross-sectional area by its velocity.

The largest river in North America, the Mississippi, discharges an average of 17,715 cubic meters (625,340 cubic feet) per second. Although this is a huge quantity of water, it is nevertheless dwarfed by the mighty Amazon, the world's largest river. Draining an area nearly three-quarters the size of conterminous United States and averaging about 200 centimeters of rain per year, the Amazon discharges 10 times more water than the Mississippi. In fact, the flow of the Amazon accounts for about 15 percent of all the fresh water discharged into the ocean by all of the world's rivers. Just one day's discharge would supply the water needs of New York City for approximately 9 years!

The discharges of rivers are far from constant. This is true because of such variables as rainfall and snowmelt. If discharge changes, the factors noted earlier must also change. When discharge increases, the width or depth of the channel must increase or the water must flow faster, or some combination of these factors must change. Indeed, measurements show that when the amount of water in a stream increases, the width, depth, and velocity all increase in an orderly fashion. In order to handle the additional water, the stream will increase the size of its channel by widening and deepening it. As we saw earlier, when the size of the channel increases, proportionally less of the water is in contact with the bed and banks of the channel. This means that friction, which acts to retard the flow, is relatively decreased. The less friction, the more swiftly the water will flow.

Changes Downstream

One useful way of studying a stream is to examine its **longitudinal profile**. Such a profile is simply a cross-sectional view of a stream from its source area (called the *head* or *headwaters*) to its *mouth*, the point downstream where the river empties into another water body. By examining Figure 6.11, you can see that the most obvious feature of a typical longitudinal profile is a constantly decreasing gradient from the head to the mouth. Although many local irregularities may exist, the overall profile is a smooth concave-upward curve.

FIGURE 6.11 A longitudinal profile is a cross section along the length of a stream. Note the concave-upward curve of the profile, with a steeper gradient upstream and a gentler gradient downstream.

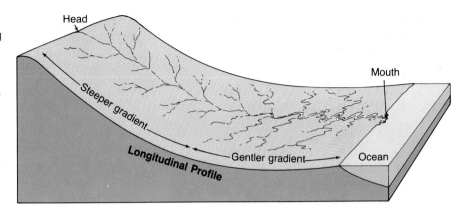

The longitudinal profile shows that the gradient decreases downstream. To see how other factors change in a downstream direction, observations and measurements must be made. When data are collected from successive gauging stations along a river, they show that discharge increases toward the mouth. This should come as no surprise since, as we move downstream, more and more tributaries contribute water to the main channel. Furthermore, in most humid regions, additional water is continually being added from the groundwater supply. Since this is the case, the width, depth, and velocity must change in response to the increased volume of water carried by the stream. Indeed, the downstream changes in these variables have been shown to vary in a manner similar to what occurs when discharge increases at one place; that is, width, depth, and velocity all increase systematically.

The observed increase in velocity that occurs downstream contradicts our impressions about wild, rushing mountain streams and wide, placid rivers. The mental picture that we may have of "old man river just rollin' along" is just not so. Although a mountain stream may have the appearance of a raging torrent, its average velocity is often less than for the river near its mouth. The difference is primarily attributable to the greater efficiency of the larger channel in a downstream direction.

In the headwaters region where the gradient is steep, the water must flow in a relatively small and often boulder-strewn channel. The small channel and rough bed create great drag and inhibit movement by sending water in all directions with almost as much backward motion as forward motion. However, downstream, the material on the bed of the stream becomes much smaller, offering less resist-

ance to flow, and the width and depth of the channel increase to accommodate the greater discharge. These factors, especially the wider and deeper channel, permit the water to flow more freely and hence more rapidly.

In summary, we have seen that there is an inverse relationship between gradient and discharge. Where gradient is high, discharge is small, and where discharge is great, gradient is small. Stated another way, a stream can maintain a higher velocity near its mouth even though it has a lower gradient than upstream because of the greater discharge, larger channel, and smoother bed.

The Effect of Urbanization on Discharge

When rains occur, stream discharge increases. If the rains are sufficiently heavy, the ability of the channel to contain the discharge is exceeded and water spills over the banks as a flood. Floods are natural events that should be expected. However, when cities are built, the magnitude and frequency of flooding often increase.

Figure 6.12A is a hypothetical hydrograph that shows the time relationship between a rainstorm and the occurrence of flooding. Notice that the water level in the stream does not rise at the onset of precipitation because time is needed for water to move from the place where it fell to the stream. This time difference is called the *lag time*.

When an area changes from being predominantly rural to largely urban, streamflow is affected. The effect of urbanization on streamflow is illustrated by the hydrograph in Figure 6.12B. Notice that after urbanization the peak discharge during a flood is

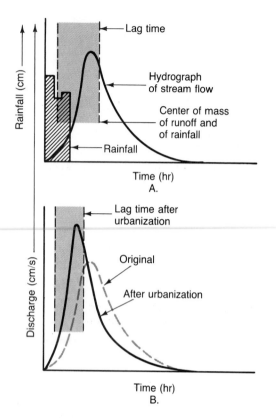

FIGURE 6.12 Generalized hydrographs. **A.** The typical lag time between the time when most of the rainfall occurs and when flooding occurs. **B.** Decreased lag time and higher floodstage due to urbanization. (After L. B. Leopold, U.S. Geological Survey Circular 559, 1968)

greater and the lag time between precipitation and flood peak is shorter than before urbanization. The explanation for this effect is relatively simple. The construction of streets, parking lots, and buildings covers the ground that once soaked up water. Thus, less water infiltrates and the rate and amount of runoff increase. Further, since much less water soaks into the ground, the low-water (dry-season) flow in urban streams, which is maintained by the seepage of groundwater into the channel, is often significantly reduced. As one might expect, the magnitude of these effects is a function of the percentage of land surface that is covered by impermeable surfaces.

Urbanization is just one example of human interference with streams. There are many other ways that land use inadvertently influences the flow of streams and the work they carry out. Moreover, there are also many ways by which people intentionally attempt to manipulate and control streams. Some of these will be discussed later in this chapter.

BASE LEVEL AND GRADED STREAMS

An important control over streamflow is base level. **Base level** is the lowest point to which a stream can erode its channel. Two general types of base level exist. Sea level is considered the **ultimate base level**, since it represents the lowest level to which stream erosion could lower the land. **Temporary**, or **local, base levels** include lakes, resistant rock, and main streams which act as base levels for their tributaries. All have the capacity to limit a stream at a certain level. For example, when a stream enters a lake, its velocity quickly approaches zero and its ability to erode decreases. Thus the lake prevents the stream from eroding below its level at any point upstream from the lake (Figure 6.13A). However, since the outlet of the lake can cut downward and drain the lake, the lake is only a temporary hindrance to the stream's ability to downcut its channel.

Any change in base level will cause a corresponding readjustment of stream activities. When a dam is built along a stream course, the reservoir which forms behind it raises the base level of the stream (Figure 6.13B). Upstream from the dam the stream gradient is reduced, lowering its velocity and, hence, its sediment-transporting ability. The stream, now unable to transport all of its load, will deposit material, thereby building up its channel. This process continues until the stream again has a gradient sufficient to carry its load. The profile of the new channel would be similar to the old, except that it would be somewhat higher.

If, on the other hand, the base level were lowered, either by an uplift of the land or by a drop in base level, the stream would readjust. The stream, now above base level, would have excess energy and would downcut its channel to establish a balance with its new base level (Figure 6.13C). Erosion would first progress near the mouth, then work upstream until the stream profile was adjusted along its full length.

The observation that streams adjust their profile for changes in base level led to the concept of a

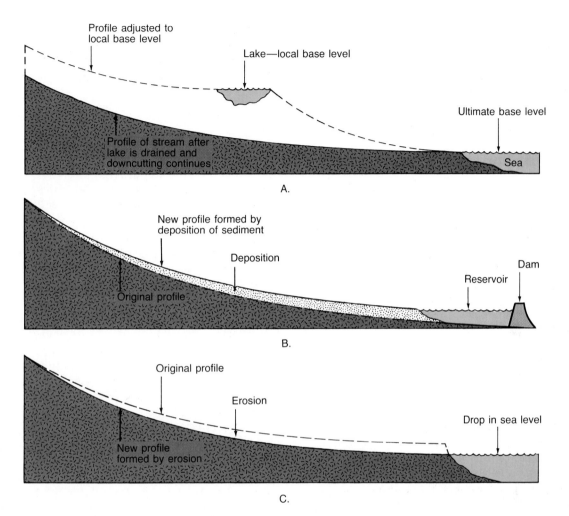

FIGURE 6.13 A. Effect of a local base level on a stream profile. **B.** Adjustment of a stream profile to a change in base level. **C.** Readjustment of a stream to a lower base level.

graded stream. A **graded stream** has the correct slope and other channel characteristics necessary to maintain just the velocity required to transport the material supplied to it. On the average, a graded system is not eroding or depositing material but is simply transporting it. Once a stream has reached this state of equilibrium, it becomes a self-regulating system in which a change in one characteristic causes an adjustment in the others to counteract the effect. Referring again to our example of a stream adjusting to a lowering of its base level, the stream would not be graded while cutting its new channel but would achieve this state after downcutting had ceased.

WORK OF STREAMS

The work of streams includes erosion, transportation, and deposition. These activities go on simultaneously in all stream channels, even though they are presented individually here.

Erosion

Although much of the material carried by streams has been brought in by underground water, overland flow, and mass wasting, streams do contribute to their load by eroding their own channels. If a channel is composed of bedrock, most of the erosion is accomplished by the abrasive action of water armed with sediment, a process analogous to sand-blasting. Pebbles caught in eddies serve as cutting tools and bore circular holes called **potholes** into the channel floor. In channels consisting of uncon-solidated material, considerable lifting can be ac-complished by the impact of water alone.

Transportation

Once streams acquire their load of sediment, they transport it in three ways: (1) in solution **(dissolved load)**; (2) in suspension **(suspended load)**; and (3) along the bottom **(bed load)**.

The dissolved load is brought to the stream by groundwater and to a lesser degree is acquired di-rectly from soluble rock along the stream's course. The quantity of material carried in solution is highly variable and depends upon such factors as climate and the geologic setting. Usually the dissolved load is expressed as parts of dissolved material per million parts of water (parts per million, or ppm). Although some rivers may have a dissolved load of 1000 ppm or more, the average figure for the world's rivers is estimated to be between 115 and 120 ppm. Almost 4 billion metric tons of dissolved mineral matter are supplied to the oceans each year by streams.

Most streams, but not all, carry the bulk of their load as suspended load. Usually only fine sand-, silt-, and clay-sized particles can be carried this way, but during floodstage larger particles are carried as well. Also during floodstage, the total quantity of material carried in suspension increases dramatically, as can be verified by persons whose homes have been sites for the deposition of this material. During floodstage the Hwang Ho (Yellow River) of China is reported to carry an amount of sediment equal in weight to the water that carries it. Rivers like this are often appro-priately described as "too thick to drink, but too thin to cultivate."

A portion of a stream's load of solid material con-sists of sediment that is too large to be carried in suspension. These coarser particles move along the bottom of the stream and constitute the bed load. In terms of the erosional work accomplished by a downcutting stream, the grinding action of the bed load is of great importance.

The particles composing the bed load move along the bottom by rolling, sliding, and saltation. Sedi-ment moving by **saltation** appears to jump or skip along the stream bed. This occurs as particles are propelled upward by collisions or sucked upward by the current and then carried downstream a short dis-tance until gravity pulls them back to the bed of the stream. Particles that are too large or heavy to move by saltation either roll or slide along the bottom, depending upon their shape.

Unlike the suspended and dissolved loads, which are constantly in motion, the bed load is in motion only intermittently, when the force of the water is sufficient to move the larger particles. Although the bed load may constitute up to 50 percent of the total load of a few streams, it usually does not ex-ceed 10 percent of a stream's total load. For exam-ple, consider the distribution of the 750 million tons of material carried to the Gulf of Mexico by the Mis-sissippi River each year. Of this total, it is estimated that approximately 500 million tons are carried in suspension, 200 million tons in solution, and the remaining 50 million tons as bed load. Estimates of a stream's bed load, however, should be viewed cau-tiously because this fraction of the load is very diffi-cult to measure accurately. Not only is the bed load more inaccessible than the suspended and dissolved loads, but it moves primarily during periods of flood-ing when the bottom of a stream channel is most difficult to study.

A stream's ability to carry its load is established using two criteria. First, the **competence** of a stream is a measure of the maximum size of particles it is capable of transporting. The stream's velocity determines its competence. If the velocity of a stream doubles, the impact force of water increases four times; if the velocity triples, the force increases nine times; and so forth. Hence, the huge boulders which are visible during the low-water stage and seem immovable can, in fact, be transported during floodstage because of the stream's increased veloc-ity. Second, the maximum load a stream can carry is termed its **capacity**. The capacity of a stream is di-rectly related to its discharge. The greater the amount of water flowing in the stream, the greater the stream's capacity for hauling sediment.

Deposition

Whenever a stream's velocity is reduced, its compe-tence is also lowered. Consequently, some of the suspended particles begin to settle out. Sediment deposited in this manner is termed **alluvium**. Although some material temporarily settles in the channel, most eventually reaches the ocean. When a stream enters the relatively still waters of an ocean or lake, its forward motion is quickly lost, and the resulting deposits form a **delta**. The finer silts and clays will settle out some distance from the mouth into nearly horizontal layers called *bottomset beds* (Figure 6.14A). Prior to the accumulation of bottomset beds, *foreset beds* begin to form. These beds are composed of coarse sediment, which is dropped almost immediately upon entering a lake or ocean, forming sloping layers. The foreset beds are usually covered by thin, horizontal *topset beds* deposited during floodstage. As the delta grows outward, the gradient of the river is continually lowering, causing the stream to search for a shorter route to base level, a process illustrated in Figure 6.14B. This figure shows how a simple delta grows into the idealized triangular shape of the Greek letter delta (Δ), for which it was named.

Large rivers such as the Nile and Mississippi have deltas extending over thousands of square kilometers. The Mississippi delta began forming millions of years ago near the present-day town of Cairo, Illinois, and has since advanced nearly 1600 kilometers (1000 miles) to the south. New Orleans rests where there was ocean less than 5000 years ago, and the present bird foot delta shown in Figure 6.15 has been built in the last 500 years. Figure 6.15 also shows the main channel dividing into several smaller ones called **distributaries**, which are found on most large deltas.

Although deltas are deposited by many large rivers, not all rivers create these features. Even streams that transport large loads of sediment may lack deltas because powerful currents and waves quickly

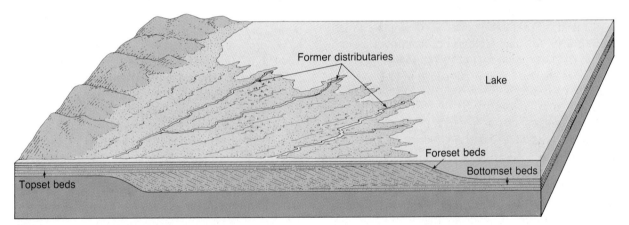

A.

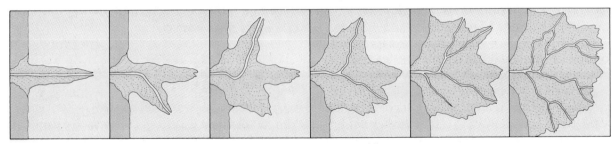

B.

FIGURE 6.14 A. Structure of a simple delta. **B.** Growth of a simple delta. Once a stream has extended its channel, the reduced gradient causes it to find a shorter distance to its base level. (After Ward's Natural Science Establishment, Inc., Rochester, N.Y.)

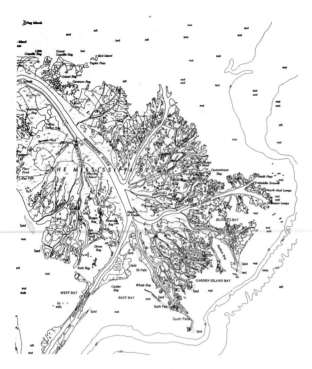

FIGURE 6.15 The bird foot delta of the Mississippi River as depicted on the Breton Sound, Louisiana, 1:250,000 topographic map. Over the past 400 years, the delta has extended the shoreline by more than 32 kilometers (20 miles). The numerous channels on the delta are distributaries. (Courtesy of U.S. Geological Survey)

redistribute the material as soon as it is deposited. The Columbia River in the Pacific Northwest is one such situation. In other cases, rivers do not carry sufficient quantities of sediment to build up a delta. The St. Lawrence River, for example, has little opportunity to pick up much sediment between Lake Ontario and its mouth in the Gulf of St. Lawrence.

Alluvial fans are features similar to deltas which form on land (Figure 6.16). When mountain streams reach a plain, their gradient is abruptly lowered and they immediately dump much of their load. Usually the coarse material is dropped near the base of the slope, while finer material is carried farther out on the plain.

Rivers that occupy valleys with broad, flat valley floors on occasion build a landform called a **natural levee** that parallels its channel. Natural levees are built by successive floods over a period of many years. When a stream overflows its banks, its velocity immediately diminishes, leaving coarse sediment deposited in strips bordering the channel (Figure 6.17). As the water spreads out over the valley a lesser amount of fine sediment is deposited over the valley floor. This uneven distribution of material produces the very gentle slope of the natural levee. The natural levees of the lower Mississippi rise 6 meters (20 feet) above the valley floor. The area behind the levee is characteristically poorly drained for the obvious reason that water cannot flow up the levee and into the river. Marshes called **back swamps** result. A tributary stream that attempts to enter a river with natural levees often has to flow parallel to the main stream until it can breach the levee. Such streams are called **yazoo tributaries** after the Yazoo River, which parallels the Mississippi for over 300 kilometers.

Sometimes artificial levees are built along rivers as a means of flood control. These are usually easy to distinguish from natural levees because their slopes are much steeper. When a river is confined by levees during periods of high water, it deposits material in its channel as the discharge diminishes. This is sediment that otherwise would have been dropped on the floodplain. Thus, each time there is a high flow, deposits are left on the river bed and the bottom of the channel is built up. With the buildup of the bed, less water is required to overflow the original levee. As a result, the height of the levee must be raised to protect the floodplain. For this reason, many levees along the lower Mississippi River have had to be raised to cope with the increasing height of the water in the channel. As you can see, artificial levees are not a permanent solution to the problem of flooding. If protection is to be maintained, the structures must be heightened periodically, a process that cannot go on indefinitely.

STREAM VALLEYS

Stream valleys can be divided into two general types. Narrow V-shaped valleys and wide valleys with flat floors exist as the ideal forms, with many gradations between. In arid regions narrow valleys often have nearly vertical walls, while in humid regions the effect of mass wasting and slope erosion caused by heavy rainfall produce the typical V-shaped valley. The type of valley gives an indication of what the

FIGURE 6.16 A large alluvial fan in Death Valley, California. These structures develop where the gradient of a stream changes abruptly, such as at the foot of a mountain. (Photo by John S. Shelton)

stream has been doing. A narrow V-shaped valley indicates that the primary work of the stream has been downcutting toward base level. On the other hand, streams with flat-floored valleys have been widened by lateral (side-to-side) erosion.

The most prominent features of a narrow valley are **rapids** and **waterfalls**. Both occur where the stream profile drops rapidly, a situation usually caused by variations in the erodibility of the bedrock into which the stream channel is cutting. Figure 6.18 shows how a resistant bed produces a rapids by acting as a temporary base level upstream while allowing downcutting to continue downstream. Once erosion has eliminated the resistant rock, the stream profile smoothes out again. Waterfalls are places where the stream profile makes a vertical drop. One type of waterfall is exemplified by Niagara Falls. The falls are supported by a resistant bed of dolomite

that is underlain by a layer of less resistant shale, similar to the situation illustrated in Figure 6.18. As water plunges over the lip of the falls it erodes the less resistant shale, undermining a section of dolomite, which eventually breaks off. In this manner the waterfall retains its vertical cliff while slowly but continually retreating upstream. Since its formation Niagara Falls has retreated approximately 11 kilometers (7 miles) upstream.

Once a stream has cut its channel closer to base level it begins to reach a graded condition, and downward erosion becomes less dominant. At this point more of the stream's energy is directed from side to side. The reason for this change is not fully understood, but the reduced gradient probably is an important factor. Nevertheless it does occur, and the result is a widening of the valley as the river cuts away first at one bank and then at the other (Fig-

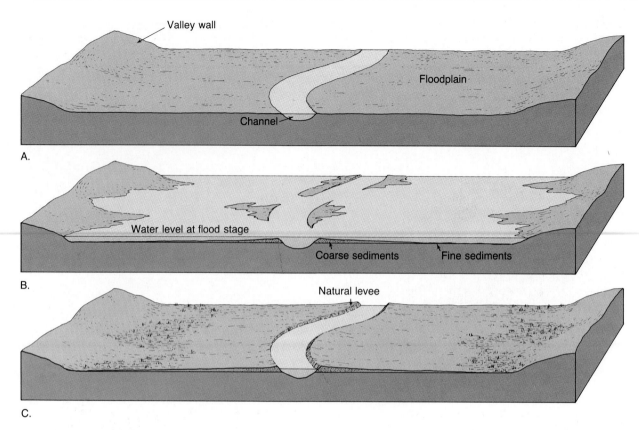

FIGURE 6.17 Formation of natural levees. After repeated flooding, streams may build very gently sloping levees.

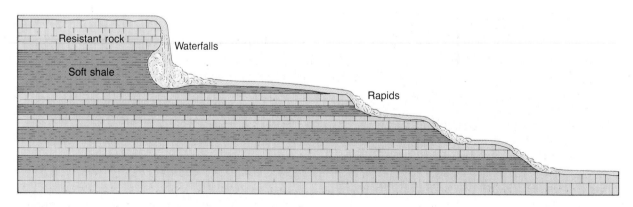

FIGURE 6.18 Formation of rapids and falls on resistant rock.

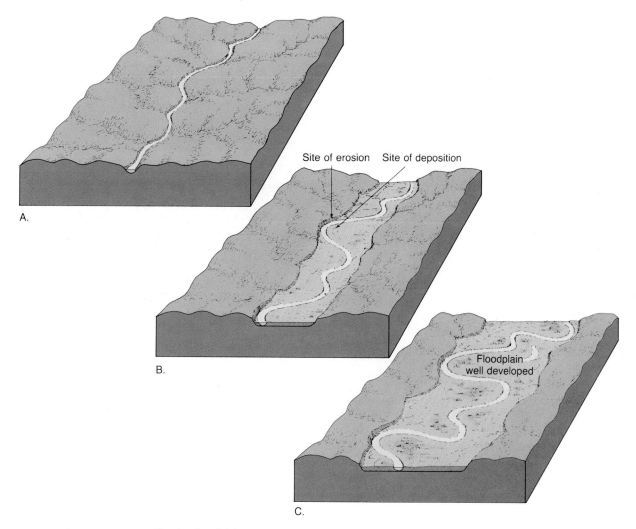

Site of erosion   Site of deposition

A.

B.

Floodplain
well developed

C.

FIGURE 6.19 Stream eroding its floodplain.

ure 6.19). In this manner the flat valley floor, or **floodplain**, is produced. It is appropriately named because the river is confined to its channel, except during floodstage, when it overflows its banks and inundates the floodplain.

When a river erodes laterally, creating a floodplain as just described, it is called an *erosional floodplain*. Floodplains can be depositional in nature as well. *Depositional floodplains* are produced by a major fluctuation in conditions, such as a change in base level. The floodplain in Yosemite Valley is one such feature, and was produced when a glacier

gouged the former stream valley about 300 meters (985 feet) deeper than it had been. After the glacial ice melted, the stream readjusted to its former base level by refilling the valley with alluvium.

Streams that flow upon floodplains, whether erosional or depositional, move in sweeping bends called **meanders**. Meanders continually change position by eroding sideways and slightly downstream. The sideways movement occurs because the maximum velocity of the stream shifts toward the outside of the bend, causing erosion of the outer bank (Figure 6.20). At the same time the reduced

FIGURE 6.20 Lateral movement of meanders. By eroding its outer bank and depositing material on the inside of the bend, a stream is able to shift its channel.

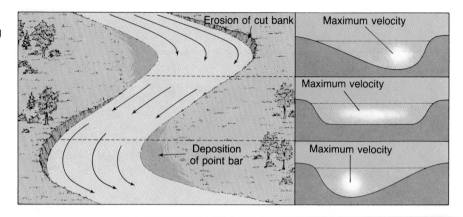

current at the inside of the meander results in the deposition of coarse sediment, especially sand. Thus by eroding its outer bank and depositing material along its inner bank, a stream moves sideways without changing its channel size. Due to the slope of the channel, erosion is more effective on the downstream side of a meander. Therefore, in addition to growing laterally, the bends also gradually migrate down the valley. Sometimes the downstream migration of a meander is slowed when it reaches a more resistant portion of the floodplain. This allows the next meander upstream to "catch up." Gradually the neck of land between the meanders is narrowed. When they get close enough, the river may erode through the narrow neck of land to the next loop (Figure 6.21). The new, shorter channel segment is called a **cutoff** and, because of its shape, the abandoned bend is called an **oxbow lake**.

One method of flood control is to straighten a channel by creating artificial cutoffs. The idea is that by shortening the stream, the gradient and hence the velocity are increased. By increasing the velocity, the larger discharge associated with flooding can be dispersed more rapidly. Since 1932 about 16 artificial cutoffs have been constructed on the lower Mississippi for the purpose of increasing the channel efficiency and reducing the threat of flooding. The program has been somewhat successful in reducing the height of the river in flood. However, since the river's tendency to meander still exists, preventing the river from returning to its previous condition has been difficult.

Artificial cutoffs increase a stream's velocity and may also accelerate erosion of the bed and banks of the channel. A case in point is the Blackwater River in Missouri, whose meandering course was shortened in 1910. Among the many effects of this project was a dramatic increase in the width of the channel caused by the increased velocity of the stream. One bridge collapsed because of bank erosion in

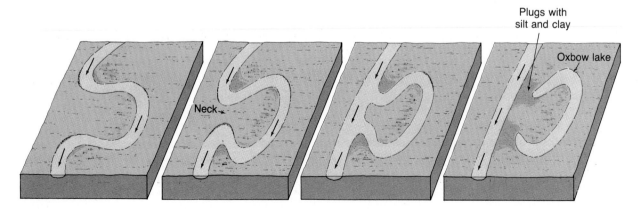

FIGURE 6.21 Formation of a cutoff and oxbow lake.

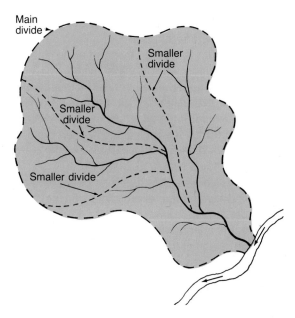

FIGURE 6.22 Drainage basins and divides. Divides separate the drainage basins of each stream. Drainage basins and divides also exist for the smallest tributaries but are not shown.

1930. Over the next 17 years the same bridge was replaced on three more occasions, each time with a wider span.

DRAINAGE SYSTEMS AND PATTERNS

A stream is just one small component in a much larger system. Each system consists of a **drainage basin**, the land area that contributes water to the stream. The drainage basin of one stream is separated from another by an imaginary line called a **divide** (Figure 6.22). Divides range from a ridge separating two small gullies to *continental divides,* which split continents into enormous drainage basins. For example, the continental divide that runs somewhat north-south through the Rocky Mountains separates the drainage which flows west to the Pacific Ocean from that which flows to the Gulf of Mexico. Although divides separate the drainage of two streams, if they are tributaries of the same river, they are both a part of that larger drainage system.

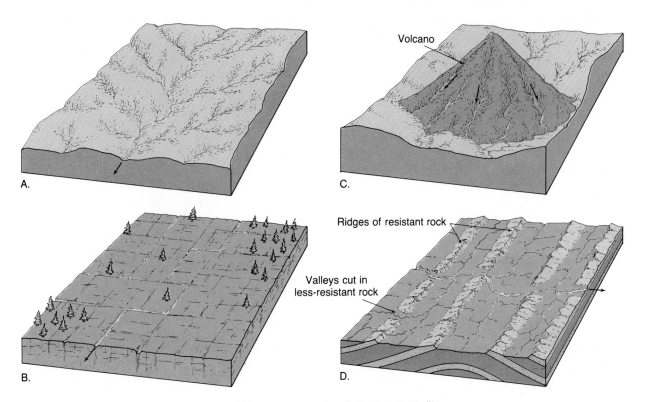

FIGURE 6.23 Drainage patterns. **A.** Dendritic. **B.** Rectangular. **C.** Radial. **D.** Trellis.

All drainage systems are made up of an interconnected network of streams which together form particular patterns. The nature of a drainage pattern can vary greatly from one type of terrain to another, primarily in response to the kinds of rock on which the streams developed or the structural pattern of faults and folds.

Certainly the most commonly encountered drainage pattern is the **dendritic** pattern (Figure 6.23A, previous page). This pattern is characterized by an irregular branching of tributary streams that resembles the branching pattern of a deciduous tree. In fact, the word *dendritic* means "treelike." The dendritic pattern forms where the underlying material is relatively uniform. Since the surface material is essentially uniform in its resistance to erosion, it does not control the pattern of streamflow. Rather, the pattern is determined chiefly by the direction of slope of the land.

Figure 6.23B illustrates a **rectangular** pattern, in which many right-angle bends can be seen. This pattern develops when the bedrock is crisscrossed by a series of joints and/or faults. Since these structures are eroded more easily than unbroken rock, their geometric pattern guides the directions of valleys.

When streams diverge from a central area like spokes from the hub of a wheel, the pattern is said to be **radial** (Figure 6.23C). This pattern typically develops on isolated volcanic cones and domal uplifts.

Figure 6.23D illustrates a **trellis** drainage pattern, a rectangular pattern in which tributary streams are nearly parallel to one another and have the appearance of a garden trellis. This pattern forms in areas underlain by alternating bands of resistant and less resistant rock and is particularly well displayed in the folded Appalachians, where both weak and strong strata outcrop in nearly parallel belts.

STAGES OF VALLEY DEVELOPMENT

Contrary to the popular belief of his day, James Hutton proposed that streams were responsible for cutting the valleys in which they flowed. Later geologic work conducted in stream valleys substantiated Hutton's pronouncement and further revealed that the development of stream valleys progresses in a somewhat orderly fashion. The evolution of a valley has been arbitrarily divided into three sequential stages: youth, maturity, and old age.

As long as the stream is downcutting to establish a graded condition with its base level, it is considered youthful. Rapids, an occasional waterfall, and a narrow V-shaped valley are all visible signs of the vigorous downcutting that is going on. Other features of a youthful stream include a steep gradient, little or no floodplain, and a rather straight course without meanders (Figure 6.24A).

When a stream reaches maturity, downward erosion diminishes and lateral erosion dominates. Thus the mature stream begins actively cutting its floodplain and meandering upon it (Figure 6.24B). During the mature stage cutoffs occur, producing oxbows, and a few streams may even produce low natural levees (Figure 6.24C). In contrast to the gradient of a youthful stream, the gradient of a mature stream is much lower and the profile is much smoother because all rapids and waterfalls have been eliminated.

A stream enters old age after it has cut its floodplain several times wider than its *meander belt,* which is the width of the meander (Figure 6.24D). When this stage is reached the stream is rarely near the valley walls; hence it ceases to significantly enlarge the floodplain. For that reason, the primary work of a river in an old-age valley is the reworking of unconsolidated floodplain deposits. Because this task is easier than cutting bedrock, a stream in an old-age valley shifts more rapidly than a stream in a mature valley. For example, some meanders of the lower Mississippi move 20 meters (66 feet) a year, and its large floodplain is dotted with oxbow lakes and old cutoffs. Natural levees are also common features of old-age valleys and, when present, are accompanied by back swamps and yazoo tributaries.

Thus far we have assumed that the base level of a stream remains constant as a river progresses from youth to old age. On many occasions, however, the land is uplifted or the base level is lowered. The effect of uplifting on a youthful stream is to increase its gradient and accelerate its rate of downcutting. However, uplifting of a mature stream would cause it to abandon lateral erosion and revert to downcutting. Rivers of this type are said to be **rejuvenated**, and the meanders are known as **entrenched meanders** (Figure 6.25). Mature streams may eventually readjust to uplift by cutting a new floodplain at a level below the old one. The remnants of the old

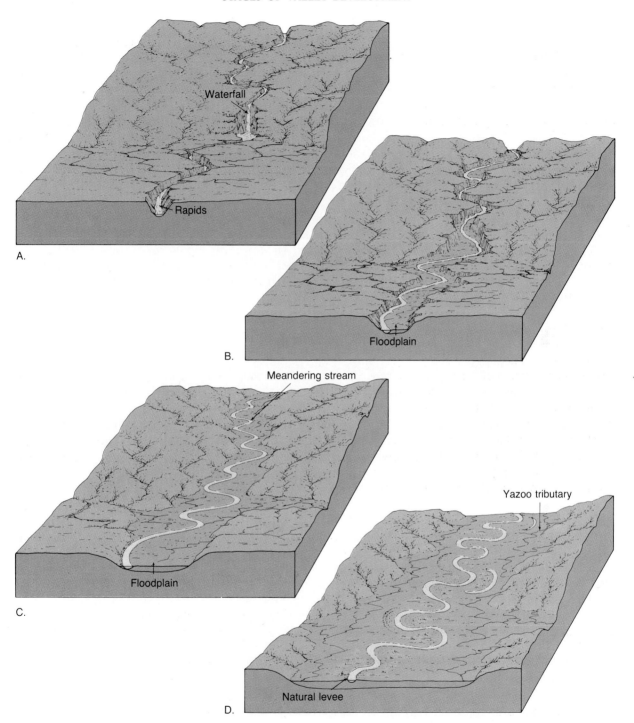

FIGURE 6.24 Stages of valley development. **A.** Youth. The youthful stage is characterized by downcutting and a V-shaped valley. **B.** and **C.** Maturity. Once a stream
has sufficiently lowered its gradient, it begins to erode laterally, producing a wide
valley. **D.** Old age. After the valley has been cut several times wider than the width
of the meander belt, it has entered old age. (After Ward's Natural Science Establishment, Inc., Rochester, N.Y.)

FIGURE 6.25 Entrenched meanders of the San Juan River, Utah. (Photo by John S. Shelton)

higher floodplain are often present in the form of flat surfaces called *terraces*.

Two additional points concerning valley development should be made. First, the time required for a stream to reach any given stage depends on several factors, including the erosive ability of the stream, the nature of the material through which the stream must cut, and the stream's height above base level. Consequently, a stream which starts out very near base level and only has to cut through unconsolidated sediments may reach maturity in a matter of a few hundred years. On the other hand, the Colorado River, where it is actively cutting the Grand Canyon, has retained its youthful nature for an estimated 15 million years. Second, individual portions of a stream reach each stage at different times. Often the lower reaches of a stream attain old age while the headwaters are still youthful in character.

CYCLE OF LANDSCAPE EVOLUTION

While streams are cutting their valleys they simultaneously sculpture the land. To describe this unending process, we need a starting point. For this reason only, we will assume the existence of a relatively flat upland area in a humid region. Until a well-established drainage system forms, lakes and ponds will occupy any depressions that exist (Figure 6.26A). As streams form and begin to downcut and erode headward, they will drain the lakes. During the youthful stage the landscape retains its relatively flat surface, interrupted only by narrow stream valleys (Figure 6.26B). As downcutting continues, relief increases, and the flat, youthful landscape is transformed into one consisting of the hills and valleys that characterize the mature stage (Figure 6.26D).

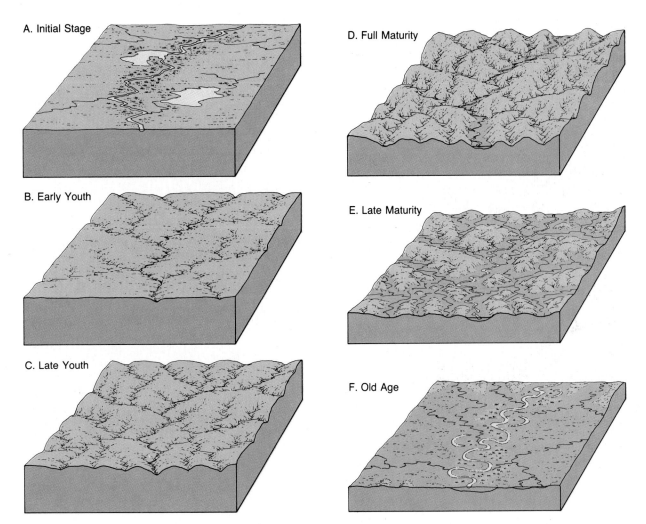

FIGURE 6.26 Idealized cycle of landscape evolution. **A.** Initial stage. The land is poorly drained and situated well above base level. **B.** Youthful stage. Streams have cut downward and drained the lakes. The landscape is still relatively flat between stream valleys. **C.** Late youth. **D.** Mature stage. All of the area between the initial streams has been eroded by running water. Most of the landscape is in slope, and maximum relief exists. **E.** Late maturity. **F.** Old-age stage. Lateral erosion, mass wasting, and slope wash have lowered most of the hills to the level of the floodplain. The entire surface has become an undulating plain near base level. (After Ward's Natural Science Establishment, Inc., Rochester, N.Y.)

Eventually some of the streams will approach base level, and downcutting will give way to lateral erosion. As the cycle nears the old-age stage, the effects of overland flow and mass wasting, coupled with the lateral erosion by streams, will reduce the land to a **peneplain** ("near plain"), an undulating plain near base level (Figure 6.26F). Although no peneplains are known to exist today, there is evidence that they formed in the past and have since been uplifted. Once a peneplain has formed, uplifting starts the cycle again. Most often, uplifting interrupts the cycle before it reaches old age.

REVIEW QUESTIONS

1 What is the controlling force of mass wasting? What other factors are important?

2 Distinguish among fall, slide, and flow.

3 Why can rock avalanches move at such great speeds?

4 What factors led to the massive rockslide at Gros Ventre, Wyoming (Figure 6.4)?

5 Compare and contrast mudflow and earthflow.

6 Describe the mechanism that leads to the slow down-slope movement called creep (Figure 6.7).

7 Why is solifluction only a summertime phenomenon?

8 Describe the movement of water through the hydrologic cycle. Is there more than one path which precipitation may take after it has fallen?

9 A stream starts out 2000 meters above sea level and travels 250 kilometers to the ocean. What is its average gradient in meters per kilometer?

10 Suppose that the stream mentioned in Question 9 developed extensive meanders so that its course was lengthened to 500 kilometers. Calculate its new gradient. How does meandering affect gradient?

11 When the discharge of a stream increases, what happens to the stream's velocity?

12 Why does the downstream portion of a river have a gentle gradient when compared to the headwater region?

13 When an area changes from predominantly rural to largely urban, how is streamflow affected?

14 Define *base level*. Name the main river in your area. For what streams does it act as base level? What is the base level for the Mississippi River?

15 In what three ways does a stream transport its load?

16 If you collect a jar of water from a stream, what part of its load will settle to the bottom of the jar? What portion will remain in the water?

17 Differentiate between competency and capacity.

18 In what way is a delta similar to an alluvial fan? In what way are they different?

19 Why must the height of many artificial levees be increased periodically?

20 What is the purpose of artificial cutoffs?

21 What is a divide?

22 Why is it possible for a youthful valley to be older (in years) than a mature valley?

23 Do streams flowing in mature and old-age valleys make good political boundaries? Explain.

CHAPTER SEVEN

GROUNDWATER

Of all the world's water only about six-tenths of one percent is found underground. Nevertheless, the amount of water stored in the rocks and sediments beneath the earth's surface is vast. The U.S. Geological Survey estimates that the quantity of water in the upper 800 meters (2600 feet) of the continental crust is about 3000 times greater than the volume of water in all rivers at any one time, and nearly 20 times greater than the combined volume in all lakes and rivers.

In many parts of the world, wells and springs provide the water needs not only for great numbers of people but also for crops, livestock, and industry. In the United States, subsurface water supplies about 20 percent of our country's freshwater requirements. In addition, underground water is important as an equalizer of streamflow and as an agent of erosion. The erosional work of subsurface water is responsible for the creation of caverns and many other related features.

DISTRIBUTION OF UNDERGROUND WATER

When rain falls, some of the water runs off, some evaporates, and the remainder soaks into the ground. This last path is the primary source of practically all subsurface water. The amount of water that takes each of these paths, however, varies greatly both in time and space. Several influential factors include steepness of slope, nature of surface material, intensity of rainfall, and type and amount of vegetation. Heavy rains falling upon steep slopes

---

Entrance to the Big Room, Carlsbad Caverns, New Mexico. Groundwater is responsible for creating and decorating the caverns. (Photo courtesy of the National Park Service)

underlain by impervious materials will obviously result in a high percentage of runoff. On the other hand, if rain falls steadily and gently upon more gradual slopes composed of materials that are easily penetrated by the water, a much larger percentage of water soaks into the ground.

Some of the water that soaks in does not travel far, because it is held by molecular attraction as a surface film on soil particles. A portion of this moisture evaporates back into the atmosphere, while much of the remainder serves as a source of water for use by plants between rains. Water that is not held near the surface penetrates downward until it reaches a zone where all of the open spaces in sediment and rock are completely filled with water. The water in this **zone of saturation** is called **groundwater**. The upper limit of this zone is known as the **water table**. The area above the water table where the soil, sediment, and rock are not saturated is known as the **zone of aeration** (Figure 7.1). The open spaces here are filled mainly with air.

THE WATER TABLE

The water table, the upper limit of the zone of saturation, is a very significant feature of the groundwater system. The water table level is important in predicting the productivity of wells, explaining the changes in the flow of springs and streams, and accounting for fluctuations in the levels of lakes.

Although we cannot observe the water table directly, its position can be mapped and studied in detail in areas where wells are numerous, because the water level in wells coincides with the upper boundary of the saturated zone. Such maps reveal that the water table is rarely level as we might expect a table to be. Instead, its shape is usually a subdued replica of the surface topography, reaching its highest elevations beneath hills and then descending toward

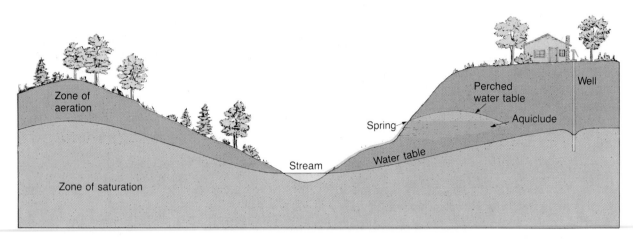

FIGURE 7.1 This diagram illustrates the relative positions of many features associated with subsurface water.

valleys (Figure 7.1). Where a swamp is encountered, the water table is right at the surface, whereas lakes and streams generally occupy areas where the water table is above the land surface.

A number of factors contribute to the irregular surface of the water table. For example, variations in rainfall and permeability from place to place can lead to uneven infiltration and thus to differences in the water table level. However, the most important cause is simply the fact that groundwater moves very slowly and at varying rates under different conditions. Because of this, water tends to "pile up" beneath high areas between stream valleys. If rainfall were to cease completely, these water table "hills" would slowly subside and gradually approach the level of the valleys. However, new supplies of rainwater are usually added frequently enough to prevent this. Nevertheless, in times of extended drought, the water table may drop enough to dry up shallow wells.

The relationship between the water table and a stream in a humid region is illustrated in Figure 7.2A. Even during dry periods, the movement of groundwater into the channel maintains a flow in the stream. In situations such as this, streams are said to be **effluent**. By contrast, in arid regions, where the water table is far below the surface, groundwater does not contribute to streamflow. Therefore, the only permanent streams in such areas are those that originate in wet regions and then happen to traverse the desert. Under these conditions the zone of satu-

ration below the valley floor is supplied by downward seepage from the stream channel which, in turn, produces an upward bulge in the water table. Streams that provide water to the water table in this manner are called **influent streams** (Figure 7.2B).

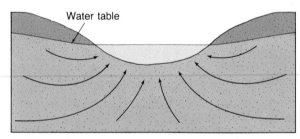

A. Effluent Stream

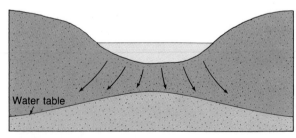

B. Influent Stream

FIGURE 7.2 **A.** Effluent streams are characteristic of humid areas and are supplied by water from the zone of saturation. **B.** Influent streams are found in deserts. Seepage from such streams produces an upward bulge in the water table.

MOVEMENT OF GROUNDWATER

The energy responsible for groundwater movement is provided by the force of gravity. In response to gravity, water moves from areas where the water table is high to zones where the water table is lower; that is, toward a stream channel, lake, or spring. Although some water takes the most direct path down the slope of the water table, much of the water follows long curving paths toward the zone of discharge. As illustrated in Figure 7.3, water percolates into the stream from all possible directions, with some paths turning upward, apparently against the force of gravity, and entering through the bottom of the channel. Such paths are followed because differences in the height of the water table create differences in the groundwater pressure at a particular height. Stated another way, the water at any given height is under greater pressure beneath a hill than beneath a stream channel and the water tends to migrate toward points of lower pressure. Thus, the looping curves followed by water in the saturated zone may be thought of as a compromise between the downward pull of gravity and the tendency of water to move toward areas of reduced pressure.

Depending upon the nature of the subsurface material, the flow of groundwater and the amount of water that can be stored are highly variable. Water soaks into the ground because bedrock, sediment, and soil contain voids or openings. These openings are similar to those of a sponge and are often called *pore spaces*. The quantity of groundwater that can be stored depends on the **porosity** of the material; that is, the percentage of the total volume of rock or sediment that consists of pore spaces. Although these openings often consist of spaces between particles of sediment or sedimentary rock, such features as vesicles (voids left by gases escaping from lava), joints and faults, and cavities formed by the solution of soluble rocks such as limestone are also common.

As one might expect, variations in porosity can be great. Sediment is commonly quite porous, and open spaces may occupy from 10 to 50 percent of the sediment's total volume. The amount of pore space depends on the size and shape of the grains, as well as the packing, degree of sorting, and in the case of sedimentary rocks, the amount of cementing material. For example, clay may have a porosity as high as 50 percent, whereas in some gravels, voids make up only 20 percent of the material's volume. Where sediments of various sizes are mixed, the porosity is reduced because the finer particles tend to fill the openings between the larger grains. Most igneous and metamorphic rocks, as well as some sedimentary rocks, are composed of tightly interlocking crystals. The amount of open space between the grains may be negligible, perhaps as little as one or two percent of the rock's volume. Therefore, if these rocks are to have greater porosity, fractures must provide a significant proportion of the open space.

Porosity alone is not a satisfactory measure of a material's ability or capacity to yield groundwater. Rock or sediment may be very porous and still not allow water to move through it. The **permeability** of a material, its ability to transmit a fluid, is also very important. Groundwater moves by twisting and turning through small openings. The smaller the pore spaces, the slower the water moves. If the spaces between particles are very small, the films of water clinging to the grains will come in contact or overlap. As a result, the force of molecular attraction binding the water to the particles extends across the opening and the water is held firmly in place. Clay exemplifies this circumstance. Although clay's ability to store water is often high, its pore spaces are so small that water is unable to move. Impermeable layers composed of materials such as clay that hinder or prevent water movement are termed **aquicludes**. On the other hand, larger particles, such as sand or

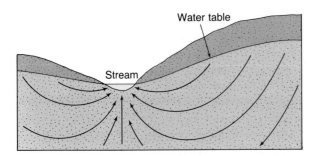

FIGURE 7.3 Arrows indicate groundwater movement through uniformly permeable material. The looping curves may be thought of as a compromise between the downward pull of gravity and the tendency of water to move toward areas of reduced pressure.

gravel, have larger pore spaces. Therefore, the water in the centers of the openings is not bound to the particles by molecular attraction and can move with relative ease. Permeable rock strata or sediment that transmit groundwater freely are called **aquifers**.

In summary, we have seen that porosity is not always a reliable guide to the amount of groundwater that can be produced and the property of permeability is a significant factor in determining the rate of groundwater movement and the quantity of water that might be pumped from a well.

SPRINGS

Springs have aroused the curiosity and wonder of people for thousands of years. The fact that springs were, and to some people still are, rather mysterious phenomena is not difficult to understand, for here is water flowing freely from the ground in all kinds of weather in seemingly inexhaustible supply, but with no obvious source. As a result, some rather interesting (although incorrect) explanations for the source of springs were proposed. Some of these have managed to live on to the present. One such explanation is that springs draw their water from the ocean. However, just how the salt is removed and how the water is elevated to the great heights it reaches in mountain springs remain unanswered questions. Another explanation for springs, one supported by Aristotle, suggested that the water originated in cold subterranean caverns where water vapor condensed from the air that had penetrated the earth and produced the needed water supply.

Not until the middle of the 17th century did the French physicist Pierre Perrault invalidate the age-old assumption that precipitation could not adequately account for the amount of water emanating from springs and flowing in rivers. Over a period of years, Perrault computed the quantity of water that fell on the Seine River basin. He then calculated the mean annual runoff by measuring the river's discharge, and after allowing for the loss of water by evaporation, he showed that there was sufficient water remaining to feed the springs. Thanks to Perrault's pioneering efforts and the measurements by many afterward, we now know that the source of springs is water from the zone of saturation and that the ultimate source of this water is precipitation.

When the water table intersects the earth's surface, a natural flow of groundwater results, which we call a **spring**. One situation leading to the formation of a spring is illustrated in Figure 7.1. Here an aquiclude is situated above the main water table. As water percolates downward, a portion of it is intercepted by the aquiclude, thereby creating a localized zone of saturation called a **perched water table**. Springs, however, are not confined to places where a perched water table creates a flow at the surface. Indeed, there are a wide variety of spring types because subsurface conditions vary greatly from place to place. Even in areas underlain by impermeable crystalline rocks, permeable zones may exist in the form of fractures or solution channels. If these openings fill with water and intersect the ground surface along a slope, a spring will result.

WELLS

The most common device used by people for removing groundwater is the **well**, an opening bored into the zone of saturation (Figure 7.1). Wells serve as reservoirs into which groundwater moves and from which it can be pumped to the surface. Digging for water dates back many centuries and continues to be an important method of obtaining water. Today in many parts of the world, including the United States, water from wells irrigates more land than does water from streams.

The level of the water table may fluctuate considerably during the course of a year, dropping during dry seasons and rising following periods of rain. Therefore, to ensure a continuous supply of water, a well should penetrate many meters below the water table. When water is pumped from a well, it produces a conical depression in the water table that is known as a **cone of depression**. If pumping is heavy, the water table may not only be lowered immediately around the well but may also be lowered over a large area (Figure 7.4). This has been the case in portions of the western United States. Under these circumstances, it can be said that the groundwater is literally being "mined," for even if pumping were to cease immediately, it could take hundreds of years for the groundwater to be replenished. The following example illustrates this point:

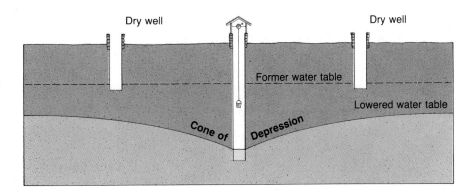

FIGURE 7.4 A cone of depression in the water table often forms around a pumping well. If heavy pumping lowers the water table, some wells may be left dry.

Consider for example, a location in the dry southwestern United States where annual recharge to an aquifer is on the order of only two-tenths of an inch of water. In such areas it is not uncommon to pump two feet or more of water per year for irrigation or other uses. In this oversimplified example, if the entire aquifer were pumped at that rate, yearly pumpage would be equivalent to 120 years' recharge, and ten years of pumping would remove a 1200-year accumulation of water. New recharge during the pumping period would be negligible. Mechanical problems and economic factors prevent complete dewatering of an aquifer, but the example is valid in principle.*

ARTESIAN WELLS

The term **artesian** is applied to any situation in which groundwater rises in a well above the level

*U.S. Geological Survey, *Water of the World*, 1968, p.16.

where it was initially encountered. For such a situation to occur, two conditions must exist (Figure 7.5): (1) Water must be confined to an aquifer that is inclined so that one end is exposed at the surface, where it can receive water; and (2) Impermeable layers, both above and below the aquifer, must be present to prevent the water from escaping. When such a layer is tapped, the pressure created by the weight of the water above will cause the water to rise. If there were no friction the water in the well would rise to the level of the water at the top of the aquifer. However, friction reduces the height of this pressure surface. The greater the distance from the recharge area (area where water enters the inclined aquifer), the greater the friction and the less the rise of water. In Figure 7.5, Well 1 is a *nonflowing artesian well,* because at this location the pressure surface is below ground level. When the pressure surface is above the ground and a well is drilled into the aquifer, a *flowing artesian well* is created (Well 2, Figure 7.5).

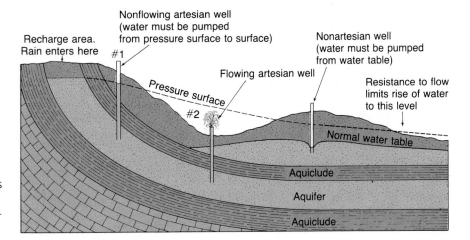

FIGURE 7.5 Artesian systems occur when an inclined aquifer is surrounded by impermeable beds.

Artesian systems act as conduits, transmitting water from remote areas of recharge great distances to the points of discharge. In this manner water which fell in central Wisconsin years ago is now taken from the ground and used by communities many kilometers away in Illinois. In South Dakota such a system has brought water from the Black Hills in the west, eastward across the state. On a different scale, city water systems may be considered as examples of artificial artesian systems. The water tower, into which water is pumped, may be considered the area of recharge, the pipes the confined aquifer, and the faucets in homes the flowing artesian wells.

ENVIRONMENTAL PROBLEMS ASSOCIATED WITH GROUNDWATER

As with many of our valuable natural resources, groundwater is being exploited at an ever-increasing rate. As we have already seen, overuse threatens the groundwater supply in some areas. In other places, groundwater withdrawal has caused the ground and everything resting upon it to sink. Still other localities are concerned with the possible contamination of their groundwater supply.

As we shall see later in this chapter, surface subsidence can result from natural processes related to groundwater. However, the ground may also sink when water is pumped from wells faster than natural recharge processes can replace it. This effect is particularly pronounced in areas underlain by thick layers of unconsolidated sediments. As water is withdrawn, water pressure drops and the weight of the overburden is transferred to the sediment. The greater pressure packs the sediment grains tightly together and the ground subsides.

Many areas may be used to illustrate land subsidence resulting from the excessive pumping of groundwater from relatively loose sediment. A classic example in the United States occurred in the San Joaquin Valley of California. Here, in a region of extensive irrigation, the water table beneath the valley has gradually been drawn down by as much as 30 meters (100 feet). As a consequence, the land has subsided by 3 meters in some places. Similar effects have occurred in the Houston and Panhandle areas

of Texas, as well as in Las Vegas and New Orleans. Damage to buildings, water and sewer lines, and roads can be extensive and costly. In addition, subsidence in coastal areas may require the construction of levees to keep out the encroaching sea.

The pollution of groundwater is a serious matter, particularly in areas where aquifers supply a large part of the water supply. A very common source of groundwater pollution is sewage which results from an ever-increasing number of septic tanks, as well as inadequate or broken sewer systems, and barnyard wastes.

If water contaminated with bacteria from sewage enters the groundwater system, it may become purified through natural processes. The harmful bacteria may be mechanically filtered out by the sediment through which the water percolates, destroyed by chemical oxidation, and/or assimilated by other organisms. In order for purification to occur, however, the aquifer must be of the correct composition. For example, extremely permeable aquifers such as highly fractured crystalline rock, coarse gravel, or cavernous limestone have such large openings that contaminated groundwater may travel long distances without being cleansed. In this case, the water flows too rapidly and is not in contact with the surrounding material long enough for purification to occur. This is the problem at Well 1 in Figure 7.6A. On the other hand, when the aquifer is composed of sand or permeable sandstone, the water can sometimes be purified within distances as short as a few tens of meters. The openings between sand grains are large enough to permit water movement, yet the movement of the water is slow enough to allow ample time for its purification (Well 2, Figure 7.6B).

Sanitary landfills and garbage dumps are another source of pollutants that may endanger the groundwater supply of an area. As rainwater oozes through the refuse, it may dissolve a variety of organic and inorganic materials, some of which may be harmful. If water containing material leached from the landfill reaches the water table, it will mix with the groundwater and contaminate the supply. Since groundwater movement is usually slow, the polluted water may go undetected for a considerable time. When the problem is finally discovered, the volume of contaminated water may be very large. Thus, even if the source of pollution is eliminated immediately (which

FIGURE 7.6 A. Although the contaminated water has traveled more than 100 meters before reaching Well 1, the water moves too rapidly through the cavernous limestone to be purified. **B.** Since the aquifer is sandstone, the water is purified in a relatively short distance.

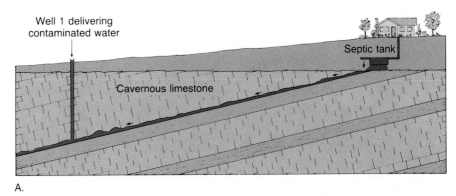

A.

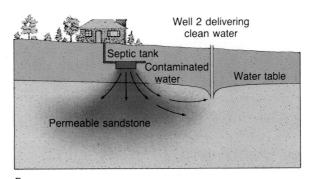

B.

is most unlikely), the problem could linger for many years until the contaminated water has migrated from the area of use.

HOT SPRINGS, GEYSERS, AND GEOTHERMAL ENERGY

Hot Springs

By definition, the water in **hot springs** is 6–9°C (11–16°F) warmer than the mean annual air temperature for the localities where they occur. In the United States alone, there are in excess of 1000 such springs.

Mineral explorations over the world have shown that temperatures in deep mines and oil wells usually rise with an increase in depth below the surface. Temperatures in such situations increase an average of about 2°C per 100 meters (1°F per 100 feet). Therefore when groundwater circulates at great depths, it becomes heated, and if it rises to the surface, the water may emerge as a hot spring. The water of some hot springs in the United States, par-

ticularly in the East, is heated in this manner. The great majority (over 95 percent) of the hot springs (and geysers) in the United States are found in the West. The reason for such a distribution is that the source of heat for most hot springs is cooling igneous rock, and it is in the West that igneous activity has been most recent.

Geysers

Geysers are intermittent hot springs or fountains where columns of water are ejected with great force at various intervals, often rising 30–60 meters (100–200 feet). After the jet of water ceases, a column of steam rushes out, usually with a thundering roar. Perhaps the most famous geyser in the world is Old Faithful in Yellowstone National Park, which erupts about once each hour (Figure 7.7). Geysers are also found in other parts of the world, including New Zealand and Iceland, where the term *geyser*, meaning "spouter" or "gusher," was coined.

Geysers occur when groundwater is heated in underground chambers. At the bottom of the chamber, the water is under great pressure because of the

FIGURE 7.7 Old Faithful, one of the most famous geysers in the world, emits as much as 45,000 liters (11,700 gallons) of hot water and steam about once each hour. (Photo by James E. Patterson)

Geothermal Energy

Many natural geyser areas around the world are potential sites for tapping **geothermal energy**, that is, natural steam used for power generation. In New Zealand, Italy, Mexico, the Soviet Union, and the United States, underground supplies of superheated steam are now being used to provide power for generating electricity. In the United States the first commercial geothermal power plant was built in 1960 at "The Geysers," north of San Francisco. The most favorable geologic factors for a geothermal reservoir of commercial value include:

1 A potent source of heat, such as a large magma chamber. The chamber should be deep enough to ensure adequate pressure and a slow rate of cooling, and yet not be so deep that the natural circulation of water is inhibited. Magma chambers of this type are most likely to occur in regions of recent volcanic activity;
2 Large and porous reservoirs with channels connected to the heat source, near which water can circulate and then be stored in the reservoir;
3 Capping rocks of low permeability that inhibit the flow of water and heat to the surface. A deep, well-insulated reservoir is likely to contain much more stored energy than an uninsulated, but otherwise similar, reservoir.*

It is too early to judge whether natural steam has the potential to satisfy an important part of the world's requirements for electrical power, but with the need to develop new sources of energy, the possibilities are definitely worth exploring.

THE GEOLOGIC WORK OF GROUNDWATER

The primary erosional work carried out by groundwater is that of dissolving rock. Since soluble rocks, especially limestone, underlie millions of square kilometers of the earth's surface, it is here that groundwater carries on its rather unique and important role as an erosional agent. Although nearly insoluble in pure water, limestone is quite easily dissolved by water containing small quantities of carbonic acid.

weight of the overlying water. Consequently, a temperature above 100°C (212°F) is required before it will boil. For example, at the bottom of a 300-meter (1000-foot) chamber water must attain a temperature of nearly 230°C (450°F) before it will boil. The heating causes the water to expand, with the result that some flows out at the top. This decreases the pressure, and the water quickly turns to steam and causes the geyser to erupt (Figure 7.8).

Groundwater from hot springs and geysers usually contains more material in solution than groundwater from other sources because hot water is a more effective dissolver than cold. When the water contains much dissolved silica, *geyserite* is deposited around the spring. *Travertine,* a form of calcite, is a characteristic deposit at hot springs in limestone regions. Some hot springs contain sulfur. In addition to making the water taste bad, sulfur-rich springs emit an unpleasant odor. Undoubtedly Rotten Egg Spring, Nevada, is such a situation.

*Adapted from U.S. Geological Survey, *Natural Steam for Power,* 1968.

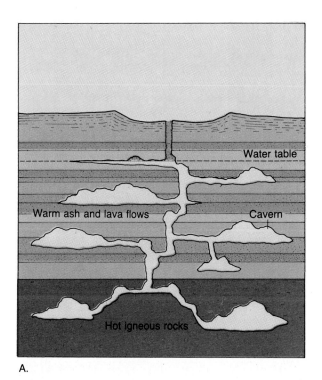

A.

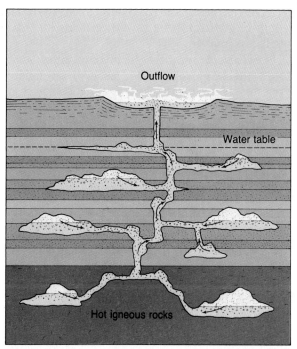

B.

FIGURE 7.8 Idealized diagrams of a geyser. A geyser can form if the heat is not distributed by convection. **A.** In this figure the water near the bottom is heated to near its boiling point. The boiling point is higher there than at the surface because the weight of the water above increases the pressure. **B.** The water higher in the geyser system is also heated and so expands and flows out at the top, reducing the weight of the water on the bottom. **C.** At the reduced pressure on the bottom, boiling occurs. The bottom water flashes into steam, and the expanding steam causes an eruption.

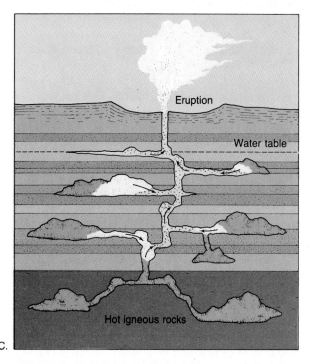

C.

Most natural water contains this weak acid because rainwater readily dissolves carbon dioxide from the air and from decaying plants. Therefore, when groundwater comes in contact with limestone, the carbonic acid reacts with calcite in the rocks to form calcium bicarbonate, a soluble material that is then carried away in solution.

Caverns

Among the most spectacular results of groundwater's erosional handiwork is the creation of limestone **caverns**. Most are relatively small, yet some have spectacular dimensions. In the United States, Carlsbad Caverns in southeastern New Mexico and Mammoth Cave in Kentucky are famous examples. One chamber in Carlsbad Caverns has an area equivalent to fourteen football fields and enough height to accommodate the U.S. Capitol Building (see chapter-opening photo). At Mammoth Cave, the total length of interconnected caverns extends for hundreds of kilometers.

Most caverns are believed to be created at or below the water table in the zone of saturation. Here the groundwater follows lines of weakness in the rock, such as joints and bedding planes. As time passes, the dissolving process slowly creates cavities and gradually enlarges them into caverns. The material dissolved by groundwater is carried away and discharged into streams.

Certainly the features that arouse the greatest curiosity for most cavern visitors are the stone formations that often exhibit quite bizarre patterns and give some caverns a wonderland appearance. These features are created by the seemingly endless dripping of water over great spans of time. The calcite that is left behind produces the limestone we call travertine. These cave deposits, however, are also commonly called *dripstone*, an obvious reference to their mode of origin.

Although the formation of caverns takes place in the zone of saturation, the deposition of dripstone is not possible until the caverns are above the water table in the zone of aeration. This commonly occurs as nearby streams cut their valleys deeper, causing the water table to drop as the elevation of the river drops. As soon as the chamber is filled with air, the stage is set for the decoration phase of cavern building to begin.

The various dripstone features found in caverns are collectively called **speleothems**, no two of which are exactly alike. Perhaps the most familiar speleothems are **stalactites**. These icicle-like pendents hang from the ceiling of the cavern and form where water seeps through cracks above. When the water reaches the air in the cave, some of the carbon dioxide in solution escapes and a residue of calcium carbonate remains. Deposition occurs as a ring around the edge of the water drop. As drop after drop follows, each leaves an infinitesimal trace of calcite behind, and a hollow limestone tube is created. Water then moves through the tube, remains suspended momentarily at the end, contributes a tiny ring of calcite, and falls to the cavern floor. The stalactite just described is appropriately called a *soda straw* (Figure 7.9A). Often the hollow tube of the soda straw becomes plugged or its supply of water increases. In either case, the water is forced to flow,

A.

B.

FIGURE 7.9 A. A "live" solitary soda straw stalactite. (Photo by Clifford Stroud, National Park Service). **B.** Stalagmites grow upward from the floor of a cavern. (Photo by E. J. Tarbuck)

and hence deposit, along the outside of the tube. As deposition continues, the stalactite takes on the more common conical shape.

Speleothems that form on the floor of a cavern and reach upward toward the ceiling are called **stalagmites** (Figure 7.9B). The water supplying the calcite for stalagmite growth falls from the ceiling and splatters over the surface. As a result, stalagmites do not have a central tube, and they are usually more massive in appearance and rounded on their upper ends than stalactites.

Karst Topography

Many areas of the world have landscapes that to a large extent have been shaped by the dissolving power of groundwater. Such areas are said to exhibit **karst topography**. The term is derived from a plateau region located along the northeastern shore of the Adriatic Sea in the border area between Yugoslavia and Italy where such topography is strikingly developed. In the United States, karst landscapes occur in areas of Kentucky, Tennessee, southern Indiana, and central and northern Florida. Generally, arid and semiarid areas do not develop karst topog-

raphy. When solution features exist in such regions, they are likely to be remnants of a time when more humid conditions prevailed.

Karst areas characteristically exhibit an irregular terrain punctuated with many depressions, called **sinkholes** or **sinks**. In the limestone areas of Kentucky and southern Indiana, there are literally tens of thousands of these depressions varying in depth from just a meter or two to a maximum of more than 50 meters.

Sinkholes commonly form in one of two ways. Some develop gradually over many years without any physical disturbance to the rock. In these situations, the limestone immediately below the soil is dissolved by downward-seeping rainwater that is freshly charged with carbon dioxide. These depressions are usually not deep and are characterized by relatively gentle slopes. By contrast, sinkholes can also form suddenly and without warning when the roof of a cavern collapses under its own weight. Typically, the depressions created in this manner are steep-sided and deep. When they form in populous areas, they may represent a serious geologic hazard. The craterlike sinkhole in Figure 7.10 began forming in Winter Park, Florida, on May 8, 1981, just one day

FIGURE 7.10 Aerial view of a large sinkhole that formed in Winter Park, Florida, in May, 1981. (Photo courtesy of George Remaine, *Orlando Sentinel Star*)

before this photograph was taken. Newspaper accounts were front page news and made this sinkhole one of the most publicized ever.

Sinkhole formation is not uncommon in northern and central Florida. In fact, the Winter Park event was just one of three that occurred in the area over a two-week span. In each case the collapse at the surface was probably triggered by a lowering of the water table brought about by a severe drought. As the water table dropped, the roofs of the underground cavities lost support and fell into the voids below.

In addition to a surface pockmarked by sinkholes, karst regions characteristically show a striking lack of surface drainage. Following a rainfall, the runoff is funneled below ground by way of sinks where it then flows through caverns until it finally reaches the water table. When streams do exist at the surface, their paths are usually short. The names of such streams often give a clue to their fate. In the Mammoth Cave area of Kentucky, for example, there is Sinking Creek, Little Sinking Creek, and Sinking Branch. Other sinkholes become plugged with clay and debris to create small lakes or ponds.

REVIEW QUESTIONS

1 Define groundwater and relate it to the water table.

2 How do porosity and permeability differ?

3 What is the difference between an aquiclude and an aquifer?

4 Under what circumstances can a material have a high porosity but not be a good aquifer?

5 What is an effluent stream? How does an influent stream differ?

6 When an aquiclude is situated above the main water table, a localized saturated zone may be created. What term is applied to such a situation?

7 Why is the pumping of water in some areas of the southwestern United States a serious problem?

8 What is meant by the term *artesian?* Under what circumstances do artesian wells form?

9 Briefly explain what happened in the San Joaquin Valley of California as the result of excessive groundwater withdrawal.

10 Which would be most effective in purifying polluted groundwater: an aquifer composed mainly of coarse gravel, sand, or cavernous limestone?

11 What is the source of heat for most hot springs and geysers? How is this reflected in the distribution of these features?

12 Name two common speleothems and distinguish between them.

13 Speleothems form in the zone of saturation. True or False? Briefly explain your answer.

14 Areas whose landscapes are largely a reflection of the erosional work of groundwater are said to exhibit what kind of topography?

15 Describe two ways in which sinkholes are created.

CHAPTER EIGHT

GLACIERS AND GLACIATION

A glacier is a thick mass of ice that originates on land from the compaction and recrystallization of snow and shows evidence of past or present movement. Since snow is the raw material that eventually produces glacial ice, glaciers must form in areas where more snow falls in winter than melts during the summer. A region that has such a net accumulation is termed a *snowfield* and its outer limits are defined by the *snowline.* The elevation of the snowline varies greatly. In the frigid polar realm, it may be sea level, while in tropical areas near the equator, the snowline exists only high in the mountains, often at elevations exceeding 4500 meters (15,000 feet). If the accumulation in the snowfield is great enough, the pressure of overlying layers transforms the snow below into glacial ice.

In Chapter Six we learned that the earth's water is in constant motion. Time and again the same water is transferred from the oceans to the atmosphere, dropped upon the land, and carried by rivers and underground flow back to the sea. However, when precipitation falls at high elevations or high latitudes, the water may not immediately soak in or run off. Rather, it may become part of a large mass of moving ice, that is, a glacier. Although the ice will eventually melt and continue its path to the sea, it can be tied up in a glacier for many tens, hundreds, or even thousands of years. For example, data collected from the glacier covering Greenland show that some of this ice is more than 25,000 years old.

How much water is stored as glacial ice? Estimates by the U.S. Geological Survey indicate that only slightly more than two percent of the world's water supply is accounted for by glaciers. But this figure can be misleading when the actual amounts of water are considered.

Although glaciers are found in many parts of the world today, most are located in areas remote from populous regions. Literally thousands of relatively small glaciers exist in mountainous regions. Such glaciers are generally confined to mountain valleys and are most often termed **alpine glaciers**. The total volume of all alpine glaciers today is about 210,000 cubic kilometers (50,000 cubic miles), comparable to the combined volume of the world's large saline and freshwater lakes.

On a different scale, **continental glaciers** are massive accumulations of ice that are not confined to valleys but cover all of the land surface of extensive areas. Two such major accumulations exist on earth today and cover most of the island of Greenland and the continent of Antarctica. Their combined area represents almost 10 percent of the earth's land surface. The Greenland ice sheet covers 80 percent of this large island, occupying about 1.7 million square kilometers (667,000 square miles) and averaging nearly 1500 meters (5000 feet) thick. However, when compared to the glacier covering the continent of Antarctica, the Greenland ice sheet seems quite small. Eighty percent of the world's ice is represented by Antarctica's glacier, which covers an area almost one and one-half times that of the United States. If this ice melted, sea level would rise an estimated 60–70 meters (200–230 feet), inundating many densely populated coastal areas. The hydrologic importance of the continent and its ice can be illustrated another way. If Antarctica's ice sheet melted at a suitable rate it could feed (1) the Mississippi River for more than 50,000 years, (2) all the rivers in the United States for about 17,000 years, (3) the Amazon River for approximately 5000 years, or (4) all the rivers of the world for about 750 years.

When alpine glaciers merge, their lateral moraines join to create a medial moraine. (Photo courtesy of the Glaciology Office, U.S. Geological Survey)

GLACIAL MOVEMENT

The movement of glacial ice is generally referred to as *flow.* The fact that glacial movement is described in this way would seem to constitute a paradox—ice is solid, yet it is capable of flow. The way ice flows is complex but is believed to be of two basic types. One mechanism involves internal movement within the ice. Ice behaves as a brittle solid until the pressure or load upon it is equivalent to the weight of about 50 to 60 meters (165 to 200 feet) of ice. Once that load is surpassed, the ice will behave as a plastic material and flow continuously. A second and often equally important mechanism of glacial movement consists of the whole ice mass slipping along the ground. With the exception of some glaciers in polar regions where the ice is probably frozen to the solid bedrock floor, the lowest portions of most glaciers are thought to move by this sliding process.

Figure 8.1 illustrates the effects of these two basic types of glacial motion. This vertical profile through a glacier also shows that all the ice does not flow forward at the same rate. Just as in streams, fric-

tional drag with the bedrock floor results in the lower portions of the glacier moving more slowly.

The uppermost portion of a glacier is often quite appropriately referred to as the *zone of fracture.* Since there is not enough overlying ice to cause plastic flow, this upper part of the glacier consists of brittle ice. Consequently, the ice in this zone is carried along piggyback style by the ice below. When the glacier moves over irregular terrain, the zone of fracture is subjected to tension, with cracks called **crevasses** resulting (Figure 8.2). These gaping cracks, which often make travel across glaciers dangerous, may extend to depths of 50 meters (165 feet). Beyond this depth, plastic flow seals them off.

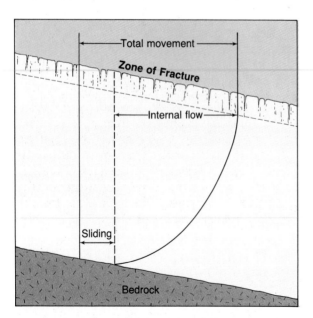

FIGURE 8.1 Glacial movement is divided into two components. Below about 50 meters, ice behaves plastically and flows. In addition, the entire ice mass may slide along the ground. The ice in the zone of fracture is carried along "piggyback" style. Notice that the rate of movement is slowest at the base of the glacier where frictional drag is greatest.

FIGURE 8.2 Crevasses form in the brittle ice of the zone of fracture. They do not continue down into the zone of flow. (Courtesy of Ward's Natural Science Establishment, Inc., Rochester, N.Y.)

Unlike streamflow, the movement of glaciers is not readily apparent to the casual observer. If we could watch a glacier in a mountain valley move, we would see that, like the water in a river, all of the ice in the valley does not move downstream at an equal rate. Just as friction with the bedrock floor slows the movement of the ice at the bottom of the glacier, the drag created by the valley walls leads to the flow being greatest in the center of the glacier.

More than 100 years ago the first measurements of glacial movement were made. In this experiment stakes were carefully placed in a straight line across the top of an alpine glacier. Periodically the positions of the stakes were noted, revealing the type of movement just described (Figure 8.3).

How rapidly does glacial ice move? Average velocities vary considerably from one glacier to another. Some move so slowly that trees and other vegetation may become well established in the debris that has accumulated on the glacier's surface, whereas others move at rates of up to several meters per day. The advance of some glaciers is characterized by periods of extremely rapid movement followed by periods during which movement is practically nonexistent. For example, Hassanabad Glacier in the Karakoram, a mountain range in Kashmir and northwestern India, advanced 10 kilometers in less than 3 months—a rate of almost 130 meters per

day. The precise cause or causes of these sporadic, short-lived advances is not well understood. One proposal is that the base of the glacier may have been frozen to the bedrock and then sudden melting released the ice. Another hypothesis suggests that a block of stagnant ice at the terminus of a glacier in an alpine valley may act as a dam until the buildup of pressure by the flowing ice behind it forces it to give way.

Glaciers are constantly gaining and losing ice. As we learned earlier, snow accumulation and ice formation occur above the snowline. Here the addition of snow thickens the glacier and promotes movement. Below the snowline, some of the glacial ice and snow from the previous winter melt. In addition, large pieces of ice may break off from the front of the glacier, a process called *calving*, sometimes creating icebergs where the glacier meets the sea. Whether the margin of a glacier is advancing, retreating, or remaining stationary depends upon the economy, or budget, of the glacier. That is, it depends upon the balance or lack of balance between accumulation on the one hand and wastage (also termed **ablation**) on the other. If ice accumulation exceeds ablation, the glacial front advances until the two factors balance. At this point, the terminus of the glacier is stationary. At a later time when ablation exceeds accumulation, the ice front will retreat until a balance is again reached. Whether the margins of a glacier are advancing, retreating, or stationary, the ice within the glacier continues to flow forward. In the case of a receding glacier, the ice simply does not flow forward rapidly enough to offset ablation. This point is illustrated rather well in Figure 8.3. While the line of stakes within Rhone Glacier continued to move downstream, the terminus of the glacier slowly retreated upstream.

GLACIAL EROSION

Glaciers are capable of carrying on great amounts of erosional work. For anyone who has observed the terminus of an alpine glacier, the evidence of its erosive force is plain. The observer can witness firsthand the melting ice unlocking rock material of all sizes. All signs lead to the conclusion that the ice has scraped, scoured, and torn rock debris from the floor and walls of the valley and carried them downslope. Indeed, as a transporter of sediment, ice has no

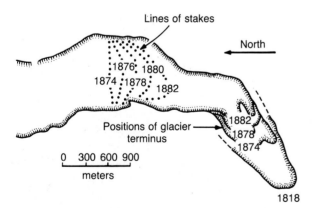

FIGURE 8.3 Map of Rhone Glacier, Switzerland. The ice along the sides of the glacier moves slowest. Also notice that even though the ice front was retreating, the ice within the glacier was advancing. (From John P. Miller and Robert Scholten, *Laboratory Studies in Geology* © 1966. W. H. Freeman and Company Publishers, San Francisco)

equal, because once the debris is acquired by the ice, it will not settle out as will the load carried by a stream or by the wind. Consequently, glaciers can carry huge blocks that no other erosional agent could possibly budge. Although glaciers are of limited importance as an erosional agent today, many landscapes that were modified by the widespread glaciers of the recent ice age still reflect to a high degree the work of ice.

Glaciers primarily erode land in two ways. First, as a glacier flows over a fractured bedrock surface, it loosens and lifts blocks of rock, incorporates them into the ice, and carries them off. This process, known as **plucking**, occurs when meltwater penetrates the cracks and joints along the rock floor of the glacier and refreezes. As the water expands, it exerts tremendous leverage that pries the rock loose. In this manner sediment of all sizes, ranging from particles as fine as flour to blocks as large as houses, becomes part of the glacier's load.

The second major erosional process is **abrasion**. As the ice with its load of rock fragments moves along, it acts as a giant rasp or file and grinds the surface below as well as the rocks within the ice. The pulverized rock produced by the glacial "grist mill" is appropriately called **rock flour**. So much rock flour may be produced that meltwater streams leaving a glacier often have the grayish appearance of skimmed milk—visible evidence of the grinding power of the ice. When the embedded material consists of large fragments, long scratches and grooves called **glacial striations** may be gouged out (Figure 8.4). These linear scratches on the bedrock surface provide clues to the direction of glacial movement. By mapping the striations over large areas, glacial flow patterns can often be reconstructed. On the other hand, not all abrasive action produces striations. When the sediment consists primarily of fine silt-sized particles, the rock surfaces over which the glacier moves may become highly polished.

The erosional effects of alpine and continental glaciers are quite different from each other. A visitor to an alpine-glaciated region is likely to see sharp and very angular topography. The reason is that as alpine glaciers move downvalley, they tend to accentuate the irregularities of the mountain landscape by creating steeper canyon walls and making bold peaks even more jagged. By contrast, continental ice sheets generally override the terrain and hence tend to subdue rather than accentuate the irregularities they encounter.

Although the erosional accomplishments of continental glaciers can be tremendous, landforms carved by these huge ice masses usually do not inspire the same degree of wonderment and awe as do the erosional features created by alpine glaciers. In regions where the erosional effects of continental ice sheets are significant, glacially scoured surfaces and subdued terrain are the rule. By contrast, in mountainous areas, erosion by alpine glaciers yields many truly spectacular features. Much of the rugged

FIGURE 8.4 Results of glacial abrasion. Scratches and grooves in limestone, St. John's Bay, Newfoundland. (Photo by Peter Kresan)

FIGURE 8.5 Prior to glaciation, a mountain valley is typically narrow and V-shaped. During glaciation, an alpine glacier widens, deepens, and straightens the valley, creating a U-shaped glacial trough. (Photo by Peter Kresan)

mountain scenery so celebrated for its majestic beauty is the product of glacial erosion.

Prior to glaciation alpine valleys are characteristically V-shaped because streams are well above base level and are therefore downcutting. However, in mountainous regions that have been glaciated, the valleys are no longer narrow. As a glacier moves down a valley once occupied by a stream, the ice modifies it in three ways: the glacier widens, deepens, and straightens the valley, so that what was once a youthful V-shaped valley is transformed in a U-shaped **glacial trough** (Figure 8.5).

Since the magnitude of the glacial erosion depends upon the thickness of the ice, main or trunk glaciers cut their valleys deeper than their tributaries are able to do. After the glaciers have receded, the tributary valleys stand high above the main trough and are termed **hanging valleys**. Hanging valleys often produce spectacular cascading waterfalls, such as those in Yosemite National Park, California (Figure 8.6).

At the head of a glacial valley is a characteristic and often imposing feature associated with an alpine glacier—a **cirque**. As Figure 8.7 illustrates, these hollowed-out, bowl-shaped depressions have precipitous walls on three sides but are open on the downvalley side. The cirque represents the focal

point of the glacier's source, that is, the area of snow accumulation and ice formation. Although the origin of cirques is still not totally clear, they are believed to begin as irregularities in the mountainside

FIGURE 8.6 Bridalveil Falls in Yosemite National Park cascades from a hanging valley into the glacial trough below. (Courtesy of Ward's Natural Science Establishment, Inc., Rochester, N.Y.)

FIGURE 8.7 Aerial view of bowl-shaped depressions called cirques in the Uinta Range, Utah. (Photo by John S. Shelton)

that are subsequently enlarged by frost wedging and plucking along the sides and bottom of the glacier. After the glacier has melted away, the cirque basin is usually occupied by a small lake.

Fiords are deep, often spectacular, steep-sided inlets of the sea that exist in many high-latitude areas of the world where mountains are adjacent to the ocean (Figure 8.8). Norway, British Columbia, Greenland, New Zealand, Chile, and Alaska all have coastlines characterized by fiords. They represent glacial troughs that were partially submerged as the ice left the valley and sea level rose following the Ice Age. The depths of fiords are often dramatic, in some instances exceeding 1000 to 1500 meters (3300 to 5000 feet). The great depths of these flooded troughs are only partly explained by the post-Ice Age rise in sea level. Unlike the situation governing the downward erosional work of rivers, sea level does not act as base level for glaciers. As a consequence, glaciers are capable of eroding their beds far below the surface of the sea. For example, an alpine glacier 300 meters (1000 feet) thick can carve its valley floor more than 250 meters (800 feet) below sea level before downward erosion ceases and the ice begins to float.

The Alps, Northern Rockies, and many other mountain landscapes carved by alpine glaciers reveal more than glacial troughs and cirques. In addition, sinuous, sharp-edged ridges called **arêtes** and sharp, pyramidlike peaks called **horns** project above the surroundings. Both features can originate from the same basic process—the enlargement of cirques produced by plucking and frost action. A group of cirques around a single high mountain creates the spires of rock called horns. As the cirques enlarge and converge, an isolated horn is produced. The most famous example is the Matterhorn in the Swiss Alps (Figure 8.9). Arêtes can form in a similar manner except that the cirques are not clustered around a point but rather exist on opposite sides of a divide. As the cirques grow, the divide separating them is reduced to a very narrow, knifelike partition. An arête, however, may also be created in another way. When two glaciers occupy parallel valleys, an arête can form when the divide separating the moving tongues of ice is progressively narrowed as the glaciers scour and widen their valleys. The landforms carved by alpine glaciers are summarized in Figure 8.10.

GLACIAL DEPOSITS

Glaciers are capable of acquiring and transporting a huge load of debris as they slowly yet steadily advance across the land. Ultimately these materials must be deposited when the ice melts. In regions where glacial sediments are deposited, the material can play a truly significant role in forming the physical landscape. For example, in many areas once cov-

ered by the continental ice sheets of the recent Ice Age, the bedrock is rarely exposed because glacial deposits that are tens or even hundreds of meters thick completely mantle the terrain. The general effect of these deposits is to reduce the local relief and

FIGURE 8.8 Like other fiords, Muir Inlet, Alaska, is a drowned glacial trough. (Photo by Bruce F. Molnia, courtesy of Terraphotographics/BPS)

FIGURE 8.9 The Matterhorn, a glacially eroded peak in the Swiss Alps. (Courtesy of Ward's Natural Science Establishment, Inc., Rochester, N.Y.)

FIGURE 8.10 Landforms created by alpine glaciers. **A.** A mountain mass not affected by glacial erosion. **B.** The mountain mass during the period of maximum glacial activity. **C.** The mountain mass shortly after the glaciers have melted from its valleys. [After William Morris Davis, "The Sculpture of Mountains by Glaciers," *Scottish Geographical Magazine* 22 (1906)]

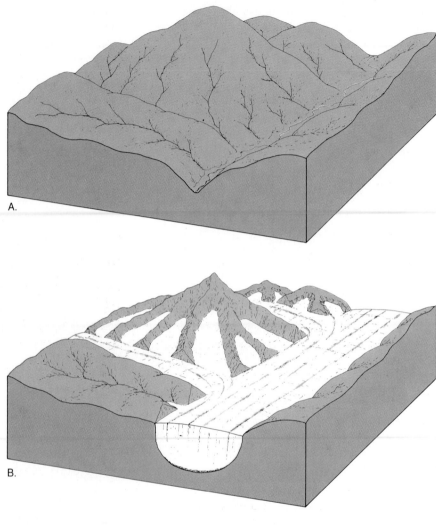

A.

B.

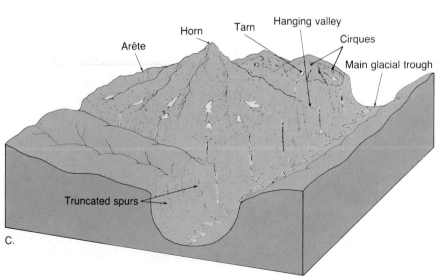

Arête

Horn

Tarn

Hanging valley

Cirques

Main glacial trough

Truncated spurs

C.

thus level the topography. Indeed, much familiar country scenery—rocky pastures in New England, wheat fields in the Dakotas, rolling farmland in the Midwest—results directly from glacial deposition.

Long before the theory of an extensive Ice Age was proposed, much of the soil and rock debris covering portions of Europe was recognized as coming from elsewhere. At the time, these foreign materials were believed to have been "drifted" into their present positions by floating ice. As a consequence, the term **drift** was applied to this sediment. Although rooted in a concept that was not correct, this term was so well established by the time the true glacial origin of the debris became widely recognized, that it remained in the glacial vocabulary. Today the word *drift* is an all-embracing term for sediments of glacial origin, no matter how, where, or in what form they were deposited.

Glacial deposits are of two types: (1) Those deposited directly by the glacier, which are known as **till**; and (2) Materials deposited by glacial meltwater, called **outwash**. Till is deposited as glacial ice melts and drops its load of rock fragments. Deposits of till are characteristically unsorted, a mixture of many sediment sizes (Figure 8.11). When boulders are found in till, they are called **glacial erratics**, indicating that they were derived from a source outside the area where they are found. In many areas erratics may be seen dotting pastures and farm fields, and sometimes these large rock fragments are cleared from the fields and piled into fences. By examining the erratics as well as the mineral composition of the remaining till, geologists are often able to trace the path taken by the glacier.

FIGURE 8.11 Glacial till is an unsorted mixture of many sediment sizes. (Photo by E. J. Tarbuck)

Outwash is sorted according to the size and weight of the fragments. Since ice is not capable of such sorting activity, these sediments are not deposited directly by the glacier as till is, but rather they reflect the sorting action of the glacial meltwater that was responsible for dropping them. Accumulations of outwash often consist largely of sand and gravel, that is, bed load material, because the finer rock flour remains suspended and is commonly carried far from the glacier by the meltwater streams. An indication that outwash consists primarily of sand and gravel can be seen in many areas where these deposits are actively mined as aggregate for road work and other construction projects.

Moraines, Outwash Plains, and Kettle Holes

Perhaps the most widespread features created by glacial deposition are *moraines,* which are simply layers or ridges of till. Several types of moraines are identified; some are common only to mountain valleys, and others are associated with areas affected by either continental or alpine glaciers. Lateral and medial moraines fall in the first category, while end moraines and ground moraines are in the second.

The sides of an alpine glacier accumulate large quantities of debris from the valley walls. When the glacier wastes away, these materials are left as ridges, called **lateral moraines**, along the sides of the valley (Figure 8.12). **Medial moraines** are formed when two alpine glaciers coalesce to form a single ice stream (see chapter-opening photo). The till that was once carried along the edges of each glacier joins to form a single dark stripe of debris within the newly enlarged glacier. The creation of these dark stripes within the ice stream is one obvious proof that glacial ice moves, because the medial moraine could not form if the ice did not flow downvalley.

As the name implies, **end moraines** form at the terminus of a glacier. Here, while the ice front is stationary, the glacier continues to carry in and deposit large quantities of rock debris, creating a ridge of till tens to hundreds of meters high. The end moraine marking the farthest advance of the glacier is called the *terminal moraine,* and those moraines formed as the ice front periodically became stationary during retreat are termed *recessional moraines.* As the glacier recedes, a layer of till is laid down,

FIGURE 8.12 A well-developed lateral moraine deposited by the shrinking Athabaska Glacier in the Canadian Rockies. (Photo by James E. Patterson)

FIGURE 8.13 These ponds occupy depressions called kettles which form when a block of ice that was buried in drift melts and leaves a pit. (Photo by Bruce F. Molnia, courtesy of Terraphotographics/BPS)

forming a gently undulating surface of **ground moraine**. Ground moraine has a leveling effect, filling in low spots and clogging old stream channels, often leading to a disruption of drainage.

At the same time that an end moraine is forming, water from the melting glacier cascades over the till, sweeping some of it out in front of the growing ridge of unsorted debris. Meltwater generally emerges from the ice in rapidly moving streams that are often choked with suspended material and carry a substantial bed load as well. As the water leaves the glacier, it moves onto the relatively flat surface beyond and rapidly loses velocity. As a consequence, much of its bed load is dropped and the meltwater begins weaving a complex pattern of braided channels. In this way a broad, ramplike surface called an

outwash plain is built adjacent to the downstream edge of most end moraines.

Often outwash plains are pockmarked with basins or depressions known as **kettles** (Figure 8.13). Kettles also occur in deposits of till. Kettles form when a block of stagnant ice becomes wholly or partially buried in drift and ultimately melts, leaving a pit in the glacial sediment. Although most kettles do not exceed two kilometers in diameter, some with diameters exceeding 10 kilometers (6 miles) occur in Minnesota. Likewise, the typical depth of most kettles is less than 10 meters (33 feet), although the vertical dimensions of some approach 50 meters. In many cases, water eventually fills the depression and forms a pond or lake. Figure 8.14 depicts a hypothetical area which shows some of the depositional features

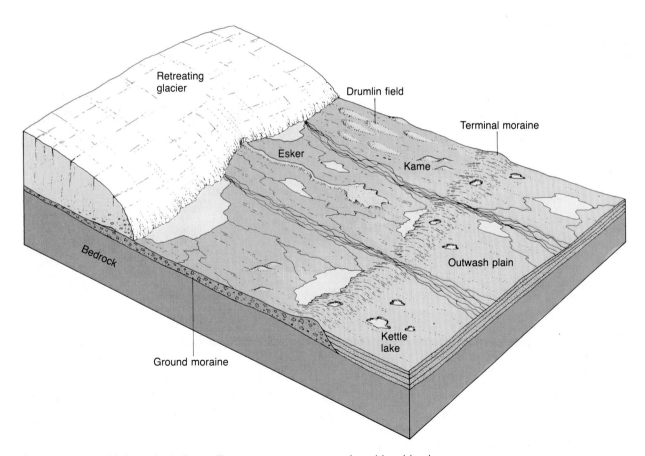

FIGURE 8.14 This hypothetical area illustrates many common depositional landforms.

described in this section as they would appear in relation to one another.

Other Depositional Features

Drumlins are streamlined asymmetrical hills composed of till (Figure 8.15). They range in height from 15 to 60 meters (50 to 200 feet) and average 0.4—0.8 kilometer (0.25–0.50 mile) in length. The steep side of the hill faces the direction from which the ice advanced, whereas the gentler slope points in the direction the ice moved (Figure 8.16). Drumlins are not found singly, but rather occur in clusters, sometimes called drumlin fields (see Figure 8.14). Although drumlin formation is not well understood, their streamlined shape seems to indicate that they were molded in the zone of flow within an active glacier. Some geologists believe drumlin fields were created when a glacier advanced and reshaped an end moraine.

In areas once occupied by continental ice sheets, sinuous ridges composed largely of sand and gravel may be found. These ridges, called **eskers**, are thought to be deposits made by streams flowing in tunnels beneath the ice, near the terminus of a gla-

FIGURE 8.15 Drumlins, such as this one in upstate New York, are depositional features associated with continental glaciers. (Courtesy of Ward's Natural Science Establishment, Inc., Rochester, N.Y.)

cier. They may be several meters high and extend for many kilometers. In many areas they are mined for sand and gravel, and for this reason, eskers are disappearing in some localities. **Kames** are steep-sided hills that, like eskers, are composed of sand and gravel (Figure 8.17). Kames are believed to have originated when sediment collected in openings in stagnant ice.

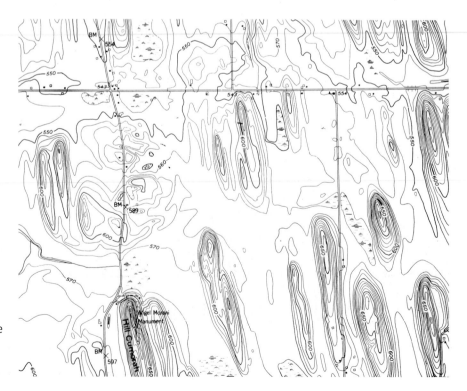

FIGURE 8.16 Portion of a drumlin field shown on the Palmyra, New York, 7.5 minute topographic map. North is at the top. The drumlins are steepest on the north side, indicating that the ice advanced from this direction.

GLACIERS IN THE GEOLOGIC PAST

At various points in the preceding pages mention was made of the Ice Age, a time when ice sheets and alpine glaciers were far more extensive than they are today. As was noted earlier, there was a time when the most popular explanation for what we now know to be glacial deposits was that the materials had been drifted in by means of icebergs or perhaps simply swept across the landscape by a catastrophic flood. However, during the nineteenth century, field investigations by many scientists provided convincing proof that an extensive ice age was responsible for these deposits as well as for many other features.

By the beginning of the twentieth century, geologists had largely determined the areal extent of Ice Age glaciation. Further, during the course of their investigations they discovered that many glaciated regions had not one layer of drift, but several. Moreover, close examination of these older deposits showed well-developed zones of chemical weathering and soil formation as well as the remains of plants that require warm temperatures. The evidence was clear; there had not been just one glacial advance but several, each separated by long periods when climates were as warm or warmer than at present. The Ice Age had not simply been a time when the ice advanced over the land, lingered for a while, and then receded. Rather, the period was a very complex event characterized by a number of advances and withdrawals of glacial ice. In North America four major stages of glaciation have been identified (Figure 8.18). Each was named for the midwestern state where the deposits of that ice sheet are well exposed and/or were first studied. These are, in order of occurrence, the Nebraskan, Kansan, Illinoian, and Wisconsinan. Additional evidence from Europe, Alaska, and elsewhere further indicates that other glacial advances probably pre-

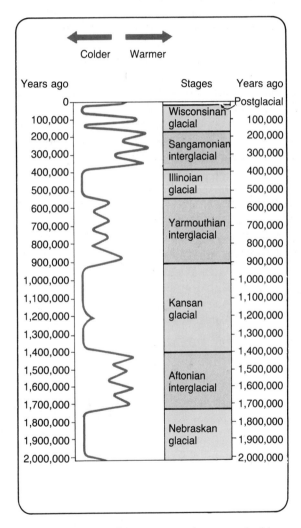

FIGURE 8.18 The Pleistocene epoch was marked by fluctuating climatic conditions that led to alternating glacial and interglacial periods. Four separate stages of glaciation are recognized for North America. (After D. B. Ericson and G. Wollin, "Pleistocene Climates and Chronology in Deep-Sea Sediments," *Science* 162 (1968): 1233. Copyright © 1968 by the American Association for the Advancement of Science)

FIGURE 8.17 White Kame in Kettle Moraine State Forest, Wisconsin. (Photo by G. J. Knudson, Wisconsin Department of Natural Resources)

ceded the Nebraskan. During the glacial age, ice left its imprint on almost 30 percent of the earth's land area, including about 10 million square kilometers of North America, 5 million square kilometers of Europe, and 4 million square kilometers of Siberia (Figure 8.19). The amount of glacial ice in the Northern Hemisphere was roughly twice that of the Southern Hemisphere. The primary reason is that the southern polar ice could not spread far beyond the margins of Antarctica. By contrast, North America and Eurasia provided great expanses of land for the spread of ice sheets.

Today we know that the Ice Age began between two and three million years ago. This means that most of the major glacial stages occurred during a division of the geologic calendar called the **Pleistocene epoch**. Although the Pleistocene is commonly used as a synonym for the Ice Age, note that this epoch does not encompass all of the last glacial period. The Antarctic ice sheet, for example, is believed to have formed about 10 million years ago.

Glaciers have not been ever-present features throughout the earth's long history. In fact, for most of geologic time glaciers have been absent. Evidence does indicate that in addition to the Pleistocene epoch there were at least three other periods of glacial activity: 2 billion, 600 million, and 300 million years ago. However, the most recent period of glaciation is of greatest interest, for the features of many present-day landscapes are a reflection of the work of Pleistocene glaciers.

SOME INDIRECT EFFECTS OF ICE AGE GLACIERS

In addition to the massive erosional and depositional work carried on by Pleistocene glaciers, the ice sheets had other, sometimes profound, effects upon the landscape. For example, as the ice advanced and retreated, animals and plants were forced to migrate. This led to stresses that some organisms could not tolerate. Furthermore, many present-day stream courses bear little resemblance to their preglacial routes. The Missouri River once flowed northward toward Hudson Bay, while the Mississippi River followed a path through central Illinois, and the head of the Ohio River reached only as far as Indiana. Other rivers that today carry only a trickle of water but nevertheless occupy broad channels are testimony to the fact that they once carried torrents of glacial meltwater.

In areas that were centers of ice accumulation, such as Scandinavia and the Canadian Shield, the land has been slowly rising for the past several thousand years. The land is rising because the added weight of the three-kilometer-thick mass of ice downwarped the earth's crust. Following the removal of this immense load, the crust has been adjusting by gradually rebounding upward ever since.

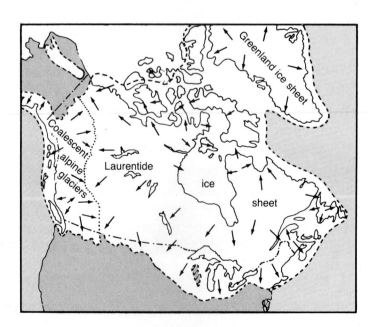

FIGURE 8.19 At their maximum extent, Pleistocene glaciers covered about 10 million square kilometers (4 million square miles) of North America.

Certainly one of the most interesting and perhaps dramatic effects of the Ice Age was the fall and rise of sea level that accompanied the advance and retreat of the glaciers. Since we know that the snow from which glaciers are made ultimately comes from the evaporation of ocean water, the growth of ice sheets must have caused a worldwide drop in sea level. Indeed, estimates suggest that sea level was as much as 130 meters (425 feet) lower than today. Thus, some land that is presently flooded by the oceans was dry. The Atlantic Coast of the United States was located more than 100 kilometers (60 miles) to the east of New York City; France and Britain were joined where the English Channel is today; Alaska and Siberia were connected across the Bering Strait; and Southeast Asia was tied by dry land to the islands of Indonesia.

While the formation and growth of ice sheets was an obvious response to significant changes in climate, the existence of the glaciers themselves triggered important climatic changes in the regions beyond their margins. In arid and semiarid areas on all continents, temperatures, and thus evaporation rates, were lower. At the same time, precipitation was moderate. This cooler, wetter climate resulted in the formation of many lakes called **pluvial lakes**, from the Latin term *pluvia* meaning *rain*. In North America, the greatest concentration of pluvial lakes occurred in the vast Basin and Range region of Nevada and Utah. Although most of the lakes completely disappeared, there are a few small remnants, the Great Salt Lake being the largest and best known.

CAUSES OF GLACIATION

A great deal is known about glaciers and glaciation. Much has been learned about glacier formation and movement, the extent of glaciers past and present, and the features created by glaciers, both erosional and depositional. However, scientists have not yet developed a completely satisfactory explanation for the causes of ice ages.

Any theory that attempts to explain the causes of glacial ages must successfully address two basic questions. First, what causes the onset of glacial conditions? For continental ice sheets to have formed, average temperatures must have been somewhat lower than at present and perhaps substantially lower than throughout much of geologic time. For that reason, a successful explanation would have to account for the gradual cooling that finally leads to glacial conditions. The second question is: What caused the alternation of glacial and interglacial stages that have been documented for the Pleistocene epoch? While the first question deals with long-term trends in temperature that occur on a scale of millions of years, this second question relates to much shorter-term changes.

Although the literature of science contains a vast array of hypotheses that attempt to explain the possible causes of glacial periods, we will discuss only a few major ideas in an effort to give the current thought on this problem.

Probably the most attractive theory for explaining why extensive glaciations have occurred only a few times in the geologic past comes from the theory of plate tectonics. Not only does this theory provide geologists with explanations about many previously misunderstood processes and features, it also provides a possible explanation for some previously unexplainable climatic changes including the onset of glacial conditions. Since glaciers can form only on the continents, we know that landmasses must exist somewhere in the higher latitudes before an ice age can commence. Many believe that ice ages have only occurred when the earth's shifting crustal plates carried the continents from tropical latitudes to more poleward positions.

Glacial features in present-day Africa, Australia, South America, and India indicate that these regions experienced an ice age near the end of the Paleozoic era, about 250 million years ago. For many years this puzzled scientists. Was the climate in these relatively tropical latitudes once like it is today in Greenland and Antarctica? Until the plate tectonics theory was formulated and proven, there was no reasonable explanation. Today scientists realize that the areas containing these ancient glacial features were joined together as a single supercontinent called Pangaea that was located at high latitudes far to the south of their present positions. Later, this landmass broke apart and its pieces, each moving on a different plate, drifted toward their present locations (Figure 8.20). It is now believed that during the geologic past, continental drift accounted for many dramatic climatic changes as landmasses shifted in relation to one another and moved to different latitudinal posi-

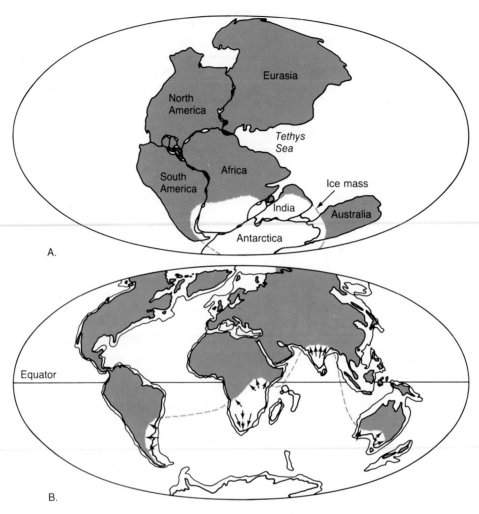

FIGURE 8.20 A. The supercontinent Pangaea showing the area covered by glacial ice 300 million years ago. **B.** The continents as they are today. The shading outlines areas where evidence of the old ice sheets exists. The dashed line joining the glaciated regions indicates how large the ice sheet would have had to have been if the continents had been in their present positions at the time of glaciation. (After R. F. Flint and B. J. Skinner, *Physical Geology*, 2nd ed., p. 418, New York: Wiley, 1977)

tions. Changes in oceanic circulation also must have occurred, altering the transport of heat and moisture, and consequently, the climate as well. Since the rate of plate movement is very slow, on the order of a few centimeters per year, appreciable changes in the positions of the continents occur only over great spans of geologic time. Thus, climatic changes brought about by continental drift are extremely gradual and happen on a scale of millions of years.

Since climatic changes brought about by moving plates are extremely gradual, the plate tectonics theory cannot explain the alternation of glacial and interglacial climates that occurred during the Pleistocene epoch. Therefore we must look to some other triggering mechanism that may cause climatic change on a scale of thousands rather than millions of years. Today many scientists believe or strongly suspect that the climatic oscillations that characterized the Pleistocene may be linked to variations in the earth's orbit. This hypothesis was first developed and strongly advocated by the Yugoslavian scientist Milutin Milankovitch and is based on the premise that variations in incoming solar radiation are a principal factor in controlling the earth's climate. Milankovitch formulated a comprehensive mathematical model based on the following elements:

1 Variations in the shape (*eccentricity*) of the earth's orbit around the sun;
2 Changes in *obliquity;* that is, changes in the angle that the earth's axis makes with the plane of the earth's orbit; and
3 The wobbling of the earth's axis, called *precession.*

Using these factors Milankovitch calculated variations in the receipt of solar energy and the corresponding surface temperature of earth back into time in an attempt to correlate these changes with the climatic fluctuations of the Pleistocene. In explaining climatic changes that result from these three variables, it should be noted that they cause little or no variation in the total amount of solar energy reaching the ground. Instead, their impact is felt because they change the degree of contrast between seasons. Somewhat milder winters in the middle to high latitudes means greater snowfall totals while cooler summers would bring a reduction in snowmelt.

Over the years the astronomical theory of Milankovitch has been widely accepted, then largely rejected, and now, in light of recent investigations, is once again very popular. Among recent studies that have added credibility and support to the astronomical theory is one in which deep-sea sediments containing certain climatically sensitive microorganisms

were analyzed in order to establish a chronology of temperature changes going back nearly one-half million years.* This time scale of climatic change was then compared to astronomical calculations of eccentricity, obliquity, and precession to determine if a correlation did indeed exist. Although the study was very involved and mathematically complex, the conclusions were straightforward. The authors found that major variations in climate over the past several hundred thousand years were closely associated with changes in the geometry of the earth's orbit; that is, cycles of climatic change were shown to correspond closely with the periods of obliquity, precession, and orbital eccentricity. More specifically, they stated, "It is concluded that changes in the earth's orbital geometry are the fundamental cause of the succession of Quaternary ice ages."**

Let us briefly summarize the theories that were just described. The theory of plate tectonics provides an explanation for the widely spaced and nonperiodic onset of glacial conditions at various times in the geologic past, whereas the astronomical theory proposed by Milankovitch and recently supported by the work of J. D. Hays and his colleagues furnishes an explanation for the alternating glacial and interglacial episodes of the Pleistocene.

In conclusion, it should be emphasized at this point that the ideas just discussed do not represent the only possible explanations for glacial ages. Although interesting and attractive, these proposals are certainly not without critics nor are they the only hypotheses currently under study. Other factors may, and in fact probably do, enter the picture.

*J. D. Hays, John Imbrie, and N. J. Shackleton, "Variations in the Earth's Orbit: Pacemaker of the Ice Ages," *Science,* 194 (4270): 1121–32.
**Ibid., p. 1131. *Quaternary* refers to the period on the geologic calendar that encompasses the last two million years.

REVIEW QUESTIONS

1 What is a glacier? Under what circumstances does glacial ice form?

2 Where are glaciers found today? What percentage of the earth's land area do they cover?

3 Describe glacial flow. At what rates do glaciers move? In an alpine glacier does all of the ice move at the same rate? Explain.

4 Why do crevasses form in the upper portion of a glacier but not below 50 meters?

5 Under what circumstances will the front of a glacier advance? Retreat? Remain stationary?

6 Describe the processes of glacial erosion.

7 How does a glaciated mountain valley differ from a mountain valley that was not glaciated?

8 List and describe the erosional features you might expect to see in an area where alpine glaciers exist or have recently existed.

9 What is glacial drift? What is the difference between till and outwash? What general effect do glacial deposits have on the landscape?

10 List the four basic moraine types. What do all moraines have in common? What is the significance of terminal and recessional moraines?

11 List and discuss depositional features other than moraines.

12 How does a kettle form?

13 How does the area covered by Pleistocene glaciers compare with the area presently covered by glacial ice?

14 How many glacial advances have been recognized for North America? List them in the order of their occurrence (Figure 8.18).

15 List three indirect effects of Ice Age glaciers.

16 How might plate tectonics help us understand the cause of ice ages? Can plate tectonics explain the alternation between glacial and interglacial climates during the Pleistocene?

CHAPTER NINE

DESERTS AND WIND

DESERTS

The word *desert* literally means *deserted* or *unoccupied*. Many desert areas are not truly deserted and indeed, in some cases, many people live there. Nevertheless, the world's dry regions are probably the least familiar land areas on earth outside of the polar realm. For example, one popular image of deserts is that they consist of mile after mile of drifting sand dunes. It is true that sand accumulations do exist in some areas and may be striking features, but they represent only a small percentage of the total desert area. In the Sahara, the world's largest desert, sand accumulations cover only 10 percent of the surface, while in the sandiest of all deserts, the Arabian, about 30 percent is sand covered. A more typical surface consists of barren rock or expanses of stoney ground.

Another commonly held but incorrect perception of dry lands is that they are practically lifeless. Although reduced in amount and different in character, plant life is usually present. Whereas desert plants differ widely from place to place, they all have one characteristic in common—they have developed adaptations that make them highly tolerant of drought. Such plants, called **xerophytes**, may have waxy leaves, stems, or branches that reduce water loss. Others have small leaves or none at all. Further, the roots of some species often extend to great depths in order to tap the moisture found there, while others produce a shallow but widespread root system that enables them to quickly absorb great amounts of moisture from the infrequent desert showers. Often the stems of desert plants are thickened by a spongy tissue that can store enough water to sustain the plant until the next rainfall. Thus, although they are often widely dispersed and provide little ground cover, plants of many kinds flourish in the desert.

DISTRIBUTION OF DRY LANDS

The desert (arid) and steppe (semiarid) regions of the world encompass about 48 million square kilometers, nearly one-third of the earth's land surface. No other climatic group covers so large a land area. The world map showing the distribution of desert and steppe regions reveals that dry lands are concentrated in the subtropics and in the middle latitudes (Figure 9.1).

The heart of the low-latitude dry climates lies in the vicinities of the Tropics of Cancer and Capricorn. A glance at Figure 9.1 shows a virtually unbroken desert environment stretching across North Africa to northwestern India. In addition to this single great expanse, lesser areas of subtropical desert include northern Mexico and the southwestern United States, parts of southern Africa, the west coast of South America, and a large portion of Australia. The existence of this dry subtropical realm is primarily the result of the prevailing global pattern of air pressure. That is to say, coinciding with the low-latitude dry regions are zones of high atmospheric pressure. These semipermanent pressure systems are characterized by dry subsiding air. Such conditions generally preclude cloud formation and precipitation.

Unlike their low-latitude counterparts, middle-latitude deserts and steppes are not controlled by the subsiding air masses associated with high pressure. Instead, these dry lands exist principally because of their position in the deep interiors of large land masses far removed from the oceans. In addition, the presence of high mountains across the paths of prevailing winds further acts to separate these areas from water-bearing, maritime air masses. In North

Transverse dunes in Great Sand Dunes National Monument, Colorado. (Photo by Stephen A. Trimble)

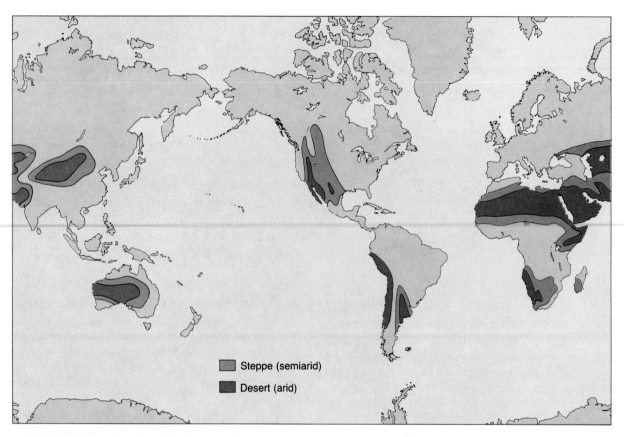

FIGURE 9.1 Arid and semiarid climates cover about 30 percent of the earth's land surface. No other climatic group covers so large an area.

America the Coast Ranges, Sierra Nevada, and Cascades are the foremost barriers, while in Asia, the great Himalayan chain prevents the summertime monsoon flow of moist Indian Ocean air from reaching into the interior. Because the Southern Hemisphere lacks extensive land areas in the middle latitudes, only a small area of desert and steppe are found in this latitude range existing primarily near the southern tip of South America in the rainshadow of the towering Andes.

GEOLOGIC PROCESSES IN ARID CLIMATES

The angular hills, the sheer canyon walls, and the pebble or sand covered surface of the desert contrast sharply with the rounded hills and curving slopes of more humid places. Indeed, to a visitor from a humid region, a desert landscape may seem to have been shaped by forces different than those operating in well-watered areas. However, while the contrasts may be striking, they are not a reflection of different processes but merely the differing effects of the same processes operating under contrasting climatic conditions.

In humid regions relatively fine textured soils support an almost continuous cover of vegetation that mantles the surface. Here the slopes and rock edges are rounded. Such a landscape reflects the strong influence of chemical weathering in a humid climate. By contrast, much of the weathered debris in deserts consists of unaltered rock and mineral fragments—the result of mechanical weathering processes. In dry lands rock weathering of any type is greatly reduced because of the lack of moisture and the scarcity of organic acids from decaying plants. Chemical weathering, however, is not completely lacking in deserts. Over long spans of time clays and thin soils do form and many iron-bearing silicate

FIGURE 9.2 Tanque Verde wash, Tucson, Arizona.
A. During a dry period.
B. Following a rain. (Photos by Tad Nichols)

A.

B.

minerals oxidize, producing the rust-colored stain found tinting some desert landscapes.

Most of the time, desert stream courses (called **washes** in the western United States) are dry (Figure 9.2A). This fact is often quite obvious even to the casual observer who, while traveling, notices the number of bridges with no streams beneath them or the number of dips in the road where dry washes cross. However, when the rare heavy showers do come, so much rain falls in such a short time that all of it cannot soak in. Since the vegetative cover is sparse, runoff is largely unhindered and consequently rapid, often creating flash floods along valley floors (Figure 9.2B). Such floods, however, are quite unlike floods in humid regions. A flood on a river such as the Mississippi may take several days to reach its crest and then to subside again, whereas desert floods arrive suddenly and likewise subside in a short time. Because much of the surface material is not anchored by vegetation, the amount of erosional work that occurs during one of these short-lived events is impressive.

Unlike the drainage in humid regions, stream courses in arid regions are seldom well integrated. That is, desert streams lack an extensive system of tributaries. In fact, a basic characteristic of deserts is that most of the streams that originate in them are small and die out before reaching the sea. Since the water table is usually far below the surface, few desert streams can draw upon it. Without a steady supply of water, evaporation soon depletes the stream and the remaining water sinks into the ground. The few permanent streams that do cross arid regions, such as the Colorado and Nile rivers, originate outside the desert, often in well-watered mountains. Here the water supply must be great to compensate for the losses occurring as the stream crosses the desert. For example, after the Nile leaves the rainy regions that are its source, it traverses almost 2000 kilometers of the Sahara without the contribution of a single tributary.

It should be emphasized that running water, although an infrequent occurrence, nevertheless does most of the erosional work in deserts. This is contrary to a commonly held belief that wind is the most important erosional agent sculpturing desert landscapes. Although wind erosion is more significant in dry areas than elsewhere, most desert landforms are nevertheless carved by running water. As we shall see shortly, the main role of wind is in the transportation and deposition of sediment, creating and shaping the ridges and mounds we call dunes.

TRANSPORTATION OF SEDIMENT BY WIND

Moving air, like moving water, is turbulent and able to pick up loose debris and transport it to other locations. Just as in a stream, the velocity of wind increases with height above the surface. Also like a stream, wind transports fine particles in suspension while heavier ones are carried as bed load. However, the transport of sediment by wind differs from that of running water in two significant ways. First, wind has a low density compared to water; thus, it is not capable of picking up and transporting coarse materials. Second, because wind is not confined to channels, it can spread sediment over large areas, as well as high into the atmosphere.

Bed Load

The **bed load** carried by wind consists of sand grains. Observations in the field and experiments using wind tunnels indicate that windblown sand moves by skipping and bouncing along the surface—a process termed **saltation**. The term is not a reference to salt, but instead derives from the Latin word meaning "to jump." The movement of sand grains begins when wind reaches a velocity sufficient to overcome the inertia of the resting particles. At first the sand rolls along the surface. Upon striking another grain, one or both of the grains may jump into the air. Once in the air, the sand is carried forward by the wind until gravity pulls the grain back toward the surface. When the sand hits the surface, it either bounces back into the air or dislodges other grains which then jump upward. In this manner a chain reaction is established, filling the air near the ground with saltating sand grains in a short period of time (Figure 9.3).

Bouncing sand grains never travel far from the surface. Even when winds are very strong, the height of the saltating sand seldom exceeds one meter and under less extreme conditions is usually confined to heights no greater than one-half meter. Some sand grains are too large to be thrown into the air by impact from other particles. When this is the case, the energy provided by the impact of the smaller saltating grains drives the larger grains forward. Estimates indicate that between 20 and 25 percent of the sand transported in a sandstorm is moved in this way.

Suspended Load

Unlike sand, dust can be swept high into the atmosphere by the wind. Since dust is often composed of rather flat particles that have large surface areas when compared to the weight of the particle, it is relatively easy for turbulent air to counterbalance the pull of gravity and keep these fine particles suspended for extended periods. Although both silt and

FIGURE 9.3 A cloud of saltating sand grains moving up the gentle slope of a dune. (Photo by Stephen Trimble)

clay can be carried in suspension, silt commonly makes up the bulk of the **suspended load** because the reduced level of chemical weathering in deserts produces only small amounts of clay.

Fine particles are easily carried by the wind, but they are not easily acquired by the turbulent air. The reason is that the wind velocity is practically zero within a very thin layer close to the ground. Thus, the wind cannot lift the sediment by itself. Instead, the dust must be ejected or spattered into the moving currents of air by bouncing sand grains or other disturbances. This idea is illustrated nicely by a dry country road on a windy day. Left undisturbed, little dust is raised by the wind. However, as a car or truck moves over the road, the previously smooth layer of silt is disturbed, creating a thick cloud of dust.

Although the suspended load is usually deposited relatively near its source, high winds are capable of carrying large quantities of dust great distances (Figure 9.4). In the 1930s, silt picked up in Kansas was

FIGURE 9.4 Dust storms like this one in Colorado were relatively common during the 1930s in the Dust Bowl region of the Great Plains. (Courtesy of the U.S. Department of Agriculture)

transported to New England and beyond into the North Atlantic. Similarly, dust blown from the Sahara has been traced as far as the West Indies.

WIND EROSION

Compared to running water and moving ice, wind is a relatively insignificant erosional agent. Recall that even in deserts, few major erosional landforms are created by the wind. Although wind erosion is not restricted to arid and semiarid regions, it does its most effective work in these areas. In humid places moisture binds particles together and vegetation anchors the soil so that wind erosion is negligible. For wind to be effective, dryness and scanty vegetation are important prerequisites. When such circumstances exist, wind may pick up, transport, and deposit great quantities of fine sediment. During the 1930s parts of the Great Plains experienced great dust storms. The plowing under of the natural vegetative cover for farming, followed by severe drought, made the land ripe for wind erosion, and led to the area being labeled the Dust Bowl.

One way that winds erode is by **deflation**, the lifting and removal of loose material. Although the effects of deflation are sometimes difficult to notice because the entire surface is being lowered at the same time, they can be significant. In portions of the 1930s Dust Bowl, vast areas of land were lowered by as much as one meter in only a few years.

The most noticeable results of deflation in some places are shallow depressions which are quite appropriately called **blowouts** (Figure 9.5). In the Great Plains region, from Texas north to Montana, thousands of blowouts are visible on the landscape. They range in size from small dimples less than one meter deep and three meters wide to depressions that approach 50 meters in depth and several kilometers across. The factor that controls the depths of these basins (that is, acts as base level) is the local water table. When blowouts are lowered to the water table, damp ground and vegetation prevent further deflation.

In portions of many deserts the surface is characterized by a layer of coarse pebbles and gravels that are too large to be moved by the wind. Such a layer, called **desert pavement**, is created as the wind lowers the surface by removing fine material until eventually only a continuous cover of coarse sedi-

FIGURE 9.5 A blowout, Sioux County, Nebraska. The remnant behind which the horse is standing indicates the level of the land prior to the formation of the blowout. (Photo by N. H. Darton, U.S. Geological Survey)

ment remains (Figure 9.6). Once desert pavement becomes established, a process that may take hundreds of years, the surface is effectively protected from further deflation.

Like glaciers and streams, wind erodes by abrasion. In dry regions as well as along some beaches, windblown sand cuts and polishes exposed rock surfaces. However, abrasion is often credited for accomplishments beyond its actual capabilities. Such features as balanced rocks that stand high atop narrow pedestals, and intricate detailing on tall pinnacles are not the results of abrasion. Since sand seldom travels more than a meter above the surface, the wind's sandblasting effect is obviously quite limited in vertical extent. Abrasion by windblown sand, however, does create interestingly shaped stones called **ventifacts** (Figure 9.7). The side of the stone exposed to the prevailing wind is abraded, leaving it polished, pitted, and with sharp edges. If the wind is not consistently from one direction, or if the pebble becomes reoriented, it may have several faceted surfaces.

WIND DEPOSITS

Although wind is relatively unimportant as a producer of erosional landforms, wind deposits are significant features in some regions. Accumulations of

FIGURE 9.6 Desert pavement composed of angular rock fragments. (Photo by Stephen Trimble)

windblown sediment are particularly conspicuous landscape elements in the world's dry lands and along many sandy coasts. Wind deposits are of two distinctive types: (1) mounds and ridges of sand from the wind's bed load and (2) extensive blankets of silt that once were carried in suspension.

Sand Deposits

As is the case with running water, wind drops its load of sediment when wind velocity falls and the energy available for transport diminishes. Thus sand begins to accumulate wherever an obstruction across the path of the wind slows the movement of the air. Unlike many deposits of silt, which form blanket-like layers over large areas, winds commonly deposit sand in mounds or ridges called *dunes*.

As moving air encounters an object, such as a clump of vegetation or a rock, the wind sweeps around and over it leaving a shadow of slower moving air behind the obstacle as well as a smaller zone of quieter air just in front of the obstacle. Some of the saltating sand grains moving with the wind come to rest in these wind shadows. As the accumulation of sand continues, it becomes a more imposing barrier to the wind and thus a more efficient trap for even more sand. If there is a sufficient supply of sand and the wind blows steadily for a long enough time, the mound of sand grows into a dune.

A profile of a dune shows an asymmetrical shape with the leeward slope being steep and the windward slope more gently inclined. Sand moves up the gentle slope on the windward side by saltation. Just beyond the crest of the dune, where the wind velocity is reduced, the sand accumulates. As more sand collects, the slope steepens and eventually some of it slides or slumps under the pull of gravity. In this way the leeward slope of the dune, called the **slip face**, maintains an angle of about 34 degrees, the angle

FIGURE 9.7 Ventifacts. (Photo by M. R. Campbell, U.S. Geological Survey)

of repose for loose dry sand.* Continued sand accumulation coupled with periodic slides down the slip face result in the slow migration of the dune in the direction of air movement.

As sand is deposited on the slip face, layers form which are inclined in the direction the wind is blowing. These sloping layers are called **cross beds** (see Figure 4.15C, p. 88). When the dunes are eventually buried under other layers of sediment and become part of the sedimentary rock record, their asymmetrical shape is destroyed, but the cross beds remain. By studying the orientation of these beds, geologists can determine the average direction of ancient winds. This information, together with other data, is then used in reconstructing climates of the geologic past. This knowledge of past climatic conditions, in turn, aids in determining earlier positions of the earth's moving lithospheric plates.

Types of Sand Dunes

Although often complex, dunes are not just random heaps of sediment. Rather they are accumulations that usually assume patterns that are surprisingly consistent. Addressing this point, a leading early investigator of dunes, the British engineer R. A. Bagnold, observed:

> Instead of finding chaos and disorder, the observer never fails to be amazed at a simplicity of form, an exactitude of repetition, and a geometric order unknown in nature on a scale larger than of crystalline structure. In places, vast accumulations of sand weighing millions of tons, move inexorably, in regular formation, over the surface of the country retaining their shape. . . .

Barchan Dunes. Solitary sand dunes shaped like crescents and with their tips pointing downwind are called **barchan dunes** (Figure 9.8A). These dunes form where supplies of sand are limited and the surface is relatively flat, hard, and lacking vegetation. They migrate slowly with the wind at a rate of up to 15 meters per year. Their size is usually modest with the largest barchans reaching heights of about 30 meters while the maximum spread between their

---

*Recall from Chapter Six that the angle of repose is the steepest angle at which material remains stable.

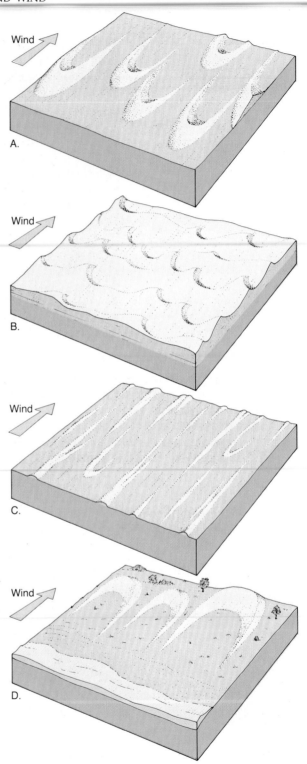

FIGURE 9.8 Sand dune types. **A.** Barchan dunes. **B.** Transverse dunes. **C.** Longitudinal dunes. **D.** Parabolic dunes.

horns approaches 300 meters. When the wind direction is nearly constant, the crescent form of these dunes is nearly symmetrical. However, when the wind direction is not perfectly fixed, one tip becomes longer than the other.

Transverse Dunes. In regions where vegetation is sparse or absent and sand is very plentiful, the dunes form a series of long ridges that are separated by troughs and oriented at right angles to the prevailing wind. Due to this orientation, they are termed **transverse dunes** (Figure 9.8B). Typically, many coastal dunes are of this type. In addition, they are common in many arid regions where the extensive surface of wavy sand is sometimes called a sand sea.

Longitudinal Dunes. **Longitudinal dunes** are long ridges of sand that form more or less parallel to the prevailing wind and where sand supplies are limited (Figure 9.8C). Apparently the prevailing wind direction must vary somewhat, yet still remain in the same quadrant of the compass. Although the smaller types are only three or four meters high and several tens of meters long, in some large deserts longitudinal dunes can reach great size. For example, in portions of North Africa, Arabia, and central Australia, these dunes may approach a height of 100 meters and extend for distances of more than 100 kilometers (62 miles).

Parabolic Dunes. Unlike the other dunes that have been described thus far, **parabolic dunes** form where vegetation partially covers the sand. The shape of these dunes resembles the shape of barchans except that their tips point into the wind rather than downwind (Figure 9.8D). Parabolic dunes often form along coasts where there are strong onshore winds and abundant sand. If the sand's sparse vegetative cover is disturbed at some spot, deflation creates a blowout. Sand is then transported out of the depression and deposited as a curved rim which grows higher as deflation enlarges the blowout.

Loess

In some parts of the world the surface topography is mantled with deposits of windblown silt. Over periods of perhaps thousands of years dust storms deposited this material which is called **loess**. As can be seen in Figure 9.9, when loess is breached by streams or road cuts it tends to maintain vertical cliffs and lacks any visible layers. The distribution of loess indicates that there are two primary sources for this sediment: deserts and glacial outwash deposits. The thickest and most extensive deposits of loess in the world occur in western and northern China, where accumulations of 30 meters are not uncommon and thicknesses of more than 100 meters have been measured. It is this fine, buff-colored sediment that gives the Yellow River (Hwang Ho) and the adjacent Yellow Sea their names. The source of China's 800,000 square kilometers of loess are the extensive desert basins of central Asia.

In the United States, deposits of loess are significant in many areas, including South Dakota, Nebraska, Iowa, Missouri, and Illinois as well as portions of the Columbia Plateau in the Pacific Northwest. The correlation between the distribution of loess and important farming regions in the Midwest and eastern Washington is not just a coincidence, because soils derived from this wind-deposited sediment are among the most fertile in the

FIGURE 9.9 A vertical loess bluff near the Mississippi River in southern Illinois. (Photo by James E. Patterson)

world. Unlike the deposits in China, the loess in the United States, as well as in Europe, is an indirect product of glaciation, for its source was deposits of outwash. During the retreat of the glacial ice, many river valleys were choked with sediment that was provided by meltwater. Strong westerly winds sweeping across the barren floodplains picked up the finer sediment and dropped it as a blanket on the east sides of the valleys. Such an origin is confirmed by the fact that loess deposits are thickest and coarsest on the lee sides of such major glacier drainage outlets as the Mississippi and Illinois rivers and rapidly thin with increasing distance from the valleys. Furthermore, the angular mechanically weathered particles composing the loess are essentially the same as the rock flour produced by the grinding action of glaciers.

THE EVOLUTION OF A DESERT LANDSCAPE

Since arid regions typically lack permanent streams, they are characterized as having **interior drainage**; that is, a discontinuous pattern of intermittent streams that do not flow out of the desert to the ocean. In the United States, the dry Basin and Range region provides an excellent example. The region includes southern Oregon, all of Nevada, western Utah, southeastern California, as well as southern Arizona and New Mexico. The name Basin and Range is an apt description for this almost 800,000-square-kilometer region, since it is characterized by more than 200 relatively small mountain ranges which rise 900–1500 meters above the basins that separate them. In this region, as in others like it around the world, erosion is carried out for the most part without reference to the ocean (ultimate base level) because drainage is in the form of local interior systems. Even in areas where permanent streams flow to the ocean, few tributaries exist, and thus only a relatively narrow strip of land adjacent to the stream has sea level as the ultimate level of land reduction.

The block models shown in Figure 9.10 depict the stages of landscape evolution in a mountainous desert such as the Basin and Range region. During and following the uplift of the mountains, running water begins carving the elevated mass and depositing large quantities of debris in the basin. During this early stage relief is greatest, for as erosion lowers the mountains and sediment fills the basins, elevation differences diminish.

When the occasional torrents of water produced by sporadic rains move down the mountain canyons, they are heavily loaded with sediment. Emerging from the confines of the canyon, the runoff spreads over the gentler slopes at the base of the mountains and quickly loses velocity. Consequently most of its load is dumped within a short distance. The result is a cone of debris at the mouth of a canyon known as an **alluvial fan**. Since the coarsest material is dropped first, the head of the fan is steepest, having a slope of perhaps 10 to 15 degrees. Moving down the fan, the size of the sediment and the steepness of the slope decrease and merge imperceptibly with the basin floor. An examination of the fan's surface would likely reveal a braided channel pattern because of the water shifting its course as successive channels became choked with sediment. Over the years, a fan enlarges, eventually coalescing with fans from adjacent canyons to produce an apron of sediment called a **bajada** along the mountain front.

On the rare occasions when rainfall is abundant, streams may flow across the bajada to the center of the basin, converting the basin floor into a shallow **playa lake**. Playa lakes are temporary features that last only a few days or at best a few weeks before evaporation and infiltration remove the water. The dry, flat lake bed that remains is termed a **playa**. Playas are typically composed of fine silts and clays, and occasionally encrusted with salts precipitated during evaporation. These precipitated salts may be unusual. A case in point is the sodium borate (better known as borax) mined from ancient playa lake deposits in Death Valley, California.

With time the mountain front is worn back and a sloping bedrock platform, called a **pediment**, is created adjacent to the steep mountain front. A pediment is an erosional surface, usually covered by a thin veneer of debris, that is formed by the action of running water. Just how the water carves the pediment, however, is unclear and still a matter for debate.

With the ongoing dissection of the mountain mass into an intricate series of valleys and sharp

THE EVOLUTION OF A DESERT LANDSCAPE

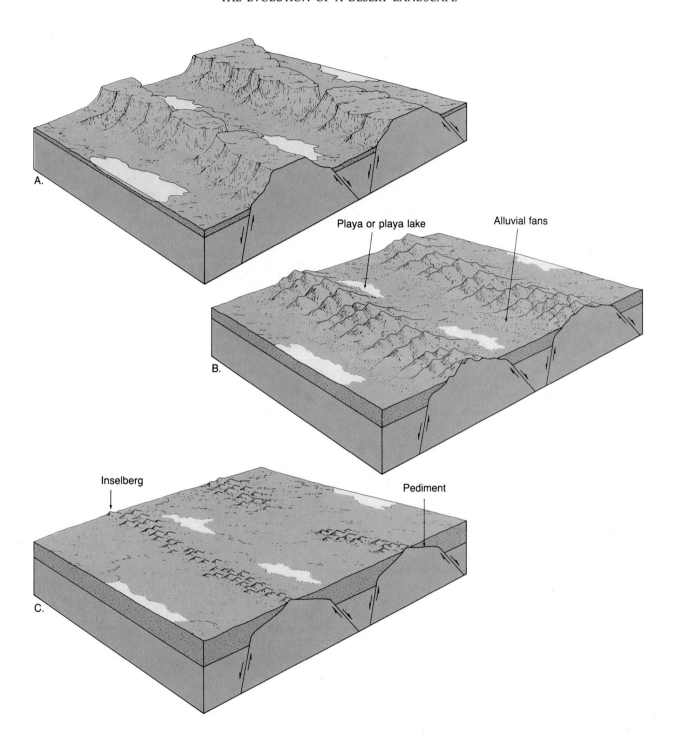

FIGURE 9.10 Stages of landscape evolution in a mountainous desert. As erosion of the mountains and deposition in the basins continue, relief diminishes. **A.** Early stage. **B.** Middle stage. **C.** Late stage.

divides as well as the accompanying sedimentation, the local relief continues to diminish. After more time passes, the steady retreat of the mountain front enlarges the pediment. Eventually this pediment growth results in nearly the entire mountain mass being consumed. Thus, by the late stages of erosion, the mountain areas are reduced to a few large bedrock knobs projecting above the surrounding pediment and sediment-filled basin. These isolated erosional remnants on an old-age desert landscape are called **inselbergs**, a German word meaning "island mountains."

Each of the stages of landscape evolution in an arid climate depicted in Figure 9.10 can be observed in the Basin and Range region. Recently uplifted mountains in an early stage of erosion are found in southern Oregon and northern Nevada. Death Valley, California, and southern Nevada fit into the more advanced middle stage, while the late stage, with its inselbergs and extensive pediments, can be seen in southern Arizona.

REVIEW QUESTIONS

1 Most deserts consist of mile after mile of drifting sand dunes. True or False? Provide some examples to support your answer.

2 How extensive are the desert and steppe regions of the earth?

3 What is the primary cause of subtropical deserts? Of middle-latitude deserts?

4 In which hemisphere (northern or southern) are middle-latitude deserts most common?

5 Why is rock weathering reduced in deserts?

6 What is the most important erosional agent in deserts?

7 What is a wash?

8 Describe the way in which wind transports sand. During very strong winds, how high above the surface can sand be carried?

9 Why is wind erosion relatively more important in arid regions than in humid areas?

10 What factor limits the depths of blowouts?

11 How do sand dunes migrate?

12 Four major dune types are recognized. Indicate which type of dune is associated with each of the statements below.
  a Dunes whose tips point into the wind.
  b Long sand ridges oriented at right angles to the wind.
  c Often form along coasts where strong winds create a blowout.
  d Solitary dunes whose tips point downwind.
  e Long sand ridges that are oriented more or less parallel to the prevailing wind.

13 Although sand dunes are the best-known wind deposits, accumulations of loess are very significant in some parts of the world. What is loess? Where are such deposits found? What are the origins of this sediment?

14 Why is sea level (ultimate base level) not a significant factor influencing erosion in desert regions?

15 Describe the features and characteristics associated with each of the stages in the evolution of a mountainous desert. Where in the United States can these stages be observed?

CHAPTER TEN

SHORELINES

The waters of the ocean are constantly in motion. The restless nature of the water is most noticeable along the shore—the dynamic interface between land and sea. Here we can observe the rhythmic rise and fall of tides and see waves constantly rolling in and breaking. Sometimes the waves are low and gentle. At other times, they pound the shore with an awesome fury.

Although it may not be readily apparent to the occasional visitor, the shoreline is constantly being shaped and modified by the moving ocean waters. However, the nature of present-day shorelines is not just the result of the relentless attack of the land by the sea. Indeed, the shore is a complex zone whose unique character results from many geologic processes. For example, practically all coastal areas were affected by the worldwide rise in sea level that accompanied the melting of glaciers at the close of the Pleistocene epoch. As the sea edged landward, the shoreline became superimposed upon landscapes that resulted from such processes as stream erosion, glaciation, volcanic activity, and the forces of mountain building.

WAVES

Wind-generated waves provide most of the energy that shapes and modifies shorelines. Where the land and sea meet, waves that may have traveled unimpeded for hundreds or thousands of kilometers suddenly encounter a barrier that will not allow them to advance farther. Stated another way, the shore is where a practically irresistible force confronts an almost immovable object. The conflict that results is never-ending and sometimes dramatic.

Wave action modifying the coastline at Point Reyes, California. (Photo by Stephen Trimble)

The undulations of the water surface, called waves, derive their energy and motion from the wind. If a breeze of less than 3 kilometers (2 miles) per hour starts to blow across still water, small wavelets appear almost instantly. When the breeze dies, the ripples disappear as suddenly as they formed. However, if the wind exceeds 3 kilometers per hour, more stable waves gradually form and progress with the wind.

All waves are described in terms of the characteristics illustrated in Figure 10.1. The tops of the waves are the *crests*, which are separated by *troughs*. The vertical distance between trough and crest is called the **wave height**, and the horizontal distance separating successive crests is the **wavelength**. The **wave period** is the time interval between the passage of successive crests at a stationary point. The height, length, and period that are eventually achieved by a wave depend upon three factors: (1) Wind speed; (2) Length of time the wind has blown; and (3) **Fetch**, the distance that the wind has traveled across the open water. As the quantity of energy transferred from the wind to the water increases, the heights of the waves transmitting it increase. In the open ocean, wave heights of 1 to 4 meters (3 to 13 feet) are common, although storms may produce much higher waves. Winds are often gusty and turbulent; for that reason, waves covering the ocean surface are quite irregular in height and length.

When the wind stops or changes direction, or the waves leave the stormy area where they were

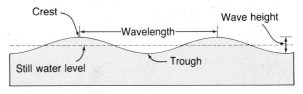

FIGURE 10.1 Characteristics of a wave.

FIGURE 10.2 Movement of water particles with the passage of a wave.

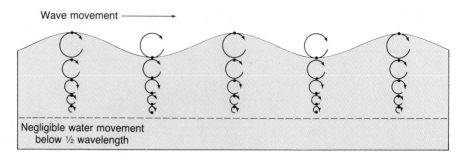

Wave movement ⟶

Negligible water movement
below ½ wavelength

created, they continue on without relation to local winds. The waves also undergo a gradual change to swells that are lower in height and longer in length and may carry the storm's energy to distant shores. Because many independent wave systems exist at the same time, the sea surface acquires a complex and irregular pattern. Hence the sea waves we watch from the shore are usually a mixture of swells from faraway storms and waves created by local winds.

It is important to realize that in the open sea the motion of the wave is different from the motion of the water particles within it. It is the wave form that moves forward, not the water itself. Each water particle moves in a circular path during the passage of a wave (Figure 10.2). As a wave passes, a water particle returns almost to its original position. This is demonstrated by observing the behavior of a floating cork as a wave passes. The cork merely seems to bob up and down and sway slightly to and fro without advancing appreciably from its original position.* For this reason, waves in the open sea are called **waves of oscillation**. The energy contributed by the wind to the water is transmitted not only along the surface of the sea but downward as well. Due to friction, there is a progressive loss of energy with an increase in depth until at a depth equal to about one-half the wavelength the movement of water particles becomes negligible. This is shown by the rapidly diminishing diameters of water-particle orbits in Figure 10.2.

As long as a wave is in deep water it is unaffected by water depth. However when a wave approaches the shore the water becomes shallower and influences wave behavior. The wave begins to "feel bottom" at a water depth equal to about one-half its

wavelength. Since some energy is used in moving small particles of sediment back and forth, the wave slows. As the wave continues to advance toward the shore, the slightly faster seaward waves catch up, decreasing the wavelength. As the speed and length of the wave diminish, the wave steadily grows higher. Finally a critical point is reached when the steep wave front is unable to support the wave, and it collapses, or *breaks* (Figure 10.3). What had been a wave of oscillation now becomes a **wave of translation** in which the water advances up the shore. The turbulent water created by breaking waves is called **surf**. Following the uprush of water onto the beach, a seaward backwash occurs. The water from expended waves most commonly moves seaward in a broad sheet that produces the undertow so often felt by swimmers. Sometimes the backwash occurs in narrow, localized channels and if strong enough, it may pull swimmers into deep water.

WAVE EROSION

During periods of calm weather, wave action is at a minimum. However just as streams do most of their work during floods, so waves do most of their work during storms. The impact of high, storm-induced waves against the shore can at times be awesome in its violence. Each breaking wave may hurl thousands of tons of water against the land, sometimes causing the earth to literally tremble. The pressures exerted by Atlantic waves, for example, average nearly 10,000 kilograms per square meter (more than 2000 pounds per square foot) in winter. The force during storms is even greater. During one such storm, for instance, a 1350-ton portion of a steel and concrete breakwater was ripped from the rest of the structure and moved to a useless position toward the shore at Wick Bay, Scotland. Five years later the 2600-ton

*The wind does drag the water slightly forward, causing the surface circulation of the oceans.

FIGURE 10.3 Changes that occur when a wave moves onto shore.

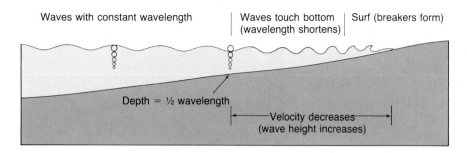

Waves with constant wavelength | Waves touch bottom (wavelength shortens) | Surf (breakers form)

Depth = ½ wavelength

Velocity decreases (wave height increases)

unit that replaced the first met a similar fate. There are many such stories that demonstrate the great force of breaking waves. It is no wonder then that cracks and crevices are quickly opened in cliffs, seawalls, breakwaters, and anything else that is subjected to these enormous shocks. Water is forced into every opening, causing air in the cracks to become highly compressed by the thrust of crashing waves. When the wave subsides, the air expands rapidly, dislodging rock fragments and enlarging and extending pre-existing fractures.

In addition to the erosion caused by wave impact and pressure, **abrasion**, the sawing and grinding action of water armed with rock fragments, is also important. In fact, abrasion is probably more intense in the surf zone than in any other environment. Smooth, rounded stones and pebbles along the shore are obvious reminders of the grinding action of rock against rock in the surf zone. Further, such fragments are used as "tools" by the waves as they cut horizontally into the land (Figure 10.4).

FIGURE 10.4 Cliff undercut by wave erosion north of Guaymas, Sonora, Mexico. (Courtesy of Ward's Natural Science Establishment, Inc., Rochester, N.Y.)

Along shorelines composed of unconsolidated material rather than hard rock, the rate of erosion by breaking waves can be extraordinary. In parts of Britain, where waves have the easy task of eroding glacial deposits of sand, gravel, and clay, the coast has been worn back 3–5 kilometers (2–3 miles) since Roman times, sweeping away many villages and ancient landmarks. A similar retreat may be seen along the cliffs of Cape Cod, which in places are retreating at a rate of up to 1 meter (3 feet) per year.

WAVE REFRACTION

Most waves approach a shoreline at an angle. When they reach the shallow water of a smoothly sloping bottom, however, they are bent and tend to become parallel to the shore. Such bending of waves is called **refraction**. The part of the wave nearest the shore touches bottom and slows first, while the end that is still in deep water continues forward at its regular speed. The net result is a wave front that may approach nearly parallel to the shore regardless of the original direction of the wave.

Due to refraction, wave impact is concentrated against the sides and ends of headlands projecting into the water, while wave attack is weakened in bays. This differential wave attack along irregular coastlines is illustrated in Figure 10.5. Since waves reach the shallow water in front of the headland sooner than they do in adjacent bays, they are bent more nearly parallel to the protruding land and strike it from all three sides. Over a period of time the effect of this process is to straighten irregular coastlines.

BEACH DRIFT AND LONGSHORE CURRENTS

Although waves are refracted, most still reach the shore at an angle, however slight. Consequently, the

FIGURE 10.5 Wave refraction along an irregular shoreline.

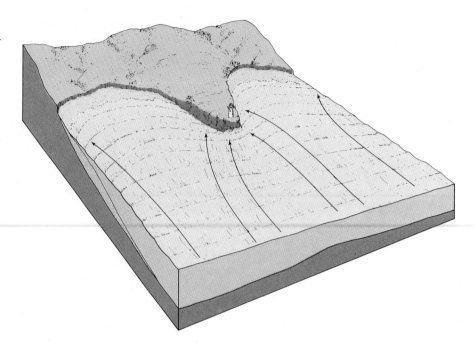

uprush of water from each breaking wave is oblique. Nevertheless, the backflow is straight down the slope of the beach. The effect of this pattern of water movement is to transport particles of sediment in a zigzag pattern along the beach (Figure 10.6). This movement, called **beach drift**, can trans-

port sand and pebbles hundreds or even thousands of meters each day.

Oblique waves also produce currents within the surf zone that flow parallel to the shore. Since the water here is turbulent, these **longshore currents** easily move the fine suspended sand as well as roll

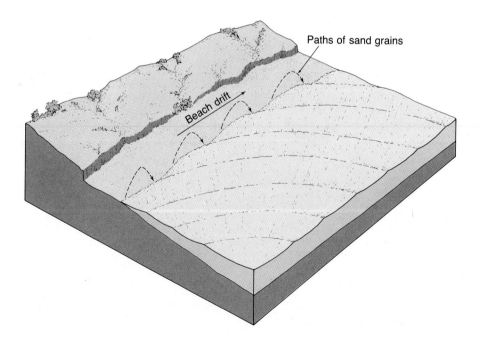

Paths of sand grains

Beach drift

FIGURE 10.6 Beach drift, caused by the uprush of water from oblique waves.

larger sand and gravel along the bottom. When the sediment transported by longshore currents is added to the quantity moved by beach drift, the total amount can be very large. At Sandy Hook, New Jersey, for example, the quantity of sand transported along the shore over a 48-year period averaged almost 750,000 tons per year. For a 10-year period at Oxnard, California, more than 1.5 million tons of sediment moved along the shore each year.

There should be little wonder that beaches have been characterized as "rivers of sand." At any point along a beach there is likely to be more sediment that was derived elsewhere than material eroded from the cliff immediately behind it. It is also worth noting that much of the sediment composing beaches is not wave-eroded debris. Rather, in many areas sediment-laden rivers that discharge into the ocean are the major source of material. For that reason, if it were not for beach drift and longshore currents, many beaches would be nearly sandless.

HUMAN INTERFERENCE WITH SHORELINE PROCESSES

The natural movement of sand by longshore currents and beach drift sometimes creates problems for those who live along the shore. Sand is either being eroded from a place where people want it to remain or it is being deposited where it is not wanted. In many coastal areas remedies for these problems include the building of such artificial structures as jetties, groins, and breakwaters. In many cases, however, these structures interfere with normal beach processes and interrupt the movement of sand. Such interference can create many new problems and result in unwanted changes that are very expensive to correct.

Jetties are usually built in pairs and extend into the ocean at the entrances to rivers and harbors. By confining the flow of water to a narrow zone, the ebb and flow caused by the rise and fall of the tides keep the sand in motion and prevent deposition in the channel. However, as illustrated in Figure 10.7, the jetty may act as a dam against which the longshore current and beach drift deposit sand. At the same time, wave activity removes sand on the other side. Since the other side is not receiving any new sand, there is soon no beach at all.

To maintain or widen beaches that lost sand from the action of longshore currents and beach drift, **groins** are often constructed. Groins are short walls built at right angles to the shore so as to trap moving sand (Figure 10.7). These structures often do their job so effectively that the longshore current beyond

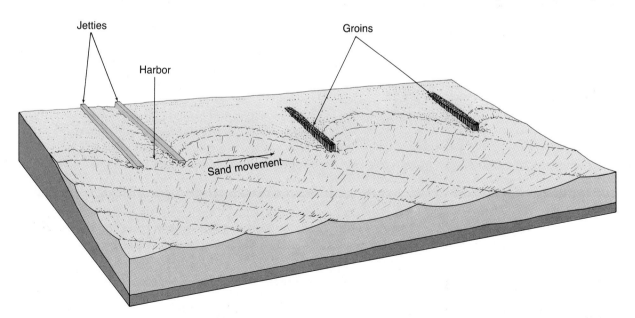

FIGURE 10.7 Jetties and groins trap sand that would otherwise be moved down the beach by the action of waves.

the groin is sand deficient. As a result, the current removes sand from the beach on the leeward side of the groin. To offset this effect, property owners downcurrent from the structure may erect a groin on their property. In this manner, the number of groins multiplies. An example of such proliferation is the shoreline of New Jersey, where more than three hundred such structures have been built. Since it has been shown that groins often do not provide a satisfactory solution, they are no longer the preferred method of keeping beach erosion in check. In some areas, a system of *beach nourishment* is used. This simply involves the periodic addition of sand to the beach system. The source of the sand may be the bottom of a nearby lagoon or inland dunes. In some cases, sand is trucked in and added to the beach. In other instances, the sand is added at an upstream location to be distributed down the coast by wave activity. Even so, it should be noted that beach nourishment can be an expensive solution. When 24 kilometers of Miami Beach were recently replenished, the cost was $64 million. Furthermore, in some instances, beach nourishment can lead to unwanted environmental effects. For example, beach replenishment at Waikiki Beach, Hawaii, involved replacing coarse calcareous sand with softer, muddier calcareous sand. Destruction of the soft beach sand by breaking waves increased the water's turbidity and killed offshore coral reefs. At Miami Beach, where quartz sand was replaced by calcareous sand, the increased turbidity damaged local coral communities.

In order to create a quiet water area near shore to protect boats from the force of large breaking waves, a **breakwater** may be constructed parallel to the shoreline. However, when this is done, the reduced wave activity along the shore behind the structure may allow sand to accumulate. If this happens, the marina will eventually fill with sand while the downstream beach erodes and retreats. At Santa Monica, California, where building a breakwater created just such a problem, the city had to install a dredge to remove sand from the protected quiet water zone and deposit it down the beach where longshore currents and beach drift could recirculate the sand.

As these examples illustrate, whenever people interfere with natural shoreline processes, the beach system responds. Any human action that does not consider the potential effects on downshore areas only results in more problems. Deposition at one site leads to erosion elsewhere and the configuration of the beach is changed.

Sometimes human activities far from the coast can also have a dramatic impact on a shoreline. Recall that the supply of sand for many beaches comes from rivers flowing into the sea. Waves then transfer this sediment along the shoreline by beach drift and longshore currents. But, if a dam is built on a river, the river's load of sediment, which once supplied the beach, is intercepted and deposited in a reservoir. Although little new sand is being added, the work of waves along the shore continues. The result is increased beach erosion, and without a supply of sand, the beach will eventually disappear.

SHORELINE FEATURES

Along the rugged and irregular New England coast or along the steep shorelines of the West Coast, the effects of wave erosion are often easily seen. **Wave-cut cliffs**, as their name implies, originate by the cutting action of the surf against the base of coastal land. As erosion progresses, rocks overhanging the notch at the base of the cliff crumble into the surf and the cliff retreats (see Figure 10.4). A relatively flat, benchlike surface, the **wave-cut platform**, is left behind by the receding cliff (Figure 10.8). The platform broadens as wave attack continues. Some debris produced by the breaking waves remains along the water's edge as part of the beach, whereas the remainder is transported farther seaward.

Headlands that extend into the sea are vigorously attacked by the waves because of refraction. The surf erodes the rock selectively, wearing away the softer or more highly fractured rock at the fastest rate. At first, sea caves may form. When two caves on opposite sides of a headland unite, a **sea arch** results (Figure 10.9). Finally the arch falls in, leaving an isolated remnant, or **sea stack**, on the wave-cut platform (Figure 10.9). Eventually it too will be consumed by the action of the waves.

Where beach drift and longshore currents are active, several features related to the movement of sediment along the shore may develop. **Spits** are elongated ridges of sand that project from the land into the mouth of an adjacent bay (Figure 10.10A). Often the end in the water hooks landward in

FIGURE 10.8 Elevated wave-cut platform along the California coast north of San Francisco. A new platform is being created at the base of the cliff. (Photo by John S. Shelton)

FIGURE 10.9 Sea arch and sea stack along the coast of Iceland. (Photo by Bruce F. Molnia, courtesy of Terraphotographics/BPS)

A.   B.   C.

FIGURE 10.10 Features created by the movement of sand by beach drift and long-shore currents. **A.** Spit. (Photo by Bruce F. Molnia, courtesy of Terraphotographics/BPS) **B.** Baymouth bar. (Photo by Robert J. Sager, Los Angeles Harbor College, © GeoPhoto Publishing Company) **C.** Tombolo and baymouth bar. (Photo by Alan L. Mayo, University of Colorado, © GeoPhoto Publishing Company)

response to wave-generated currents. The term **baymouth bar** is applied to a sand bar that completely crosses a bay, sealing it off from the open ocean (Figure 10.10B). Such a feature tends to form across bays where currents are weak, allowing a spit to extend to the other side. A **tombolo**, a ridge of sand that connects an island to the mainland or to another island (Figure 10.10C), forms in much the same manner as does a spit.

The gently sloping coastline found along the Gulf Coast and much of the eastern shore of the United States south of New York City is frequently characterized by **barrier islands**, which are low offshore ridges of sand that parallel the coast. The lagoons that separate these narrow islands from the shore represent zones of relatively quiet water that allow small craft traveling between New York and northern Florida to avoid the rough waters of the North Atlantic.

How barrier islands originate is still not certain. Quite possibly they form in three or more ways. Some are thought to have originated as spits that were subsequently severed from the mainland by wave erosion or by the general rise in sea level following the last episode of glaciation. It is also possible that some barrier islands are created when turbulent waters in the line of breakers heap up sand that has been scoured from the bottom. Since these sand barriers rise above normal sea level, the piling up of sand likely resulted from the work of storm waves at high tide. Finally, some studies suggest that barrier islands may be former sand dune ridges that originated along the shore during the last glacial

period, when sea level was lower. As the ice sheets melted, sea level rose and flooded the area behind the beach-dune complex.

There is little question that a shoreline soon undergoes modification regardless of its initial configuration. At first most coastlines are irregular, although the degree of and reason for the irregularity may vary considerably from place to place. Along a coastline that is characterized by varied geology, the pounding surf may at first increase its irregularity because the waves will erode the weaker rocks more easily than the stronger ones. Be that as it may, it is commonly agreed that if a shoreline remains stable, marine erosion and deposition will eventually produce a more regular coast. Figure 10.11 illustrates the evolution of an initially irregular coast. As waves erode the headlands, creating cliffs and a wave-cut platform, sediment is carried along the shore. Some material is deposited in the bays, while other debris is formed into spits and baymouth bars. At the same time rivers fill the bays with sediment; ultimately a smooth coast results.

EMERGENT AND SUBMERGENT COASTS

The great variety of present-day shorelines suggests that they are complex areas. Indeed, to understand the nature of any particular coastal area, many factors must be considered, including rock types, size and direction of waves, number of storms, tidal range, and submarine profile. Moreover, recent tectonic events and changes in sea level must also be

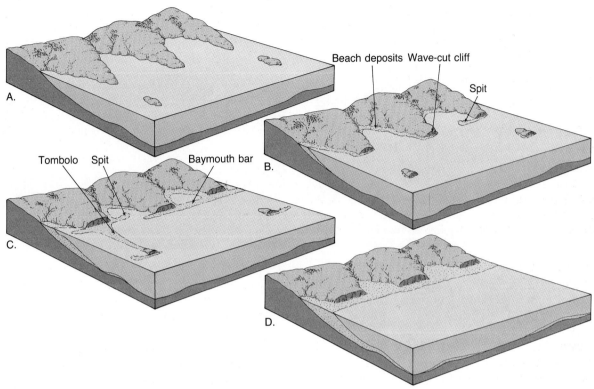

FIGURE 10.11 Development of an initially irregular coastline.

taken into account. These variables make shoreline classification difficult.

One way that many geologists classify coasts is based upon changes that have occurred with respect to sea level. This commonly used, although incomplete, classification divides coasts into two categories: emergent and submergent. **Emergent coasts** develop either because an area has been uplifted or as a result of a drop in sea level. Conversely, **submergent coasts** are created when sea level rises or the land adjacent to the sea subsides.

In some areas the coast is clearly emergent because rising land or a falling water level expose wave-cut cliffs and platforms above sea level. Excellent examples include portions of coastal California where uplift has occurred in the recent geological past. The elevated wave-cut platform shown in Figure 10.8 illustrates this. In the case of the Palos Verdes Hills, south of Los Angeles, seven different terrace levels exist, indicating seven episodes of uplift. The ever-persistent sea is now cutting a new platform at the base of the cliff. If uplift follows, it too will become an elevated marine terrace.

Other examples of emergent coasts include regions that were once buried beneath great ice sheets. When glaciers were present, their weight depressed the crust and when the ice melted, the crust began to gradually spring back. As a result, prehistoric shoreline features today are found high above sea level. The Hudson Bay region of Canada is one such area, portions of which are still rising at a rate of more than one centimeter per year.

In contrast to the preceding examples, other coastal areas show definite signs of submergence. The shoreline of a coast that has been submerged in the relatively recent past is often highly irregular because the sea typically floods the lower reaches of river valleys flowing into the ocean. The ridges separating the valleys, however, remain above sea level and project into the sea as headlands. These drowned river mouths, which are often called **estuaries**, characterize many coasts today. Along the Atlantic coast, the Chesapeake and Delaware bays are examples of estuaries created by submergence. The picturesque coast of Maine, particularly in the vicinity of Acadia National Park, is another excellent

A.

B.

FIGURE 10.12 A. High tide and **B.** low tide in the Bay of Fundy at Parrsboro, Nova Scotia. (Photos by G. Blouin, Information Canada Photothèque)

example of an area that was flooded by the post-glacial rise in sea level and transformed into a highly irregular submerged coastline.

It should be kept in mind that most coasts have a rather complicated geologic history. With respect to sea level, many have at various times emerged and then submerged again. Each time they retain some of the features created during the previous situation.

TIDES

Tides are periodic changes in the elevation of the ocean surface at a specific location. Their rhythmic rise and fall along coastlines have been known since antiquity, and other than waves, they are the easiest ocean movements to observe (Figure 10.12). Although known for centuries, tides were not explained satisfactorily until Sir Isaac Newton applied the law of gravitation to them. Newton showed that there is a mutual attractive force between two bodies, and that since oceans are free to move, they are deformed by this force. Hence tides result from the gravitational attraction exerted upon the earth by the moon, and to a lesser extent by the sun.

To illustrate how tides are produced, we will assume that the earth is a rotating sphere covered to a uniform depth with water (Figure 10.13). It is easy to see how the moon's gravitational force can cause the water to bulge on the side of the earth nearest the moon. In addition, however, an equally large

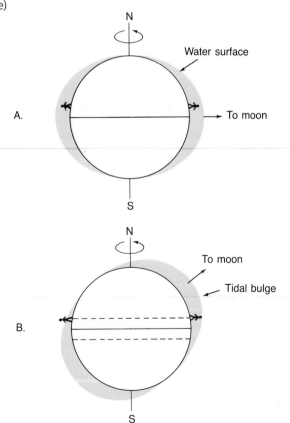

FIGURE 10.13 Tidal bulges on an ocean-covered earth with the moon **A.** in the plane of the earth's equator and **B.** above the plane of the equator. In the latter situation the observer experiences unequal high tides.

tidal bulge is produced on the side of the earth directly opposite the moon. Both tidal bulges are caused, as Newton discovered, by the pull of gravity, a force inversely proportional to the square of the distance between two objects. In this case the two objects are the moon and the earth. Because the force of gravity decreases as distance increases, the moon's gravitational pull on the earth is slightly greater on the near side of the earth than on the far side. The result of this differential pulling is to stretch (elongate) the earth. Although the solid earth is stretched by the moon's gravitational pull, the amount of elongation is very slight. However, the world's oceans, which are mobile, are deformed quite dramatically by this effect to produce the two opposing tidal bulges.

Since the position of the moon changes slightly in a single day, it is the tidal bulges that remain in place while the earth rotates beneath them. For that reason, the earth will carry an observer at any given location alternately into areas of deeper and shallower water. When the person is carried into regions of deeper water, the tide rises, and as the person is carried away, the tide falls. Therefore during one day the observer would experience two high tides and two low tides. In addition to the earth rotating, the tidal bulges also move as the moon revolves about the earth every 28 days. As a result, the tides, like the time of moonrise, occur about 50 minutes later each day. After 28 days one cycle is complete and a new one begins.

There may be an inequality between the high tides during a given day. Depending upon the moon's position, the tidal bulges may be inclined to the equator as in Figure 10.13B. This figure illustrates that an observer in the Northern Hemisphere experiences a high tide on the side of the earth under the moon that is considerably higher than the high tide half a day later. On the other hand, a Southern Hemisphere observer would experience the opposite effect.

The sun also influences the tides, but because it is so far away, the effect is considerably less than that of the moon. In fact, the tide-generating potential of the sun is slightly less than half that of the moon. Near the times of new and full moons, the sun and moon are aligned and their forces are added together (Figure 10.14A). Accordingly, the two tide-producing bodies cause higher tidal bulges (high tides) and lower tidal troughs (low tides). These are called the **spring tides**. Spring tides create the largest daily tidal range, that is, the largest variation between high and low tides. Conversely, at about the time of the first and third quarters of the moon,

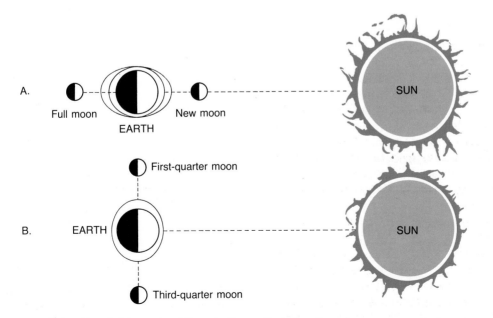

FIGURE 10.14 Relationship of the moon and sun to the earth during **A.** spring tides and **B.** neap tides.

the gravitational forces of the moon and sun on the earth are at right angles, and each partially offsets the influence of the other (Figure 10.14B). As a result, the daily tidal range is less. These are the **neap tides**.

The discussion thus far explains the basic causes and patterns of tides but keep in mind these theoretical considerations cannot be used to predict either the height or the time of actual tides at a particular place. The shape of coastlines and the configuration of ocean basins greatly influence the tide. Consequently, tides at various locations respond differently to the tide-producing forces. This being the case, the nature of the tide at any place can be determined most accurately by actual observation. The predictions in tidal tables and the tidal data on nautical charts are based upon such observations.

Tidal current is the term used to denote the horizontal flow of water accompanying the rise and fall of the tide. As the tide rises water flows in toward the shore as a **flood tide**, submerging the low-lying coastal zone. When the tide falls, a reverse flow, the **ebb tide**, again exposes the drowned portion of the shore. The areas affected by these alternating tidal currents are called **tidal flats**. Depending upon the nature of the coastal zone, tidal flats vary in size from narrow strips lying seaward of the beach to zones that may extend for several kilometers.

Although tidal currents are not important in the open sea, tides may create rapid currents in bays, river estuaries, straits, and other narrow places. Off the coast of Brittany, for example, tidal currents which accompany a high tide of 12 meters (40 feet) may attain a speed of 20 kilometers (12 miles) per hour. While it is generally believed that tidal currents are not major agents of erosion and sediment transport, notable exceptions occur where tides move through narrow inlets. Here they constantly scour the small entrances to many good harbors that would otherwise be blocked.

TIDES AND THE EARTH'S ROTATION

The tidal drag of the moon is steadily slowing the earth's rotation. The rate of slowing, however, is not great. It has been estimated that the weak but steady braking action of the tide is slowing the earth's rotation at a rate between one second per day per 120,000 years and one second per day per 100,000 years. Although this may seem inconsequential, over millions of years this small effect will become very large. Eventually, billions of years into the future, the earth will no longer have alternating days and nights.

If the earth is slowing, the number of days per year must have been greater in the geologic past. By studying fossil corals and clam shells, geologists have shown that this is indeed the case. By counting the number of daily growth rings of some well-preserved fossil specimens, the number of days per year can be ascertained. Studies using this ingenious technique indicate that 350 million years ago a year had 400–410 days, while 280 million years ago there were 390 days in a year. These figures closely agree with current estimates of the earth's slowing rotation.

TIDAL POWER

With increased public interest in the rising costs and eventual depletion of petroleum, greater attention is being focused upon alternate energy sources. Although several methods of generating electrical energy from the oceans have been proposed, the ocean's energy potential remains largely untapped. The development of tidal power is the principal example of energy production from the ocean.

Tides have been used as a source of power for centuries. Beginning in the twelfth century, water wheels driven by the tides were used to power gristmills and sawmills. During the seventeenth and eighteenth centuries, much of Boston's flour was produced at a tidal mill. Today, far greater energy demands must be met and more sophisticated ways of using the force created by the perpetual rise and fall of the ocean must be employed.

Tidal power is harnessed by constructing a dam across the mouth of a bay or an estuary in a coastal area having a large tidal range. The narrow opening between the bay and the open ocean magnifies the variations in water level that occur as the tides rise and fall. The strong in-and-out flow that results at such a site is then used to drive turbines and electrical generators.

Tidal energy utilization is exemplified by the tidal power plant at the mouth of the Rance River in France. By far the largest yet constructed, this plant went into operation in 1966 and produces enough power to satisfy the needs of Brittany and also contribute to the demands of other regions. Much smaller experimental facilities near Murmansk in the Soviet Union and near Taliang in China are also being used to generate electricity. The United States has not yet tapped its tidal power potential, although a site at Passamaquoddy Bay in Maine, where the tidal range approaches 15 meters (50 feet), has been under review for more than 50 years.

Along most of the world's coasts it is not possible to harness tidal energy. If the tidal range is less than 8 meters (25 feet) or narrow, enclosed bays are absent, tidal power development is uneconomical. For this reason, the tides will never provide a very high proportion of our ever-increasing electrical energy requirements. Nevertheless, the development of tidal power may be worth pursuing as Paul R. Ryan points out:

> Although total tidal power potential represents only a relatively small proportion of world energy requirements, its realization would nevertheless save a significant amount of fossil fuels. Tidal projects worldwide have been estimated to have a potential energy output of 635,000 gigawatts, the equivalent of more than a billion barrels of oil, a year.*

In addition to the fact that electricity produced by the tides consumes no exhaustible fuels (and hence creates no noxious wastes), such facilities disturb the landscape much less than the large reservoirs that are created when rivers are dammed.

---

*"Harnessing Power from the Tides: State of the Art," *Oceanus* 22 (1980): 64.

REVIEW QUESTIONS

1 List three factors that determine the height, length, and period of a wave.

2 Describe the motion of a water particle as a wave passes.

3 Explain what happens when a wave breaks.

4 How do waves cause erosion?

5 What is wave refraction? What is the effect of this process along irregular coastlines (see Figure 10.5)?

6 Why are beaches often called "rivers of sand"?

7 For what purpose is a groin built? Why might the building of one groin lead to the building of others?

8 Describe the formation of the following features: wave-cut cliff, wave-cut platform, sea stack, spit, baymouth bar, tombolo.

9 List three possible ways that barrier islands form.

10 What observable features would lead you to classify a coastal area as emergent?

11 Are estuaries associated with submergent or emergent coasts? Why?

12 Discuss the origin of ocean tides.

13 Explain why an observer can experience two unequal high tides during one day (see Figure 10.13).

14 How does the sun influence tides?

15 What is meant by flood tide? Ebb tide?

16 How have tides affected the earth's rotation? How did geologists substantiate this idea?

17 Why will tidal power never contribute substantially to filling the world's growing energy needs? What advantages does tidal power offer?

CHAPTER ELEVEN

THE OCEAN FLOOR

A glance at a globe or a look at a photograph of the earth from space reveals a planet dominated by the oceans. Indeed, it is for this reason that the earth is often called the blue planet.

The area of the earth is nearly 510 million square kilometers (197 million square miles). Of this total, approximately 360 million square kilometers (140 million square miles), or 71 percent, are represented by the oceans and marginal seas. The remaining 29 percent, 150 million square kilometers (58 million square miles), is represented by the continents, which protrude from the water like enormous islands.

By studying a globe or world map, it is readily apparent that the continents and oceans are not evenly divided between the Northern and Southern hemispheres. When we compute the percentages of land and water in the Northern Hemisphere, we find that nearly 61 percent of the surface is water, whereas about 39 percent is land. In the Southern Hemisphere, on the other hand, almost 81 percent of the surface is water, and only 19 percent is land. It is no wonder then that the Northern Hemisphere is called the *land hemisphere,* and the Southern Hemisphere the *water hemisphere.*

The volume of the ocean basins is many times greater than the volume of the continents above sea level. In fact, the volume of all land above sea level is only 1/18 that of the ocean. Figure 11.1 helps to illustrate this point. Note that the mean elevation of the land surface is 840 meters (2755 feet), while the average depth of the ocean is more than 4.5 times this figure—3800 meters (12,465 feet). Therefore, if the solid earth were perfectly smooth (level) and round, the oceans would cover it to a depth of more than 2000 meters.

THE EARTH BENEATH THE SEA

If all the water were drained from the ocean basins, what kind of surface would be revealed? It would not be the quiet, subdued topography as was once thought, but a surface characterized by a great diversity of features—towering mountain chains, deep canyons, and flat plains. In fact, the scenery would be just as varied as that on the continents (Figure 11.2).

Although the tremendous extent of the oceans became apparent through voyages of discovery in the 15th and 16th centuries, an understanding of the ocean floor's complex topography did not unfold until much later. The beginnings of this realization can be traced to the historic 3½-year voyage of the H.M.S. *Challenger.* Beginning in December, 1872, and ending in May, 1876, the *Challenger* expedition made the first, and perhaps most comprehensive, study of the global ocean ever attempted by one agency. The 110,000-kilometer (69,000-mile) trip took the ship and its crew of scientists to every ocean except the Arctic. Throughout the voyage they sampled the total depth of the water by laboriously lowering a weighted line overboard. Not many years later the knowledge gained by the *Challenger* of the ocean's great depths and varied topography was further expanded with the laying of transatlantic cables. However, as long as ocean depth measurements had to be taken with weighted lines, our knowledge of the sea floor remained slight. Then, in the 1920s a technological breakthrough occurred with the invention of electronic depth-sounding equipment (echo sounder).

The **echo sounder** works by transmitting sound waves from the ship toward the ocean bottom. A

A scenic view of the Pacific from Cypress Cove, Point Lobos, California. (Photo by James E. Patterson)

FIGURE 11.1 Elevations of the earth's crust. (From Foster, *Physical Geology*, 4th ed., Columbus, Ohio: Charles E. Merrill, 1983, p. 302.

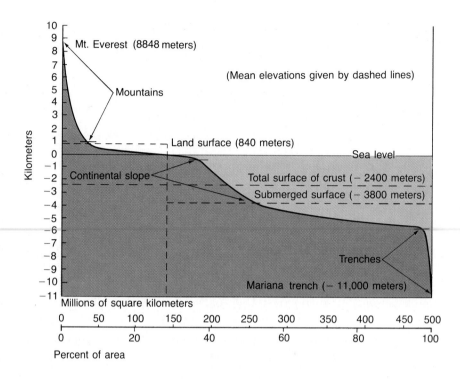

delicate receiver catches the echo from the bottom, and a highly accurate clock measures the time interval in small fractions of a second. By knowing the velocity of the sound waves in the water (about 1500 meters, or 4900 feet, per second), the depth can be calculated very precisely. Since ultrasonic waves are often used because they can be distinguished from the audible sounds made during the operation of the ship, the "sound" is not detectable by human ears (Figure 11.3).

Since the invention of the echo sounder, millions of kilometers of continuous sonic depth determinations have provided us with a more complete and detailed view of the ocean floor.

Oceanographers studying the topography of the ocean basins have delineated three major units: continental margins, the ocean basin floor, and mid-ocean ridges. The map in Figure 11.4 outlines these provinces for the North Atlantic, and the profile at the bottom of the illustration shows the varied topography. Such profiles usually have their vertical dimension exaggerated many times—40 times in this case—to make topographic features more conspicuous. Due to this, the slopes shown in the profiles of the sea floor in Figure 11.4 appear to be much steeper than they actually are.

CONTINENTAL MARGINS

The features comprising the **continental margin** include the continental shelf, continental slope, and continental rise (Figure 11.5). The first of these parts, the **continental shelf**, is a gently sloping submerged surface extending from the shoreline toward the deep-ocean basin. Since it is underlain by continental-type crust, it is clearly a flooded extension of the continents. The continental shelf varies greatly in width. Almost nonexistent along some continents, the shelf may extend seaward as far as 1500 kilometers (900 miles) along others. On the average, the continental shelf is about 80 kilometers (50 miles) wide and 130 meters (425 feet) deep at the seaward edge. The average inclination of the continental shelf is less than one-tenth of one degree, a drop of only about 2 meters per kilometer (10 feet per mile). The slope is so slight that it would appear to an observer to be a flat surface.

The continental shelves represent 7.5 percent of the total ocean area, which is equivalent to about 18 percent of the earth's total land area. These areas have taken on increased economic and political significance since they have been found to be sites of important mineral deposits, including large reser-

FIGURE 11.2 The ocean floor is characterized by a great diversity of features. This is an artist's view of what would be seen if all of the water were removed from the Atlantic Ocean basin. (From a painting by Heinrich Berann; courtesy of Aluminum Company of America)

voirs of petroleum and natural gas, as well as huge sand and gravel deposits. Of course, the waters of the continental shelf contain many important fishing grounds that are significant sources of food.

When compared with many parts of the deep-ocean floor, the surface of the continental shelf is relatively featureless. This is not to say that the shelves are completely smooth. The most profound features are long valleys running from the coastline into deeper waters. Many of these valleys are seaward extensions of river valleys on the adjacent landmass. Such valleys were excavated during the Pleis-

tocene epoch (Ice Age). During this time great quantities of water were tied up in vast ice sheets on the continents, causing sea level to drop by 90 to 120 meters (300 to 400 feet) and exposing the continental shelves. For this reason, rivers extended their courses, and land-dwelling plants and animals inhabited the newly exposed portions of the continents. Today these areas are covered by the sea and inhabited by marine organisms. Dredging along the eastern coast of North America has produced the remains of numerous land dwellers, including mammoths, mastodons, and horses. Bottom sampling

FIGURE 11.3 An echo sounder determines water depth by measuring the time required for a sonic wave to travel from a ship to the sea floor and back. The speed of sound in water is 1500 m/sec. Therefore water depth = ½(1500 m/sec × echo travel time).

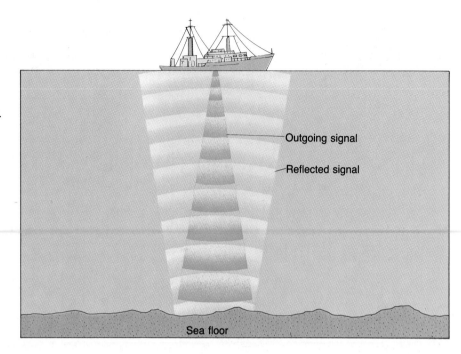

Outgoing signal

Reflected signal

Sea floor

has also revealed that freshwater peat bogs existed, adding to the evidence that the continental shelves were once land areas.

Marking the seaward edge of the continental shelf is the **continental slope**, which leads into deep water and has a steep gradient compared to the continental shelf. While the slope varies from place to place, it has an average drop of about 70 meters per kilometer (370 feet per mile). The conti-

nental slope marks the boundary between the continental crust and the oceanic crust.

Along some mountainous coasts the continental slope descends abruptly into deep-ocean trenches which intervene between the continent and ocean basin. In such cases, the shelf is very narrow or does not exist at all. The side of the trench and the continental slope are essentially the same feature and grade into the adjacent mountains which tower

FIGURE 11.4 Major topographic divisions of the North Atlantic and a profile from New England to the coast of North Africa. (After B. C. Heezen, M. Tharp, and M. Ewing. "The Floors of the Oceans," *Geological Society of America Special Paper 65,* p. 16)

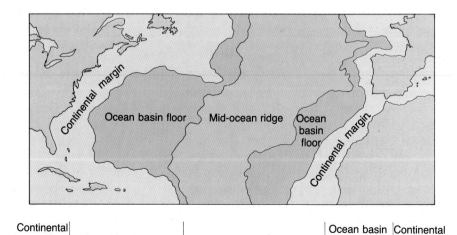

Continental margin

Ocean basin floor

Mid-ocean ridge

Ocean basin floor

Continental margin

Continental margin | Ocean basin floor | Mid-ocean ridge | Ocean basin floor | Continental margin

FIGURE 11.5 Schematic profile showing the provinces of the continental margin. (Vertical exaggeration 135:1)

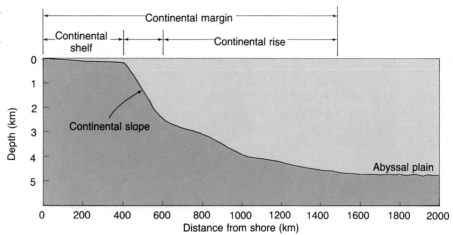

thousands of meters above sea level. This situation occurs along the west coast of South America. Here the vertical distance from the high peaks of the Andes Mountains to the floor of the deep Peru-Chile trench bordering the continent exceeds 12,200 meters (40,000 feet).

In regions where trenches do not exist, the steep continental slopes merge into a more gradual incline known as the **continental rise**. Here the gradient lessens to between 4 and 8 meters per kilometer (20 to 40 feet per mile). While the width of the continental slope averages about 20 kilometers (12 miles), the continental rise may reach for hundreds of kilometers. This feature consists of a thick accumulation of sediment that moved downslope from the continental shelf to the deep-ocean floor. Although rises are relatively featureless, their surfaces are occasionally interrupted by submarine canyons or by submarine volcanoes that have not yet been completely buried by sediments.

The deep, steep-sided valleys known as **submarine canyons** originate on the continental slope and may reach to water depths of 3 kilometers (2 miles). Although some of these canyons appear to be seaward extensions of valleys that were carved on the continental shelf during the Ice Age, others are not oriented in this manner. Furthermore, the canyons reach depths far below the maximum lowering of sea level, which indicates that they were created by some process that operates below the ocean surface (Figure 11.6). Most available information seems to favor the view that submarine canyons have been excavated by turbidity currents.

TURBIDITY CURRENTS

Turbidity currents are downslope movements of dense, sediment-laden water. They are created when sand and mud on the continental shelf and slope are dislodged, perhaps by an earthquake, and are thrown into suspension. Since the muddy water is denser than the clearer water above, it flows down the slope, eroding and accumulating more sediment

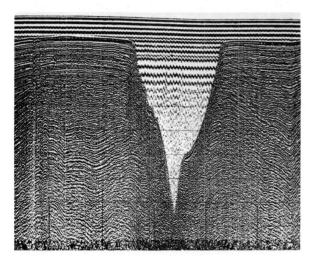

FIGURE 11.6 An echo-sounding profile of the Congo submarine canyon off the west coast of Africa. The bottom of the canyon is about 3 kilometers below the level of the sea floor of the continental shelf. Across the top, the canyon is more than 10 kilometers wide. (Courtesy of K. O. Emery, Woods Hole Oceanographic Institution)

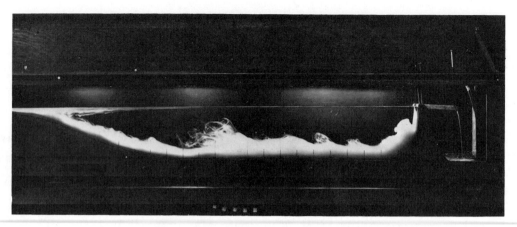

FIGURE 11.7 Turbidity current produced in a water-filled laboratory tank. (Courtesy of H. S. Bell Sedimentation Laboratory, California Institute of Technology)

as it continues to gain speed (Figure 11.7). The erosional work repeatedly carried on by these muddy torrents eventually excavates submarine canyons.

Turbidity currents usually originate along the continental slope and continue across the continental rise, still cutting channels. Eventually they lose momentum and come to rest along the ocean basin floor. As these currents slow, suspended sediments begin to settle out. First, the coarser sand is dropped, followed by successively finer deposits of silt and then clay. Consequently, these deposits, called **turbidites**, are characterized by a decrease in sediment grain size from bottom to top, a phenomenon known as **graded bedding**.

For many years the existence of turbidity currents in the ocean was a matter of considerable debate among marine geologists. Not until the 1950s did the speculation begin to subside. Two lines of evidence helped establish turbidity currents as important mechanisms of submarine erosion and sediment transportation. The first important evidence came from records of a rather severe earthquake that took place off the coast of Newfoundland in 1929 and resulted in the breakage of 13 transatlantic telephone and telegraph cables. At the time it was presumed that the tremor had caused the multiple breaks. However, when the data were examined, it appeared that this was not the case. After plotting the locations of the breaks on a map, it was seen that all the breaks had occurred along the steep continental slope and the gentler continental rise. Since the time of each break was known from information

provided by automatic recorders, a pattern of what had happened could be deduced. The breaks high up on the continental slope took place first, almost concurrently with the earthquake. The other breaks happened in succession, the last occurring 13 hours later, some 720 kilometers (450 miles) from the source of the quake (Figure 11.8). The breaks downslope had obviously taken place too long after the tremor to have been caused by the shock of the earthquake. The existence of a turbidity current, triggered by the quake, thus appeared as a plausible alternative. As the avalanche of sediment-choked water raced downslope it snapped the cables in its path. Investigators calculated that the current reached speeds approaching 80 kilometers (50 miles) per hour on the steep slopes and about 24 kilometers (15 miles) per hour on the gentler slopes below. Subsequent investigations of cable breaks in other areas revealed a similar sequence of events.

A second compelling line of evidence relating turbidity currents to submarine erosion and transportation of sediment came from the examination of deep-sea sediment samples. These cores show that extensive graded beds of sand, silt, and clay exist in the quiet waters of the deep ocean. Some samples also include fragments of plants and animals that live only in the shallower waters of the continental shelves. No mechanism other than turbidity currents could explain the existence of these deposits.

Although there is still much to be learned about the complex workings of turbidity currents, it has been well established that they are a very important

FIGURE 11.8 Profile of the sea floor showing the events of the November 18, 1929, earthquake off the shores of Newfoundland. Arrows point to cable breaks; numbers show times of breaks in hours and minutes after the earthquake. The vertical scale is greatly exaggerated. (After B. C. Heezen and M. Ewing, "Turbidity Currents and Submarine Slump and the 1929 Grand Banks Earthquake," *American Journal of Science* 250:867)

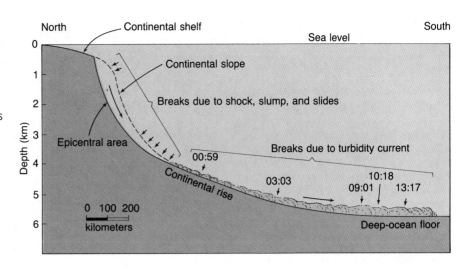

mechanism of sediment transport in the ocean. By the action of turbidity currents, submarine canyons are created and sediments are carried to the deep-ocean floor.

FEATURES OF THE OCEAN BASIN FLOOR

Between the continental margin and the oceanic ridge system lies the ocean basin floor (see Figure 11.4). The size of this region—almost 30 percent of the earth's surface—is roughly comparable to the percentage of the surface that projects above the sea as land. Here we find ocean trenches, which are dramatically deep grooves in the ocean floor; remarkably flat regions, known as abyssal plains; and steep-sided volcanic peaks, called seamounts.

Deep-Ocean Trenches

Deep-ocean trenches are long, relatively narrow features that represent the deepest parts of the ocean. Several in the western Pacific approach or exceed depths of 10,000 meters (32,800 feet), and at least a portion of one, the Challenger Deep in the Mariana trench, is more than 11,000 meters (36,000 feet) below sea level.

Although deep-ocean trenches represent only a very small portion of the ocean floor area, they are nevertheless very significant geological features. Trenches are the sites where moving crustal plates are destroyed as they plunge back into the mantle.

In addition to the earthquakes created as one plate descends beneath another, volcanic activity is also associated with trench regions. Trenches in the open ocean are paralleled by volcanic island arcs, while volcanic mountains, such as the Andes, may be found paralleling trenches that are adjacent to continents. The melting of a descending plate produces the molten rock that leads to this volcanic activity.

Abyssal Plains

Abyssal plains are incredibly flat features; in fact, these regions are likely the most level places on the earth. The abyssal plain found off the coast of Argentina, for example, has less than 3 meters (10 feet) of relief over a distance exceeding 1300 kilometers (800 miles). The monotonous topography of abyssal plains will occasionally be interrupted by the protruding summit of a buried volcanic structure.

By employing seismic profilers, instruments whose signals penetrate far below the ocean floor, researchers have shown that abyssal plains consist of thick accumulations of sediment that were deposited atop the low, rough portions of the ocean floor. The nature of the sediment indicates that these plains consist primarily of sediments transported far out to sea by turbidity currents. The turbidite deposits are interbedded with sediments composed of minute clay-sized particles that continuously settle onto the ocean floor.

Abyssal plains are found as part of the sea floor in all of the oceans. However, they are more widespread where there are no deep-ocean trenches adjacent to the continents. Since the Atlantic Ocean has fewer trenches to act as traps for the sediments carried down the continental slope, it has more extensive abyssal plains than the Pacific.

Seamounts

Dotting the ocean floors are isolated volcanic peaks called **seamounts** that may rise hundreds of meters above the surrounding topography. These steep-sided conical peaks have been discovered in all oceans, but the greatest number have been identified in the Pacific.

Many of these undersea volcanoes begin to rise near oceanic ridges, which are divergent plate boundaries where the plates of the lithosphere move apart. They continue to grow as they ride along on the moving plate. If the volcano rises fast enough, it emerges as an island. Examples in the Atlantic include the Azores, Ascension, Tristan da Cunha, and St. Helena. During the time they exist as islands, some of these volcanoes are eroded to near sea level by running water and wave action. Over a span of millions of years the islands gradually sink as the moving plate slowly carries them from the oceanic ridge area. These submerged, flat-topped seamounts are called **guyots**. In other instances, guyots may be remnants of eroded volcanic islands that were formed away from the ridge crest, possibly by hot spot activity. Here subsidence occurs after the volcanic activity ceases and the sea floor cools and contracts.

MID-OCEAN RIDGES

Mid-ocean ridges are found in all major oceans and represent more than 20 percent of the earth's surface. They are certainly the most prominent topographic features in the oceans, for they form an almost continuous mountain range which extends for about 65,000 kilometers (40,000 miles) in a manner similar to the seam on a baseball. Although ocean ridges stand high above the adjacent deep-ocean basins, they are much different than the mountains found on the continents. Rather than thick sequences of folded and faulted sedimentary rocks,

oceanic ridges consist of layer upon layer of basaltic rocks that have been faulted and uplifted. The term *ridge* may also be misleading since these features are not narrow, but have widths from 500 to 5000 kilometers and, in places, may occupy as much as one-half of the total ocean floor area. Ridge crests are marked by deep clefts, or **rifts**, and are flanked by ridges and lines of peaks that extend outward for hundreds of kilometers (see Figure 11.2). Axes of the ridges are marked by frequent earthquakes and characterized by a much higher heat flow through the crust. The rifts at the center of the ridges are sites where new magma wells up from the asthenosphere below, continually creating new oceanic crust. The rifts therefore represent divergent plate boundaries where sea-floor spreading is taking place.

The primary reason for the elevated position of a ridge system is the fact that newly created oceanic crust is hot, and therefore occupies more volume than cooler rocks of the deep-ocean basin. As the young lithosphere travels away from the spreading center, it gradually cools and contracts. This thermal contraction accounts in part for the greater ocean depths that exist away from the ridge. Almost 100 million years must pass before cooling and contraction cease completely. By this time, rock that was once a part of a majestic oceanic mountain system is located in the deep-ocean basin, where it is mantled by thick accumulations of sediment.

A CLOSE-UP VIEW OF THE OCEAN FLOOR

Although much has been (and continues to be) learned about the floor of the ocean from echo sounders and other remote sensing equipment, as well as from drilling and sampling from surface ships, oceanographers in the 1970s became aware that direct manned observation was essential to bring about a true understanding of many deep-sea phenomena. What was needed was a firsthand view of the previously unseen world below.

Today, the names and accomplishments of deep-diving manned submersibles such as the *Alvin* are common knowledge among oceanographers (Figure 11.9). Manned submersibles are now extending the coverage provided by traditional oceanographic

examine the structure of the rift valley in the Mid-Atlantic Ridge. The data collected by the three vessels proved invaluable and led to more realistic explanations of how the spreading process works in creating new ocean floor (Figure 11.10A).

More recently, dives were made by the *Alvin* to a spreading center at 21°N latitude on the East Pacific Rise near the mouth of the Gulf of California. Here, in addition to gathering large quantities of basic data, the scientists aboard the *Alvin* discovered the existence of spectacular geyserlike hot springs. They witnessed two- to five-meter-high chimney-like structures spewing dark, mineral-rich hot (350°C–400°C) water (Figure 11.10B). As the heated solutions hit the surrounding 2°C seawater, sulfides of copper, iron, and zinc precipitated immediately, forming mounds of minerals around the steaming vents. In addition to viewing firsthand the formation of massive sulfide deposits, the scientists aboard the *Alvin* found communities of exotic, bottom-dwelling animals living near cooler (20°C) hot springs. The discovery of an animal community thriving more than three kilometers below the surface where no light can reach was totally unexpected. An analysis of the sulfur-rich vent water as well as the stomach contents of some animals revealed that the base of the food chain was sulfur-oxidizing bacteria.

Dives such as the ones briefly highlighted here have demonstrated the value of deep-diving manned submersibles in detecting and studying the fine-scale features of the ocean floor. These vessels now appear to occupy a permanent and important place as tools in oceanographic research.

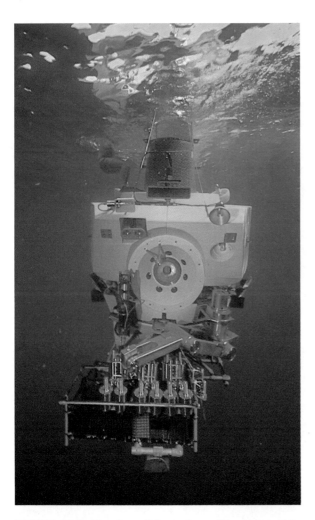

FIGURE 11.9 The deep-diving submersible *Alvin* is 7.6 meters long, weighs 16 tons, has a cruising speed of 1 knot, and can reach depths as great as 4000 meters (13,000 feet). A pilot and two scientific observers are along during a normal 6- to 10-hour dive. (Courtesy of Woods Hole Oceanographic Insitution)

research vessels by allowing scientists to investigate the fine-scale features that previously eluded detection.

One of the pioneering research projects that used deep-diving submersibles was a cooperative venture called Project FAMOUS (French-American Mid-Ocean Undersea Study). In 1974, after three years of preliminary surveying and study by surface ships, two French submersibles and one American vessel made a total of forty-four dives to the floor of the Atlantic. The primary purpose of the project was to

CORAL REEFS AND ATOLLS

Coral reefs are among the most picturesque features found in the ocean. They are constructed primarily from the calcareous (calcite-rich) skeletal remains and secretions of corals and certain algae. The term *coral reef* is somewhat misleading in that it makes no mention of the skeletons of many small animals and plants found inside the branching framework built by the corals; nor of the fact that limy secretions of algae help bind the entire structure together.

Coral reefs are confined largely to the warm waters of the Pacific and Indian oceans, although a few

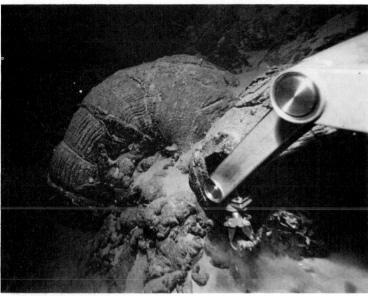

A.   B.

FIGURE 11.10   A. A photograph taken from the *Alvin* during Project FAMOUS shows lava extrusions in the rift valley of the Mid-Atlantic Ridge. Large tooth-pastelike extrusions such as the one in this photograph were common features. A mechanical arm is sampling an adjacent blisterlike extrusion. **B.** This is one of about two dozen vents called "black smokers" that were found by the *Alvin* at 21°N latitude on the East Pacific Rise in May, 1979. The "smoke" is actually hot, mineral-rich water that has circulated through the ocean crust and picked up iron, copper, and zinc. When the hot solution meets cold seawater, it precipitates sulfide ores that now coat the vents. (Courtesy of Woods Hole Oceanographic Institution)

occur elsewhere. Since reef-building corals grow best in waters with an average annual temperature of about 24°C (75°F), their location is in part the result of their need for warm water. They can survive neither sudden temperature changes nor prolonged exposure to temperatures below 18°C (65°F). In addition, these reef-builders require clear, sunlit water. For this reason, the limiting depth of active reef growth is about 45 meters (150 feet).

From 1831 to 1836 the naturalist Charles Darwin was aboard the British ship *Beagle* on a surveying expedition that circumnavigated the globe. One outcome of Darwin's studies during the five-year voyage was a theory on the formation of coral islands, or **atolls**. As Figure 11.11 illustrates, atolls consist of a continuous or broken ring of coral reef surrounding a central lagoon. From the time that Darwin first studied them until shortly after World War II, their manner of origin challenged people's curiosity.

FIGURE 11.11   View from space of a group of atolls in the Pacific Ocean. (Courtesy of NASA)

Darwin's theory explained what seemed to be a paradox; that is, how can corals, which require warm, shallow, sunlit water no deeper than 45 meters to live, create structures that reach thousands of meters to the floor of the ocean? Commenting on this in *The Voyage of the Beagle,* Darwin stated:

> . . . from the fact of the reef-building corals not living at great depths, it is absolutely certain that throughout these vast areas, wherever there is now an atoll, a foundation must have originally existed within a depth of from 20 to 30 fathoms from the surface.

The essence of Darwin's theory was that coral reefs form on the flanks of sinking volcanic islands. As the island slowly sinks, the corals continue to build the reef complex upward (Figure 11.12):

> For as mountain after mountain, and island after island slowly sank beneath the water, fresh bases would be successively afforded for the growth of the corals.

Thus atolls, like guyots, are thought to owe their existence to the gradual sinking of oceanic crust. In succeeding years there were numerous challenges to Darwin's theory. These arguments were not completely put to rest until after World War II when the United States made extensive studies of two atolls (Eniwetok and Bikini) that were going to become sites for testing atomic bombs. Drilling operations at these atolls revealed that volcanic rock did indeed underlie the thick coral reef structure. This finding was a striking confirmation of Darwin's theory.

SEA-FLOOR SEDIMENTS

Except for a few areas, such as near the crests of mid-ocean ridges, the ocean floor is mantled with sediment. Part of this material has been deposited by turbidity currents, and the rest has slowly settled to the bottom from above. The thickness of this carpet of debris varies greatly. In some trenches, which act as traps for sediments originating on the continental margin, accumulations may approach 10 kilometers

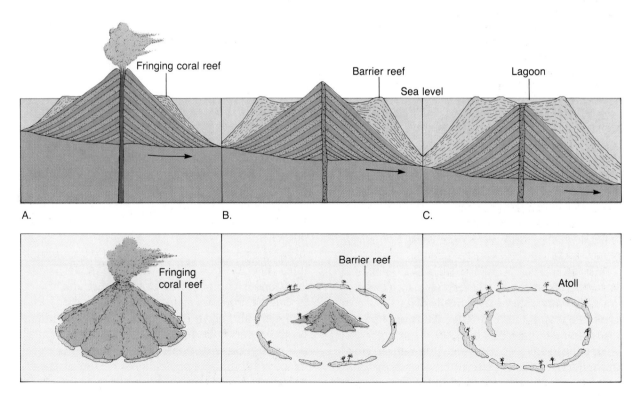

FIGURE 11.12 Cross-sectional views and map views of the formation of a coral atoll.

(6 miles). In general, however, sediment accumulations are considerably less. In the Pacific Ocean, uncompacted sediment measures about 600 meters or less, whereas on the floor of the Atlantic, the thickness varies from 500 to 1000 meters.

Although accumulations of sand-sized particles are found on the deep-ocean floor, mud is the most common sediment covering this region. Muds also predominate on the continental shelf and slopes, but the sediments in these areas are coarser overall because of greater quantities of sand. Sampling has shown that sands are generally deposited on the continental shelf, forming beaches along the shore. However, in some cases this coarse sediment, which is expected to be found near the shore, occurs in irregular patches at greater depths near the seaward limits of the continental shelves. While some of the sand may have been deposited by local currents that are capable of moving coarse sediment far from shore, the bulk of it appears to result from sand deposition on ancient beaches. Such beaches formed during the Ice Age, when sea level was much lower than it is today. These patches of sand were then submerged as sea level rose again.

Types of Sea-Floor Sediments

Sea-floor sediments can be classified according to their origin into three broad categories: (1) Lithogenous ("derived from rocks") sediment; (2) Biogenous ("derived from organisms") sediment; and (3) Hydrogenous ("derived from water") sediment. Although each category is discussed separately, it should be remembered that all sea-floor sediments are mixtures. No body of sediment comes from a single source.

Lithogenous sediment consists primarily of mineral grains which were weathered from continental rocks and transported to the ocean. The sand-sized particles settle near shore. However, since the very smallest particles take years to settle to the ocean floor, they may be carried for thousands of kilometers by ocean currents. As a consequence, virtually every area of the ocean receives some lithogenous sediment. However, the rate at which this sediment accumulates on the deep-ocean floor is indeed very slow. From 5000 to 50,000 years are necessary for a 1-centimeter layer to form. Conversely, on the continental margins near the mouths

of large rivers, lithogenous sediment accumulates rapidly. In the Gulf of Mexico, for example, the sediment has reached a depth of many kilometers.

Since fine particles remain suspended in the water for a very long time, there is ample opportunity for chemical reactions to occur. Because of this, the colors of the deep-sea sediments are often red or brown. This results when iron on the particle or in the water reacts with dissolved oxygen in the water and produces a coating of iron oxide (rust).

Biogenous sediment consists of shells and skeletons of marine animals and plants. This debris is produced mostly by microscopic organisms living in the sunlit waters near the ocean surface. The remains continually "rain" down upon the sea floor.

The most common biogenous sediments are known as calcareous ($CaCO_3$) oozes, and as their name implies, they have the consistency of thick mud. These sediments are produced by organisms that inhabit warm surface waters. When calcareous hard parts slowly sink through a cool layer of water, they begin to dissolve. This results because cold seawater contains more carbon dioxide and is thus more acidic than warm water. In seawater deeper than about 4500 meters (15,000 feet), calcareous shells will completely dissolve before they reach bottom. Consequently, calcareous ooze does not accumulate where depths are great.

Other examples of biogenous sediments are siliceous (SiO_2) oozes and phosphate-rich materials. The former is composed primarily of opaline skeletons of diatoms (single-celled algae) and radiolaria (single-celled animals), while the latter is derived from the bones, teeth, and scales of fish and other marine organisms.

Hydrogenous sediment consists of minerals that crystallize directly from seawater through various chemical reactions. For example, some limestones are formed when calcium carbonate precipitates directly from the water; however, most limestone is composed of biogenous sediment.

One of the principal examples of hydrogenous sediment, and one of the most important sediments on the ocean floor in terms of economic potential, are **manganese nodules**. These rounded blackish lumps are composed of a complex mixture of minerals that form very slowly on the floor of the ocean basins (Figure 11.13). In fact, their formation rate represents one of the slowest chemical reactions

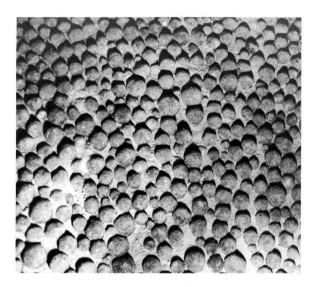

FIGURE 11.13 Manganese nodules photographed at a depth of 2909 fathoms (5323 meters) beneath the *Robert Conrad* south of Tahiti. (Courtesy of Lawrence Sullivan, Lamont-Doherty Geological Observatory)

known. By analyzing the radioactive elements continually incorporated into growing nodules, researchers have determined that the growth rates vary from 0.001 to 0.2 millimeter per 1000 years. Some portions of the sea floor are littered with these deposits whereas others lack them altogether. The presence or absence of nodules has been correlated with the sedimentation rate. If sediment accumulates too rapidly (at a rate exceeding about 7 millimeters per 1000 years), newly forming nodules are buried and growth ceases. Since nodule growth is exceedingly slow, why are nodules not buried even where sediment accumulates at less than 7 millimeters per 1000 years? Some scientists suggest that benthic animals living in and on the sea floor are responsible for keeping the nodules at the surface. By stirring the sediment, burrowing organisms are thought to produce a slight lifting effect which, in combination with small surface animals that consume newly arrived sediment from nodule surfaces, keeps nodules from being buried.

Although manganese nodules may contain more than 20 percent manganese, the interest in them as a potential resource lies in the fact that other more valuable metals may be enriched in them. In addition to manganese, nodules may contain significant quantities of iron, copper, nickel, and cobalt. All re-

gions containing nodules, however, are not equally good potential sites for mining. Possible mining locations must have abundant nodules (more than 5 kilograms per square meter) and contain the economically optimum mix of cobalt, copper, and nickel. Sites meeting these criteria are relatively limited. Furthermore, before such areas prove to be valuable commercial sources for these metals, the logistics of extracting nodules from the floor of the deep-ocean basins must be worked out.

Sea-Floor Sediments and Climatic Change

Due to the fact that instrumental climate records go back only a couple of hundred years (at best), how do scientists find out about climates and climatic changes prior to that time? The obvious answer is that they must reconstruct past climates from indirect evidence; that is, they must examine and analyze phenomena that respond to and reflect changing atmospheric conditions. One of the more interesting and important techniques for analyzing the earth's climatic history is the study of sediments from the ocean floor.

Although sea-floor sediments are of many types, most contain the remains of organisms that once lived near the sea surface (the ocean-atmosphere interface). When such near-surface organisms die, their shells slowly settle to the ocean floor where they become part of the sedimentary record. One reason that sea-floor sediments are useful recorders of worldwide climatic change is that the numbers and types of organisms living near the sea surface change as the climate changes. This principle is explained by Richard Foster Flint as follows:

> . . . we would expect that in any area of the ocean/atmosphere interface the average annual temperature of the surface water of the ocean would approximate that of the contiguous atmosphere. The temperature equilibrium established between surface seawater and the air above it should mean that . . . changes in climate should be reflected in changes in organisms living near the surface of the deep sea. . . . When we recall that the sea-floor sediments in vast areas of the ocean consist mainly of shells of pelagic foraminifers, and that these animals are sensitive to variations in water

temperature, the connection between such sediments and climatic change becomes obvious.*

*Glacial and Quaternary Geology (New York: Wiley, 1971), p. 718.

Thus, in seeking to understand climatic change as well as other environmental transformations, scientists have become increasingly interested in the huge reservoir of data in sea-floor sediments.

REVIEW QUESTIONS

1 How does the area covered by the oceans compare with that of the continents? Describe the distribution of land and water on earth.

2 How does the average depth of the ocean compare to the average elevation of the continents (see Figure 11.1)?

3 Assuming that the average speed of sound waves in water is 1500 meters per second, determine the water depth if the signal sent out by an echo sounder requires 6 seconds to strike bottom and return to the recorder (see Figure 11.3).

4 List the three major subdivisions of the continental margin. Which subdivision is considered a flooded extension of the continent? Which has the steepest slope?

5 How does the continental margin along the west coast of South America differ from the continental margin along the east coast of North America?

6 Defend or rebut the statement "Most submarine canyons were formed during the Ice Age when rivers extended their courses seaward."

7 What are turbidites? What is meant by the term graded bedding?

8 Discuss the evidence that helped confirm the existence of turbidity currents in the ocean and establish them as significant mechanisms of erosion and sediment transport.

9 Why are abyssal plains more extensive on the floor of the Atlantic than on the floor of the Pacific?

10 How are mid-ocean ridges and deep-ocean trenches related to sea-floor spreading?

11 What is an atoll? Describe Darwin's hypothesis on the origin of atolls. Was the hypothesis ever confirmed?

12 Differentiate among the three basic types of sea-floor sediment.

13 If you were to examine recently deposited biogenous sediment taken from a depth in excess of 4500 meters (15,000 feet), would it more likely be rich in calcareous materials or siliceous materials? Explain.

14 Why are sea-floor sediments useful in studying climates of the past?

CHAPTER TWELVE

EARTHQUAKES AND THE EARTH'S INTERIOR

Quake Kills Hundreds

MANAGUA, Nicaragua (AP)—A disastrous earthquake rolled through this Central American city of 300,000 early yesterday, leaving a heavy toll of death and destruction. Unofficial estimates of the dead ranged as high as 18,000 but that figure appeared to be exaggerated.

Many of those who were not injured sat on the curbstones in a daze, surrounded by what few possessions they could save from the rubble. Many others fled the city.

"It's like standing on jelly down here," radioed U.S. communications satellite technician Ray Hashberger from a station two miles outside the city.

Half the downtown section of the city lay in ruins as night fell. The quake devastated 36 blocks in the central area.

Fires burned out of control through the afternoon. The quake, which measured between 6 and 7 on the Richter scale of magnitude, struck at 12:40 A.M. yesterday following a series of lesser jolts.

All normal communications, water and electrical services were out. The U.S. Embassy was among the buildings destroyed.

Jack Burton, an information officer in the U.S. Embassy, was in Lima, Peru, for that city's major earthquake May 31, 1970.

"The Lima quake was more gradual though it was about of the same intensity," he said. "But this quake came on like gangbusters. It knocked us on our knees. We had no warning at all."

Smaller tremors continued to hit the city throughout the day and into the evening, loosening debris from already wrecked buildings.*

---

*Courtesy of The Associated Press.

Trace of the San Andreas fault in the Carrizo Plain. (Photo by John S. Shelton)

This was not the first earthquake to devastate the city of Managua, and it probably will not be the last. It is estimated that nearly one million earthquakes occur each year. Fortunately few are as devastating as the 1972 Managua earthquake. Generally only a few destructive earthquakes occur worldwide each year, but when they do, they are among the most destructive natural forces on earth. The shaking of the ground coupled with the liquefaction of some soils wreak havoc on buildings (Figure 12.1). In addition, when a quake occurs in a populated area, power and gas lines are often ruptured, causing numerous fires. In the 1906 San Francisco earthquake, most of the damage was caused by fires which ran unchecked when broken water mains left firefighters with only trickles of water (Figure 12.2).

FIGURE 12.1 These leaning apartment houses rest on unconsolidated soil, which imitated quicksand during the 1964 earthquake in Niigata, Japan. Although some of the buildings were hardly damaged, their new orientation left something to be desired. (Courtesy of NOAA)

FIGURE 12.2 San Francisco in flames after the 1906 earthquake. (Reproduced from the collection of the Library of Congress)

WHAT IS AN EARTHQUAKE?

An **earthquake** is the vibration of the earth produced by the rapid release of energy. This energy radiates in all directions from its source, the **focus**, in the form of waves analogous to those produced when a bell is struck, vibrating the air around it. During an earthquake and for many hours following, the earth could be described as "ringing like a bell." Even though the energy dissipates rapidly with increasing distance from the focus, instruments located throughout the world record the event.

The tremendous energy released by atomic explosions or by volcanic eruptions can produce an earthquake, but these events are usually weak and infrequent. What mechanism does produce a destructive earthquake? Ample evidence exists that the earth is not a static planet. Numerous ancient wave-cut benches can be found many meters above the level of the highest tides, which indicates crustal uplifting of comparable magnitude. Other regions exhibit evidence of extensive subsidence. In addition to these vertical displacements, offsets in fence lines, roads, and other structures indicate that horizontal movement is also prevalent (Figure 12.3). These movements are frequently associated with large fractures in the earth called **faults**. Most of the motion along faults can be satisfactorily explained by the plate tectonics theory. This theory proposes that large slabs of the earth are continually in motion. These mobile plates interact with neighboring plates, straining and deforming the rocks at their edges. It is along these plate boundaries that most earthquakes occur.

The actual mechanism of earthquake generation eluded geologists until H. F. Reid conducted a study following the great 1906 San Francisco earthquake.

FIGURE 12.3 This fence was offset 2.5 meters (8.2 feet) during the 1906 San Francisco earthquake. (Photo by G. K. Gilbert, U.S. Geological Survey)

The earthquake was accompanied by displacements of several meters along the northern portion of the San Andreas fault, a 950-kilometer (600-mile) fracture which runs northward through southern California. Using land surveys conducted several years apart, Reid discovered that during the 50 years prior to the 1906 earthquake the land at distant points on both sides of the San Andreas fault showed a relative displacement of slightly more than 3 meters (10 feet). The mechanism for earthquake formation which Reid deduced from this information is illustrated in Figure 12.4. Tectonic forces ever so slowly deform the crustal rocks on both sides of the fault as illustrated by the bent features. Under these conditions, rocks are bending and storing elastic energy, much like a wooden stick would if bent. Eventually, the frictional resistance holding the rocks together is overcome. As slippage occurs at the weakest point (the focus), displacement will exert stress farther along the fault where additional slippage will occur until most of the built-up strain is released. This slippage allows the deformed rock to "snap back." The vibrations we know as an earthquake occur as the rock elastically returns to its original shape. The "springing back" of the rock was termed **elastic rebound** by Reid, since the rock behaves elastically, much like a stretched rubber band does when it is released.

The intense vibrations of the 1906 San Francisco earthquake lasted about 40 seconds. Although most of the displacement along the fracture occurred in

this rather short period, additional movements and adjustments in the rocks occurred for several days following the main quake. The adjustments that follow a major earthquake often generate smaller earthquakes called **aftershocks**. Although these aftershocks are usually much weaker than the main earthquake, they can sometimes cause significant destruction to already badly weakened structures. In addition, small earthquakes called **foreshocks** often precede a major earthquake by days or in some cases by as much as several years. Monitoring of these foreshocks has been used as a means of predicting forthcoming major earthquakes. We will consider the topic of earthquake prediction in a later section of this chapter.

The tectonic forces responsible for the strain that was eventually released during the 1906 San Francisco earthquake are still active. Currently laser beams* are used to establish the relative motion between the opposite sides of this fault. These measurements have revealed a displacement of 2 centimeters per year. Although this rate of movement seems slow, it is fast enough to produce substantial movement over millions of years of geologic time. In 30 million years such a rate of displacement is sufficient to slide the western portion of California northward so that Los Angeles, on the northward-moving Pacific plate, would be adjacent to San Francisco on the North American plate. Once the strain along any segment of this active fault again reaches sufficient levels, another slippage with an accompanying earthquake can be expected. It is estimated that great earthquakes occur about every 50 to 200 years along plate boundaries such as the San Andreas fault. This repetitive process is often described as *stick-slip motion,* since elastic energy is stored over a period of time and then released through slippage.

Not all motion along the San Andreas fault is of the stick-slip type. Along certain portions of this fault the motion is a slow *creep.* Thus, while some sections of the fault are continually creeping, other "locked" segments are building up strain that could result in a major earthquake.

In addition, not all movement along faults is horizontal. Vertical displacement along faults, in which

---

*Laser beams are used in very precise surveying instruments because of their incredibly accurate straight-line qualities.

FIGURE 12.4 Elastic rebound. As rock is deformed it bends, storing elastic energy. Once the rock is strained beyond its breaking point it ruptures, releasing the stored-up energy in the form of earthquake waves.

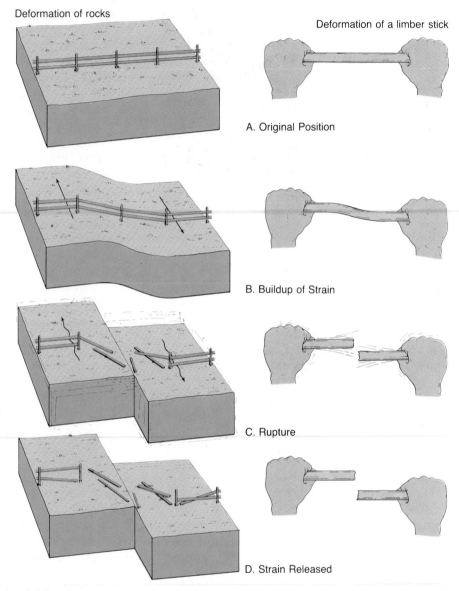

Deformation of rocks

Deformation of a limber stick

A. Original Position

B. Buildup of Strain

C. Rupture

D. Strain Released

one side is lifted higher than the other, is also common. Figure 12.5 shows a scarp (cliff) produced by such vertical displacement. In the same manner, the 1964 Good Friday earthquake in Alaska produced a 15-meter vertical offset at one location. Further, many earthquakes occur at such great depths that no displacement is evident at the surface.

EARTHQUAKE WAVES

The study of earthquake waves, **seismology**, dates back to attempts by the Chinese almost 2000 years ago to determine the direction to the source of each earthquake. The principle used in modern **seismographs**, instruments which record earthquake waves, is rather simple. A weight is freely suspended from a support that is attached to bedrock (Figure 12.6). When waves from a distant earthquake reach the instrument, the inertia* of the weight keeps it stationary, while the earth and the support vibrate.

---

*Simply stated, inertia refers to the fact that objects at rest tend to stay at rest and objects in motion tend to remain in motion unless acted upon by an outside force. You probably have experienced this phenomenon when you tried to stop your automobile quickly and your body continued to move forward.

FIGURE 12.5 Scarp resulting from vertical movement along a fault zone during the 1964 Alaskan earthquake. (Courtesy of U.S. Geological Survey)

The movement of the earth in relation to the stationary weight is recorded on a rotating drum.

The principle of a seismograph can be demonstrated by attaching a heavy mass to a string and holding the other end of the string so that the mass is just off the floor. Rapid side-to-side movements of the string represent the vibrations of an earthquake. Notice that the mass remains relatively motionless. Any noticeable movement will be small and will represent the natural oscillation of your pendulum (this is similar to the oscillation of a clock pendulum). Modern seismographs use a damping mechanism to remove the effect of the natural oscillation of the suspended masses.

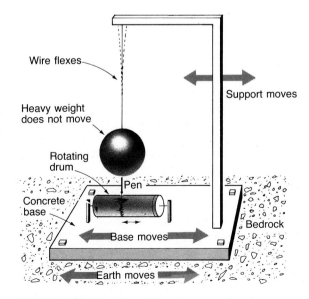

FIGURE 12.6 Principle of the seismograph.

The records of seismographs, called **seismograms**, provide a great deal of information about the behavior of seismic waves. Simply stated, seismic waves are elastic energy which radiates outward in all directions from the focus. The propagation (transmission) of this energy can be compared to the shaking of gelatin in a bowl which results as some is spooned out. Whereas the gelatin will have one mode of vibration, seismograms reveal that two main groups of seismic waves are generated by the slippage of a rock mass. One of these wave types travels along the outer layer of the earth. These are called **surface waves**. Others travel through the earth's interior and are called **body waves**. Body waves are further divided into two types called **primary**, or **P**, **waves** and **secondary**, or **S**, **waves**.

These two wave forms are divided by their method of travel (propagation) through the earth. P waves push (compress) and pull (dilate) rocks in the direction the wave is traveling. This wave motion is the same as that generated by your vocal cords as they move air to and fro in order to transmit sound. S waves on the other hand "shake" the particles at right angles to their direction of travel. This can be illustrated by tying one end of a rope to a fence post and shaking the other end while holding the rope tense.

Notice in Figure 12.7 that the propagation of P waves involves changing the volume and shape of the intervening material, whereas S waves change only the shape. Solids, liquids, and gases all resist being compressed and will elastically spring back once the force is removed. For this reason, P waves can travel through all types of matter. On the other hand, S waves change only the shape of the medium through which they travel, and because fluids do not resist changes in shape, fluids will not transmit S waves.

The motion of surface waves is somewhat more complex. As surface waves travel along the ground, they cause the ground and anything resting upon it to move, much like ocean swells toss a ship. In addition to their up-and-down motion, surface waves have a side-to-side motion similar to an S wave oriented in a horizontal plane. This latter motion causes most of the structural damage to buildings and their foundations.

By observing a "typical" seismic record, as shown in Figure 12.8, one of the differences between these

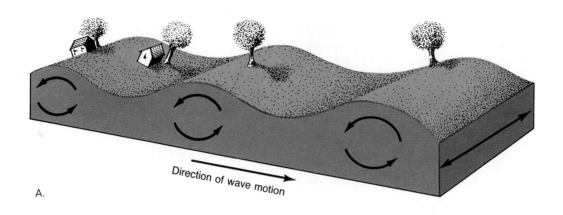

A.

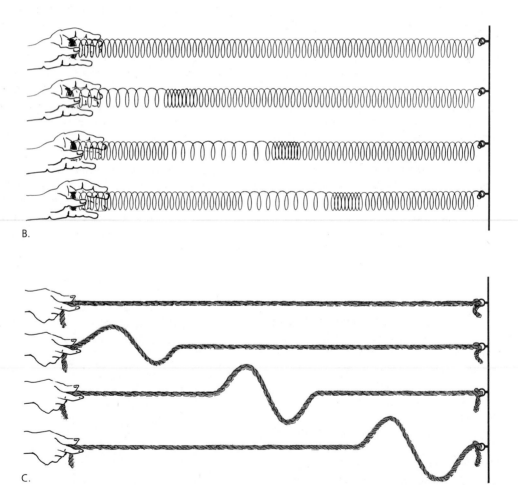

B.

C.

FIGURE 12.7 Types of seismic waves and their characteristic motion. **A.** Surface waves move the land in a circular path similar to the motion of water in ocean swells. **B.** P waves cause the particles in the material to vibrate back and forth in the same direction as the waves move. **C.** S waves cause particles to oscillate at right angles to the direction of wave motion.

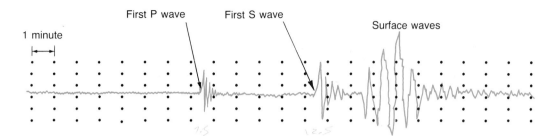

FIGURE 12.8 Typical seismic record. Note the time interval between the arrival of each wave type.

seismic waves becomes apparent: P waves arrive at the recording station before S waves, which themselves arrive before the surface waves. This is a consequence of their relative velocities. For purposes of illustration, the velocity of P waves through granite within the crust is about 6 kilometers per second, whereas S waves under the same conditions travel at 3.5 kilometers per second. Differences in density and elastic properties of the transmitting material greatly influence the velocities of these waves. However, in any solid material, P waves travel about 1.7 times faster than S waves, and surface waves can be expected to travel at 90 percent of the velocity of the S waves that are traveling in the layer directly below.

As we shall see, seismic waves allow us to determine the location and magnitude of earthquakes. In addition, seismic waves provide us with a tool for probing the earth's interior.

LOCATION OF EARTHQUAKES

Recall that the focus is the place where the earthquake originates, usually below ground. The **epicenter** is the location on the surface directly above the focus (Figure 12.9). The difference in velocities of P and S waves provides a method for determining the epicenter. The principle used is analogous to a race between two autos, one faster than the other. The greater the distance of the race, the greater will be the difference in the arrival times at the finish line. Therefore, the greater the interval between the arrival of the first P wave and the first S wave, the greater the distance to the earthquake source.

A system for locating earthquake epicenters was developed through the use of seismograms from earthquakes whose epicenters could be easily pinpointed from physical evidence. From these seismo-

grams, travel-time graphs as shown in Figure 12.10 were constructed. The first travel-time graphs were greatly improved when seismograms became available from nuclear explosions, because the location and time of detonation were well established.

Using the sample seismogram in Figure 12.8 and the travel-time curve in Figure 12.10, we can determine the distance separating the recording station from the earthquake. The time interval between the arrival of the first P wave and the first S wave is determined, then the place on the travel-time graph which exhibits an equivalent time spread between the P and S wave curves is found. From this information we can determine that this earthquake occurred 3800 kilometers (2350 miles) from the recording instrument. Although the distance to an earthquake is established in this manner, its location could be in

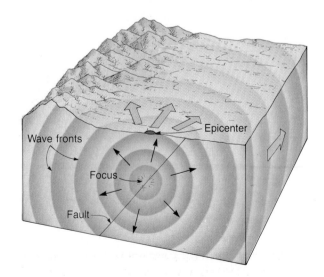

FIGURE 12.9 The focus of most earthquakes is located at depth. The surface location directly above it is called the epicenter.

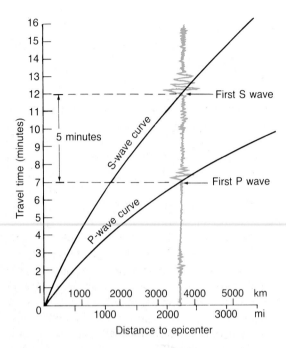

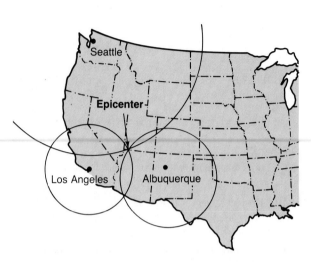

FIGURE 12.10 A travel-time graph is used to determine the distance to the epicenter. The difference in arrival times of the first P and S waves in the example is 5 minutes. Thus, the epicenter is roughly 3800 kilometers (2350 miles) away.

FIGURE 12.11 Earthquake epicenter is located using the distances obtained from three seismic stations.

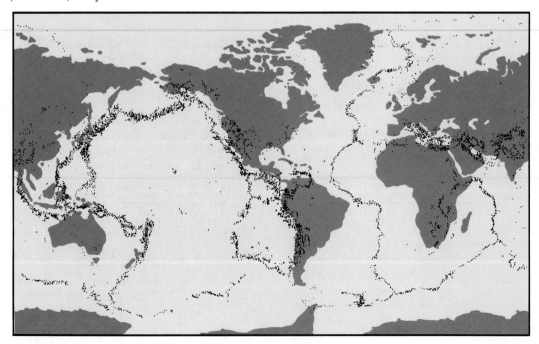

FIGURE 12.12 World distribution of earthquakes for a nine-year period. (Data from NOAA)

any direction from the observer. As shown in Figure 12.11, the precise location can be found only when the distance is known from three or more different seismic stations. By drawing circles representing the epicenter distance for each of these observatories, an accurate location is established.

About 95 percent of the energy released by earthquakes is concentrated in a few relatively narrow zones that wind around the globe (Figure 12.12). The greatest energy is released along a path near the outer edge of the Pacific Ocean known as the *circum-Pacific belt.* Included in this zone are regions of great seismic activity, such as Japan, the Philippines, Chile, and numerous volcanic island chains, as exemplified by the Aleutian Islands. Another major concentration of strong seismic activity runs through the mountainous regions that flank the Mediterranean Sea and continues through Iran and on past the Himalayan complex. Figure 12.12 indicates that yet another continuous belt extends for thousands of kilometers through the world's oceans. This zone coincides with the oceanic ridge system, an area of frequent but low-intensity seismic activity. By comparing this figure with Figure 13.5, we can see a close correlation between the location of

earthquake epicenters and plate boundaries, a phenomenon which will be explored in the next chapter.

EARTHQUAKE INTENSITY AND MAGNITUDE

Early attempts to establish the intensity of earthquakes relied heavily on subjective descriptions. There was an obvious problem with this method—people's accounts varied widely, making an accurate classification of the quake's intensity difficult. Then in 1902 a fairly reliable scale based on the amount of damage caused to various types of structures was developed by Giuseppe Mercalli. A modified form of this tool is presently used by the U.S. Coast and Geodetic Survey (Table 12.1). However, the destruction wrought by earthquakes is not an adequate means for comparison. Many factors, including distance from the epicenter, nature of surface materials, and building design, cause variations in the amount of damage. Consequently, methods were devised to determine the total amount of energy released during an earthquake, a measurement referred to as **magnitude**.

Ideally, the magnitude of an earthquake would be determined from the amount of material which

TABLE 12.1 Modified Mercalli intensity scale.

I	Not felt except by a very few under especially favorable circumstances.
II	Felt only by a few persons at rest, especially on upper floors of buildings.
III	Felt quite noticeably indoors, especially on upper floors of buildings, but many people do not recognize as an earthquake.
IV	During the day felt indoors by many, outdoors by few. Sensation like heavy truck striking building.
V	Felt by nearly everyone, many awakened. Disturbances of trees, poles, and other tall objects sometimes noticed.
VI	Felt by all; many frightened and run outdoors. Some heavy furniture moved; few instances of fallen plaster or damaged chimneys. Damage slight.
VII	Everybody runs outdoors. Damage negligible in buildings of good design and construction; slight to moderate in well-built ordinary structures; considerable in poorly built or badly designed structures.
VIII	Damage slight in specially designed structures; considerable in ordinary substantial buildings with partial collapse; great in poorly built structures. (Fall of chimneys, factory stacks, columns, monuments, and other vertically oriented features.)
IX	Damage considerable in specially designed structures. Buildings shifted off foundations. Ground cracked conspicuously.
X	Some well-built wooden structures destroyed. Most masonry and frame structures destroyed with foundations. Ground badly cracked.
XI	Few, if any, (masonry) structures remain standing. Bridges destroyed. Broad fissures in ground.
XII	Damage total. Waves seen on ground surfaces. Objects thrown upward into air.

*Source: U.S. Coast and Geodetic Survey.

slides along the fault and the distance it is displaced. Even in an ideal setting such as that of the 1906 San Francisco earthquake, where the fault trace is visible and displacement can be measured from physical evidence, this method can only provide a crude estimate of the forces involved. In most earthquakes, the fault does not penetrate the surface, therefore the amount of displacement cannot be measured directly. In 1935, Charles Richter of the California Institute of Technology attempted to rank the earthquakes of southern California into groups of large, medium, and small magnitude. The system he developed determines earthquake magnitudes from the motions measured by seismic instruments.

Today a refined **Richter scale** is used worldwide to describe earthquake magnitude. Using Richter's scale, the magnitude is determined by measuring the amplitude of the largest wave recorded on the seismogram. Although seismographs greatly magnify the ground motion, large-magnitude earthquakes will cause the recording pen to be displaced farther than small-magnitude earthquakes. In order for seismic stations worldwide to obtain the same magnitude for a given earthquake, adjustments must be made for the weakening of seismic waves as they move from the focus and for the sensitivity of the recording instrument.

The largest earthquakes ever recorded have Richter magnitudes near 8.6. These great shocks released energy roughly equivalent to the detonation of one billion tons of TNT. Conversely, earthquakes with a Richter magnitude of less than 2.5 are usually not felt by humans. Table 12.2 shows how earthquake magnitudes and their effects are related.

As we have seen, earthquakes vary enormously in strength; consequently, the wave amplitudes generated vary by factors of thousands of times as well. To accommodate this wide variation, Richter used a logarithmic scale to express magnitude. On this scale a tenfold increase in wave amplitude corresponds to an increase of one on the magnitude scale. Thus, the amplitude of the largest surface wave for a 5-magnitude earthquake is 10 times greater than the wave amplitude produced by an earthquake having a magnitude of 4. Further, each unit of magnitude increase on the Richter scale equates to roughly a 30-fold increase in the energy released. Thus, an earthquake with a magnitude of 6.5 releases 30 times more energy than one with a magnitude of 5.5, and roughly 900 times (30×30) more energy than a 4.5-magnitude quake. A major earthquake with a magnitude of 8.5 releases millions of times more energy than the smallest earthquakes felt by humans. This dispels the notion that a moderate earthquake decreases the chances for the occurrence of a major quake in the same region. Thousands of moderate tremors would be needed to release the amount of energy released by one "great" earthquake.

Some of the world's major earthquakes and their corresponding Richter magnitudes are listed in Table 12.3. Great earthquakes such as these can be ex-

TABLE 12.2 Earthquake magnitudes and expected world incidence.

Richter Magnitudes	Earthquake Effects	Estimated Number per Year
< 2.5	Generally not felt, but recorded.	900,000
2.5–5.4	Often felt, but only minor damage detected.	30,000
5.5–6.0	Slight damage to structures.	500
6.1–6.9	Can be destructive in populous regions.	100
7.0–7.9	Major earthquakes. Inflict serious damage.	20
≥ 8.0	Great earthquakes. Produce total destruction to communities near epicenter.	One every 5–10 years

Source: *Earthquake Information Bulletin* and others.

TABLE 12.3 Some notable worldwide and U.S. earthquakes.

Year	Location	Deaths (est.)	Magnitude	Comments
1290	Chihli (Hopei), China	100,000		
1556	Shensi, China	830,000		Possibly the greatest natural disaster.
1737	Calcutta, India	300,000		
1755	Lisbon, Portugal	70,000		Tsunami damage extensive.
*1811–12	New Madrid, Missouri	Few		Three major earthquakes.
*1886	Charleston, South Carolina	60		
*1906	San Francisco, California	700	8.25	Fires caused extensive damage.
1908	Messina, Italy	120,000	7.5	
1920	Kansu, China	180,000	8.5	
1923	Tokyo, Japan	150,000	8.2	Fire caused extensive destruction.
1960	Southern Chile	5700	8.5–8.7	Possibly the largest magnitude earthquake ever recorded.
*1964	Alaska	131	8.4–8.6	
1970	Peru	66,000	7.8	Great rockslide.
*1971	San Fernando, California	65	6.5	Damage exceeded one billion dollars.
1975	Liaoning Province, China	Few	7.5	First major earthquake to be predicted.
1976	Tangshan, China	650,000	7.6	Not predicted.

*U.S. earthquakes.
Source: U.S. National Oceanic and Atmospheric Administration.

pected in a tectonically active region every 50 to 200 years. The region of the San Andreas fault, along which slippage occurred to produce the 1906 San Francisco earthquake, has not generated a large earthquake in over 75 years—an alarming statistic to the residents of this region.

EARTHQUAKE DESTRUCTION: THE 1964 ALASKAN EARTHQUAKE

The most violent earthquake to jar North America this century—the Good Friday Alaskan Earthquake—occurred at 5:36 P.M. on March 27, 1964. Felt throughout that state, the earthquake had a magnitude of 8.4–8.6 on the Richter scale and reportedly lasted 3 to 4 minutes. This brief event left 131 persons dead, thousands homeless, and the economy of the state badly disrupted. Had the schools and business districts been open, the toll surely would have been higher. The total financial loss has been estimated at 300 million dollars, roughly twenty times the original 15 million dollar purchase price of the state. The location of the epicenter and the towns which were hardest hit by the quake are shown in Figure 12.13. Within 24 hours of the initial

shock, 28 aftershocks were recorded, 10 of which exceeded a magnitude of 6 on the Richter scale.

Many factors determine the amount of destruction that accompanies an earthquake. The most obvious of these are the magnitude of the earthquake and the proximity of the quake to a populated area. Fortunately, most earthquakes are small and occur in remote regions of the earth. However, about 20 major earthquakes are reported annually, one or two of which are catastrophic.

Destruction Caused by Seismic Vibrations

The 1964 Alaskan earthquake provided geologists with new insights into the role of ground shaking as a destructive force. As the energy released by an earthquake travels along the earth's surface, it causes the ground to vibrate in a complex manner by moving up and down as well as from side to side. The amount of structural damage attributable to the vibrations depends on several factors, including: (1) The intensity and duration of the vibrations; (2) The nature of the material upon which the structure rests; and (3) The design of the structure.

All of the multistory structures in Anchorage were damaged by the vibrations; the more flexible wood

FIGURE 12.13 Region most affected by the Good Friday earthquake of 1964. Note the location of the epicenter (red dot). (After U.S. Geological Survey)

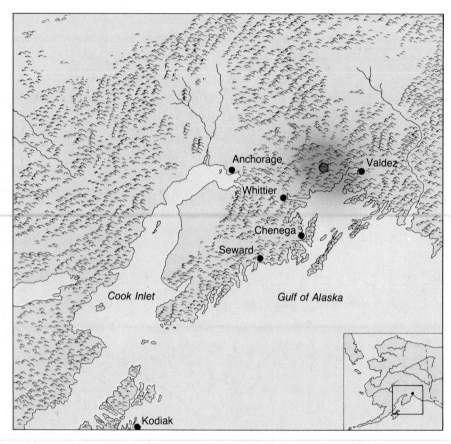

frame residential buildings fared best. However, many homes were destroyed when the ground failed. A striking example of how construction variations affect earthquake damage is shown in Figure 12.14. We can see in this photo that the steel-frame building on the left withstood the vibrations, whereas the relatively rigid concrete structure was badly damaged.

The greatest loss of life from an earthquake can be partially attributed to the type of structures that are inhabited. In 1556, in the Shensi region of China, an estimated 830,000 persons perished when an early morning earthquake struck the region. Many of these people lived in dwellings carved out of a compacted windblown sediment called loess (see Chapter Nine). The walls of these structures failed, allowing the roofs to bury the inhabitants. The dense population and present-day construction practices in this region are such that a repetition of the 1556 event is possible.

Most of the large structures in Anchorage were damaged even though they were built to conform to the earthquake provisions of the Uniform Building Code of California. Perhaps some of that destruction can be attributed to the unusually long duration of this earthquake, which was estimated at 3 to 4 minutes. Most earthquakes consist of tremors lasting from 20 seconds to one minute. The San Francisco earthquake of 1906 was felt for about 40 seconds.

Although the region within 20 to 50 kilometers of an epicenter will experience about the same degree of ground shaking, the destruction will vary considerably within this area. This difference is mainly attributable to the nature of the ground on which the structures are built. Soft sediments, for example, generally amplify the vibration more than solid bedrock. Thus, the buildings in Anchorage, which were situated on unconsolidated sediments, experienced heavy structural damage. By contrast, most of the town of Whittier, although located much nearer to the epicenter than Anchorage, rests on a firm foundation of granite and hence suffered much less damage from the seismic vibrations. However, Whittier was damaged by a seismic sea wave.

FIGURE 12.14 Damage to the five-story J. C. Penney Co. building, Anchorage, Alaska. Very little structural damage was incurred by the adjacent building. (Courtesy of NOAA)

Tsunami

Most of the deaths associated with the 1964 Alaskan quake were caused by **seismic sea waves**, or **tsunami**.* These destructive waves have popularly been called "tidal waves." However, this name is not accurate since these waves are not generated by the tidal effect of the moon or sun.

Most tsunamis result from vertical displacement of the ocean floor during an earthquake as illustrated in Figure 12.15. Once formed, a tsunami resembles the ripples created when a pebble is dropped into a pond. In contrast to ripples, tsunamis advance at speeds between 500 and 800 kilometers (300 and 500 miles) per hour. Despite this striking characteristic, a tsunami in the open ocean can pass undetected because its height is usually less than one meter and the distance between wave crests ranges from 100 to 700 kilometers. However, upon entering shallower coastal waters, these destructive waves are slowed and the water begins to pile up to heights that occasionally exceed 30 meters (100 feet). As a tsunami approaches shore, it appears as a rapid rise in sea level with a turbulent and chaotic surface.

*Seismic sea waves were given the name *tsunami* by the Japanese, who have suffered a great deal from them. The term *tsunami* is now used worldwide.

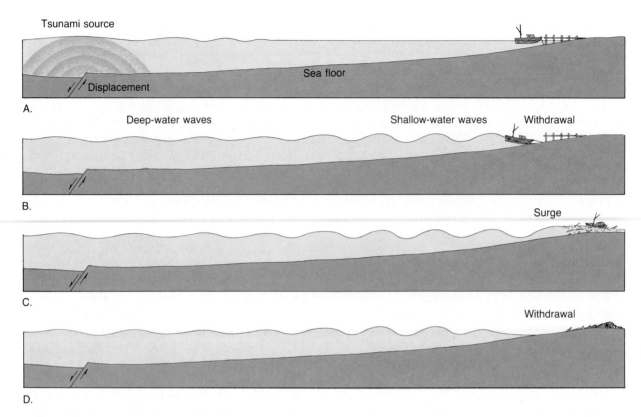

Tsunami source

Sea floor

Displacement

A.

Deep-water waves   Shallow-water waves   Withdrawal

B.

Surge

C.

Withdrawal

D.

FIGURE 12.15   Schematic drawing of a tsunami generated by displacement of the ocean floor. The size and spacing of the swells are not to scale.

Usually the first warning of a tsunami is a rather rapid withdrawal of water from beaches (Figure 12.15B). Residents of coastal areas have learned to heed this warning and move to higher ground. About 5 to 30 minutes later the retreat of water is followed by a surge capable of extending hundreds of meters inland. In a successive fashion, each surge is followed by rapid oceanward retreat of the water. These waves, separated by intervals of between 10 and 60 minutes, are able to traverse large stretches of the ocean before their energy is totally dissipated.

The tsunami generated in the 1964 Alaskan earthquake inflicted heavy damage to communities in the vicinity of the Gulf of Alaska, completely destroying the town of Chenega. The town of Seward was also heavily damaged as most of its port facilities were demolished by this seismic sea wave (Figure 12.16). The deaths of 107 persons have been attributed to this tsunami. By contrast, only 9 persons died in Anchorage as a direct result of the vibrations. Tsunami damage following the Alaskan earth-

quake extended along much of the west coast of North America, and in spite of a one-hour warning, 12 persons perished in Crescent City, California, where all of the deaths and most of the destruction

FIGURE 12.16   The effects of a tsunami at Seward, Alaska. (Courtesy of NOAA)

were caused by the fifth wave. The first wave crested about 4 meters (13 feet) above low tide and was followed by three progressively smaller waves. Believing that the tsunami had ceased, people returned to the shore, only to be met by the fifth and most devastating wave, which, superimposed upon high tide, crested about 6 meters higher than the level of low tide.

Fire

Fire was only a minor consequence in the 1964 Alaskan earthquake, but often it is the most destructive result. The 1906 earthquake centered near the city of San Francisco reminds us of the formidable threat of fire. The central city contained mostly large, older wooden structures and brick buildings. Although many of the unreinforced brick buildings were extensively damaged, the greatest destruction was caused by innumerable fires which started when gas and electrical lines were severed. The fires raged out of control for three days and devastated over 500 blocks of the city (see Figure 12.2). The problem was compounded by the initial ground shaking which broke the city's water lines into hundreds of unconnected pieces.

The fire was finally contained when buildings were dynamited along a wide boulevard to provide a fire break. Although only a few deaths were attributed to the fires, that is not always the case. An earthquake which rocked Japan in 1923 triggered an estimated 250 fires, which devastated the city of Yokohama and destroyed more than half the homes in Tokyo. Over 100,000 deaths were attributed to the fires, which were driven by unusually high winds.

Landslides and Ground Subsidence

In the 1964 Alaskan earthquake, it was not ground vibrations directly, but landslides and ground subsidence triggered by the vibrations that probably caused the greatest damage to structures. At Valdez and Seward the violent shaking caused deltaic materials to liquify; the subsequent slumping carried both waterfronts away. Because of the threat of recurrence, the entire town of Valdez was relocated about 7 kilometers away in a region of stable ground. The destruction at Valdez was compounded by the tragic loss of 31 lives. While waiting for an

incoming vessel, the 31 persons and the dock slid into the sea.

Most of the damage in the city of Anchorage was also attributed to landslides caused by the shaking and lurching ground. Many homes were destroyed in Turnagain Heights when a layer of clay lost its strength and over 200 acres of land slid toward the ocean. The destruction was so complete that this area was bulldozed over and made into a park, which was appropriately named "Earthquake Park." Downtown Anchorage was equally disrupted as blocks of earth broke loose and sections of the main business district dropped by as much as 3 meters.

EARTHQUAKE PREDICTION AND CONTROL

The vibrations that shook the San Fernando, California, area on the morning of February 9, 1971, inflicted 65 deaths and almost one billion dollars in damages (Figure 12.17)—all from an earthquake that lasted only 60 seconds and had a moderate rating of 6.5 on the Richter scale. Fortunately, because of the early hour, freeways, businesses, and schools were sparsely occupied, reducing the possible toll. Also, had the Lower Van Norman Lake Dam, badly damaged during the earthquake, actually broken, 80,000 additional lives might have been lost which would have made it the most catastrophic event ever in the United States. This quake in populous southern California re-emphasized the need for reliable methods of earthquake prediction and control.

Japan's location in an earthquake-prone region has resulted in great interest in earthquake prediction there. The Japanese have established a complex seismic network extending 200 kilometers (125 miles) out into the ocean. Here on the ocean floor where background noise is slight, the Japanese plan to monitor microearthquakes (foreshocks), which precede the main earthquake. It is hoped that by monitoring these seismic activities some pattern will emerge which can be used to accurately predict forthcoming tremors.

In California, uplift or subsidence of the land and changes in movement of a fault zone from a slow creep to a locked position have been found to precede moderate earthquakes. It therefore seems reasonable that earthquakes may be predicted by continually monitoring ground tilt, fault movement, and

FIGURE 12.17 Collapsed overpass to Golden State Freeway, San Fernando earthquake, 1971. (Photo by R. W. Wallace, U.S. Geological Survey)

seismic activity. Some monitoring networks are already operating in the earthquake-prone regions of the United States; others have been proposed.

Although no reliable method of short-range prediction has yet been devised, a few successful predictions have been made. In 1966, an earthquake in Tashkent, U.S.S.R., was predicted by monitoring the radon level in wells. Radon is an inert gas generated by radioactive decay of radium, a small amount of which is found in certain rocks. Normally this gas is locked within rock, but during the buildup of stress, the newly formed cracks allow for its release. In February, 1975, an earthquake in northeast China was predicted only hours before it occurred. By warning an estimated 3 million people to remain outdoors on a cold evening, tens of thousands of lives were believed to have been spared. Western observers confirmed Chinese reports that almost 90 percent of the structures in the city of Haicheng were heavily damaged. The rather large foreshocks which preceded this earthquake aided the prediction and also prompted people to heed the warning.

Unfortunately, the Chinese were able to predict but not pinpoint the exact date of the great Tangshan earthquake of 1976. Their long-range warning of an upcoming earthquake was not precise enough

to save as many as 650,000 persons estimated to have been killed and another 780,000 who were injured. The Chinese have also had false alarms. In a province near Hong Kong, people evacuated their dwellings for over a month, but no earthquake followed. The debate that would have to precede an evacuation order for a large American city such as Los Angeles would be considerable. The cost of the evacuation, loss of work time, and innumerable other problems associated with an evacuation would have to be weighed against the earthquake's probability. Earthquake prediction must become a more proven science before such warnings will be heeded.

The actual control of earthquakes is another matter altogether. The discovery that humans have inadvertently triggered earthquakes has given earth scientists some encouragement. The most convincing evidence that people can initiate earthquakes came between 1962 and 1966 when studies of the seismic activity at the Rocky Mountain Arsenal near Denver were conducted. For a period of 80 years prior to 1962, the U.S. Coast and Geodetic Survey reported no significant earthquake activity in the Denver region. In 1962 the arsenal began disposing of wastes from its chemical warfare production into a well over 3600 meters deep. During the period of fluid waste injection, from April, 1962 to September, 1965, about 700 microearthquakes were reported, 75 intense enough to be felt. The injection of water under pressure is believed to have "lubricated" the fault, which had been building up strain over the years. This lubricating effect is not one of making rocks along the fault zone slippery. Rather, the water exerts an outward force that is directed perpendicularly to the fault plane. This outward force opposes the natural inward force caused by the weight of rock piled above. When injection was halted for about a year, a marked drop in seismic activity was also detected. When pumping resumed, the frequency of tremors increased markedly.

Other earthquakes caused by human activity have occurred in regions adjacent to large reservoirs such as Lake Mead on the Arizona-Nevada border. Ever since Lake Mead was filled in 1936, hundreds of small tremors have been recorded. They are thought to have been caused by the added weight of the lake, and perhaps aided by the "lubricating" effect of water seeping into the rock below. Another large reservoir in India is believed responsible for triggering a disastrous earthquake in which 200 persons were killed. Underground nuclear explosions have also initiated numerous small aftershocks, although none has been as great as the explosion itself.

The hope of many scientists is that we may someday reduce the threat of earthquakes by triggering numerous small earthquakes using fluid injections or nuclear explosions. Such methods would slowly and continually release the elastic strain that might otherwise build up and be released as a high-magnitude earthquake. Recall, however, that many thousands of minor tremors are required to equal the energy released by one strong earthquake. This fact coupled with the inaccessibility of many fault zones make the possibility of earthquake control not quite as feasible as it might first appear. A favorable condition for earthquake control in California is the shallow depth of earthquake foci, making drilling operations possible. Many tests will have to be made in remote regions before we dare risk such a venture along a fault system like the San Andreas, at least in those portions located in populous regions.

THE EARTH'S INTERIOR

The earth's interior lies just below us, yet its accessibility to direct observation is very limited. Most of our knowledge of the earth's interior comes from the study of P and S waves that travel through the earth and emerge at some distant point. Simply stated, the technique involves accurately measuring the time required for seismic waves to travel from the focus of an earthquake or nuclear explosion to a seismographic station. Since the time required for P and S waves to travel through the earth depends upon the properties of the rock materials encountered, seismologists search for variations in travel times that cannot be accounted for simply by differences in the distances traveled. These variations correspond to changes in rock properties. Based upon these seismological data the earth has been divided into four major layers: (1) The **crust**, a very thin outer layer; (2) The **mantle**, a rocky layer located below the crust and having a thickness of 2885 kilometers (1789 miles); (3) The **outer core**, a layer about 2270 kilometers (1407 miles) thick which exhibits the characteristics of a mobile liquid; and (4) The **inner core**, a solid metallic sphere about 1216 kilometers (754 miles) in radius (Figure 12.18).

In 1909, a pioneering Yugoslavian seismologist, Andrija Mohorovičić, presented the first convincing evidence for layering within the earth. By studying seismic records, he found that the velocity of waves increases abruptly below a depth of 50 kilometers. The boundary that he discovered separates the crust from the underlying mantle and is known as the **Mohorovičić discontinuity** in his honor (Figure 12.18). For reasons that are obvious, the name for this boundary was quickly shortened to **Moho**.

A few years later another major boundary was discovered by the German seismologist Beno Gutenberg. This discovery was based primarily on the observation that P waves diminish and eventually die out completely about 105 degrees from an earthquake. Then, about 140 degrees away, the P waves reappear, but about two minutes later than would be expected based on the distance traveled. This belt where direct seismic waves are absent is approximately 35 degrees wide and has been named the **shadow zone*** (Figure 12.19). Gutenberg realized that the shadow zone could be explained if the earth contained a core composed of material unlike the overlying mantle and had a radius of 3420 kilometers. The core must somehow hinder the transmis-

---

*As more sensitive instruments were developed, weak and delayed P waves that enter this zone via reflection were detected.

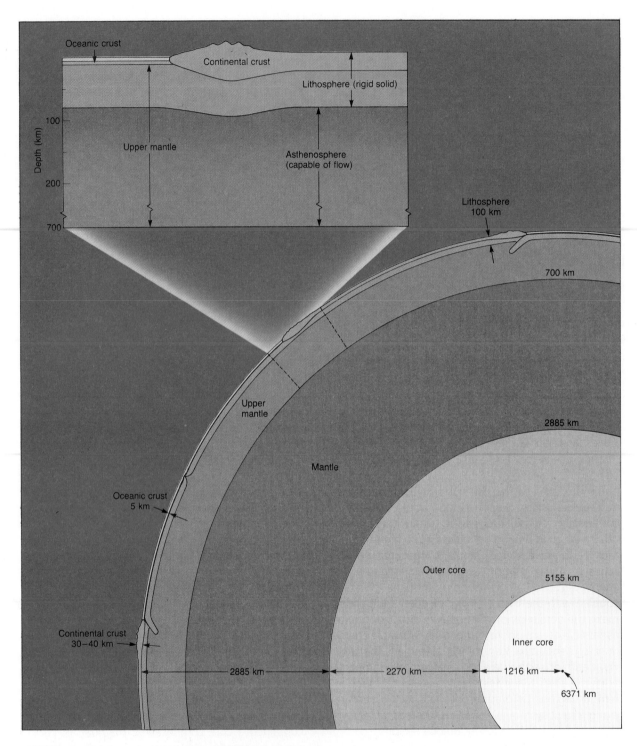

FIGURE 12.18 Cross-sectional view of the earth showing internal structure.

sion of P waves in a manner similar to the light rays blocked by an opaque object that casts a shadow. However, rather than actually stopping the P waves, the shadow zone is produced by the bending of P waves which enter the core as shown in Figure 12.19.

It was further learned that S waves could not propagate through the core; therefore, geologists concluded that at least a portion of this region is liquid. This conclusion was further supported by the observation that P-wave velocities suddenly decrease about 40 percent as they enter the core. Since melting would reduce the elasticity of rock, all evidence points to the existence of a liquid layer below the rocky mantle.

In 1936, the last major subdivision of the earth's interior was predicted by the discovery of seismic waves believed to be reflected from a boundary within the core. Hence, a core within a core was discovered. The actual size of the inner core was not accurately calculated until the early 1960s when underground nuclear tests were conducted in Nevada. Because the precise location and time of the explosions were known, echoes from seismic waves which bounced off the inner core provided an accurate means of determining its size. From these data and subsequent studies, the inner core was found to have a radius of about 1216 kilometers. Further, P waves passing through the inner core have appreciably faster travel times than those penetrating the outer core exclusively. The apparent increase in elasticity of the inner core material is considered evidence for the solid nature of the earth's innermost region.

A most important zone exists within the upper mantle and deserves special mention. This region, called the **asthenosphere**, is located between the depths of 70 kilometers and 700 kilometers (Figure 12.18). In the asthenosphere the velocity of S waves decreases, indicating to seismologists that this zone consists partly of melted rock (approximately 10 percent). Situated above the asthenosphere is the outer solid portion of the earth, called the **lithosphere**, which includes part of the upper mantle and the crust. It is believed that the "plastic" material of the asthenosphere moves and carries along the rigid

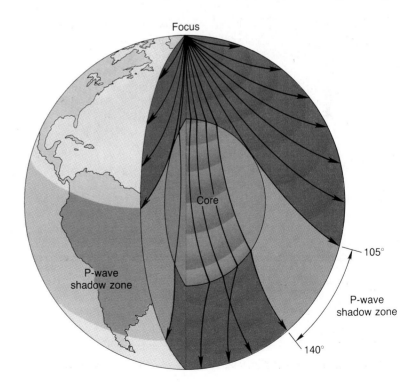

FIGURE 12.19 The abrupt change in physical properties at the mantle-core boundary causes the wave paths to bend sharply. This abrupt change in wave direction results in a shadow zone for P waves between about 105 and 140 degrees.

lithosphere. Thus, the discovery of the astheno-sphere was an important contribution to the theory which proposes that the continents ''drift'' about. This is the topic of the next chapter. It is also from the asthenosphere that some, but not all, molten material for volcanic activity is thought to originate.

COMPOSITION OF THE EARTH

The crust of the earth varies in thickness, being greater than 70 kilometers in some mountainous regions and less than 5 kilometers in some oceanic regions (see Figure 12.18). Early seismic data indicated that the continental crust, which is mostly made of granitic rocks, is quite different in composition from the oceanic crust. Until recently, however, scientists had only seismic evidence from which to determine the composition of oceanic crust, which lies beneath 3 kilometers of water as well as hundreds of meters of sediment. With the development of the deep-sea drilling ship *Glomar Challenger* (see Figure 13.14), recovery of ocean floor samples became possible. The samples were of basaltic composition—indeed different from the rocks which compose the continents.

Our knowledge of the compositions of the mantle and core is much more speculative. However, we do have some clues. Recall that some of the lava that reaches the earth's surface originates in the partially melted asthenosphere located within the mantle. In the laboratory, experiments have shown that partial melting of a rock called peridotite results in a melt that has a basaltic composition similar to those associated with the volcanic activity of oceanic islands. Rocks such as peridotite are thought to make up the mantle and provide the lava for oceanic eruptions.

Surprisingly, meteorites—''shooting stars''—which fall to the earth from space, are considered evidence of the earth's inner composition. Since meteorites are part of the solar system, they are assumed to be representative samples. Their composition ranges from metallic types made of iron and nickel to stony meteorites composed of rock similar to peridotite. Because the earth's crust contains a much smaller percentage of iron than meteorites do, geologists believe that the heavy minerals sank during the early history of the earth. By the same token, the lighter minerals may have floated to the top, creating the crust. Thus, the core of the earth is thought to be mainly iron and nickel, similar to metallic meteorites, whereas the surrounding mantle is believed to be composed of rock material similar to that found in stony meteorites.

The concept of a molten iron outer core is further supported by the existence of the earth's magnetic field, which acts as a large bar magnet. The most widely accepted mechanism explaining the magnetic field requires that the earth's core be made of a material which conducts electricity, such as iron, and which is mobile enough that circulation can occur. Both of these conditions are met by the model of the earth's core that was established on the basis of seismic data.

Not only does an iron core explain the earth's magnetic field, it also explains the high density of the inner earth, about 13.5 times that of water. Even under the extreme pressure at those depths, average crustal rocks with densities 2.8 times that of water would not have the density calculated for the core. But iron, which is 3 times more dense than crustal rocks, has the required density.

REVIEW QUESTIONS

1 What is an earthquake? Under what circumstances do earthquakes occur?

2 How are faults, foci, and epicenters related?

3 Faults that are experiencing no active creep may be considered "safe." Rebut or defend this statement.

4 Describe the principle of a seismograph.

5 Using Figure 12.10, determine the distance between an earthquake and a seismic station if the first S wave arrives 3 minutes after the first P wave.

6 List the major differences between P and S waves.

7 An earthquake measuring 7 on the Richter scale releases about ___ times more energy than an earthquake with a magnitude of 6.

8 List three factors that affect the amount of destruction caused by seismic vibrations.

9 In addition to the destruction created directly by seismic vibrations, list three other types of destruction associated with earthquakes.

10 Distinguish between the Mercalli scale and the Richter scale.

11 What is a tsunami? How is one generated?

12 Cite some reasons why an earthquake with a moderate magnitude might cause more extensive damage than a quake with a high magnitude.

13 How might earthquakes be controlled in the future?

14 What evidence do we have that the earth's outer core is molten?

15 Contrast the physical makeup of the asthenosphere and the lithosphere.

16 Earthquakes occur only in the rigid lithosphere, not in the plastic asthenosphere. Using the elastic rebound idea, explain this phenomenon.

17 Why are meteorites considered important clues to the composition of the earth's interior?

18 Describe the chemical (mineral) makeup of the following:
a Continental crust.
b Oceanic crust.
c Mantle.
d Core.

CHAPTER THIRTEEN

PLATE TECTONICS

Early in this century geologic thought was dominated by a belief in the geographic permanency of the ocean basins and continents. During the last few decades, however, vast accumulations of new data have dramatically changed our ideas about the nature and workings of the earth. Earth scientists now realize that the positions of landmasses are not fixed. Rather, the continents gradually migrate over the surface of the globe. The splitting of continental blocks has resulted in the formation of new ocean basins, while older segments of the sea floor are continually being recycled in areas where we find deep-ocean trenches. Further, because of this movement, once-disjointed segments of continental material have collided and formed the earth's great mountain ranges. In short, a revolutionary new model of the earth's tectonic* processes has emerged in marked contrast to what was accepted just a few decades ago.

This profound reversal of scientific opinion has been appropriately described as a scientific revolution. Like other scientific revolutions, an appreciable length of time elapsed between the idea's inception and its general acceptance. The revolution began in the early part of the twentieth century as a relatively straightforward proposal that the continents drift about the face of the earth. After many years of heated debate, the idea of drifting continents was rejected by the vast majority of earth scientists as being improbable. However, during the 1950s and 1960s new evidence began to rekindle interest in this abandoned proposal. By 1968 these new developments led to the unfolding of a far more encompassing theory than continental drift—a theory known as plate tectonics.

*Tectonics refers to the deformation of the earth's crust and results in the formation of structural features such as mountains.

The Sinai Peninsula is bordered by rifts believed to be caused by sea-floor spreading. (Courtesy of NASA)

CONTINENTAL DRIFT: AN IDEA BEFORE ITS TIME

The idea that continents, particularly South America and Africa, fit together like pieces of a jigsaw puzzle originated with improved world maps. However, little significance was given this idea until 1915, when Alfred Wegener,* a German climatologist and geophysicist, published an expanded version of a 1912 lecture in his book *The Origin of Continents and Oceans*. In this monograph, Wegener set forth the basic outline of his radical hypothesis of **continental drift**. One of his major tenets suggested that a supercontinent he called **Pangaea** (meaning "all land") once existed (Figure 13.1). He further hypothesized that about 200 million years ago this supercontinent began breaking into smaller continents, which then "drifted" to their present positions. Wegener and others who advocated this position collected substantial evidence to support these claims. The fit of South America and Africa, ancient climatic similarities, fossil evidence, and rock structures all seemed to support the idea that these now-separate landmasses were once joined.

Fit of the Continents

Like a few others before him, Wegener first suspected that the continents might have been joined when he noticed the remarkable similarity between the coastlines on opposite sides of the South Atlantic. However, his use of present-day shorelines to make a fit of the continents was challenged immediately by other earth scientists. These opponents correctly argued that shorelines are continually

*Wegener's ideas were actually preceded by those of an American geologist, F. B. Taylor, who in 1910 published a paper on continental drift. Taylor's paper provided little supporting evidence for continental drift, which may have been the reason that it had a relatively small impact on the geologic community.

FIGURE 13.1 Reconstruction of Pangaea as it is thought to have appeared 200 million years ago. (After Robert S. Dietz and John C. Holden. *Journal of Geophysical Research* 75: 4943. Copyright © by American Geophysical Union)

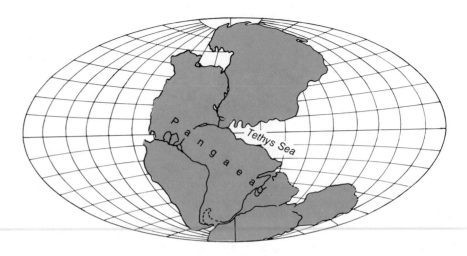

modified by erosional processes and even if continental displacement had taken place, a good fit today would be unlikely. Wegener appeared to be aware of this problem, and, in fact, his original jigsaw fit of the continents was only very crude.

A much better approximation of the outer boundary of the continents is the seaward margin of the continental shelf. Today the continental shelf's edge lies several hundred meters below sea level. In the early 1960s, Sir Edward Bullard and two associ-

ates produced a map with the aid of computers that attempted to fit the continents at a depth of 900 meters. The remarkable fit that was obtained is shown in Figure 13.2. Although the continents overlap in a few places, these are regions where streams have deposited large quantities of sediment, thus enlarging the continents. The overall fit obtained by Bullard and his associates was better than even the supporters of the continental drift theory suspected it would be.

Fossil Evidence

Although Wegener was intrigued by the remarkable similarities of the shorelines on opposite sides of the Atlantic, he at first thought the idea of a mobile earth improbable. Not until he came across an article citing fossil evidence for the existence of a land bridge connecting South America and Africa did he begin to take his own idea seriously. Through a search of the literature Wegener learned that most paleontologists were in agreement that some type of land connection was needed to explain the existence of identical fossils on the widely separated landmasses.

To add credibility to his argument for the existence of Pangaea, Wegener used the already documented evidence that several fossil organisms exist which could not have made the journey across the vast oceans presently separating the continents. In particular, the fossil fern *Glossopteris* was known to have been widely dispersed in the southern continents of Africa, Australia, and South America during the Mesozoic era. Later, fossil remains of *Glossop-*

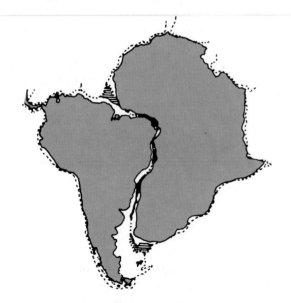

FIGURE 13.2 The best fit of South America and Africa along the continental slope at a depth of 500 fathoms. (After A. G. Smith. "Continental Drift." In *Understanding the Earth,* edited by I. G. Gass. Courtesy of Artemis Press)

teris were discovered in Antarctica as well. In addition, remains of a species of swimming reptile called *Mesosaurus* were found in eastern South America and western Africa. Although this reptile probably swam in the shallow waters of these regions, *Mesosaurus* was clearly not capable of making the long journey across the Atlantic Ocean. For Wegener, fossils provided undeniable proof that these landmasses were once joined together as the supercontinent Pangaea.

How could these fossil flora and fauna be so similar in places separated by thousands of kilometers of open ocean? The idea of land bridges (isthmian links) was the most widely accepted solution to the problem of migration. We know for example that during the recent glacial period the lowering of sea level allowed animals to cross the narrow Bering Straits between Asia and North America. Was it possible then that land bridges once connected Africa and South America? We are now quite certain that land bridges of this magnitude did not exist, for their remnants should still lie below sea level, but are nowhere to be found.

Rock Type and Structural Similarities

Anyone who has worked a picture puzzle knows that in addition to the pieces fitting together, the picture must be continuous as well. The "picture" that must match in the "Continental Drift Puzzle" is represented by the rock types and mountain belts found on the continents. If the continents were once together, the rocks found in a particular region on one continent should closely match in age and type with those found in adjacent positions on the matching continent.

Such evidence has been found in the form of several mountain belts which appear to terminate at one coastline only to reappear on a landmass across the ocean. For instance, the mountain belt that includes the Appalachians trends northeastward through the eastern United States and disappears off the coast of Newfoundland. Mountains of comparable age and structure are found in Greenland and Northern Europe. When these landmasses are reassembled, the mountain chains form a nearly continuous belt.

Paleoclimatic Evidence

Alfred Wegener was a climatologist by training. For that reason, he was keenly interested in obtaining paleoclimatic (ancient climatic) data in support of continental drift. His efforts in this area were rewarded when he found evidence for apparently dramatic climatic changes. For instance, glacial deposits indicate that near the end of the Paleozoic era (between 220 and 300 million years ago), ice sheets covered extensive areas of the Southern Hemisphere. Layers of glacial till were found in southern Africa and South America, as well as in India and Australia. Below these beds of glacial debris lay striated and grooved bedrock. In some locations the striations and grooves indicated the ice had moved from the sea onto land (Figure 13.3). Much of the land area containing evidence of this late Paleozoic glaciation presently lies within 30 degrees of the equator in a subtropical or tropical climate.

Could the earth have gone through a period sufficiently cold to have generated extensive continental glaciers in what is presently a tropical region? Wegener rejected this explanation because during the late Paleozoic, large tropical swamps existed in the Northern Hemisphere. These swamps with their lush vegetation eventually became the major coal fields of the eastern United States, Europe, and Siberia. As Wegener proposed, a better explanation is provided if the landmasses are fitted together as a supercontinent and then moved nearer the South Pole. This would account for the conditions necessary to generate extensive expanses of glacial ice over much of the Southern Hemisphere. At the same time this shift would place the northern landmasses nearer the tropics and account for their vast coal deposits (see Figure 8.20).

How does a glacier develop in hot, arid Australia? How do land animals migrate across wide expanses of open water? As compelling as this evidence may have been, fifty years passed before most of the scientific community would accept it and the logical conclusions to which it led.

THE GREAT DEBATE

Wegener's proposal did not attract much open criticism until 1924 when his book was translated into English. From this time on, until his death in 1930, his drift hypothesis encountered a great deal of hostile criticism. To quote the respected American geologist R. T. Chamberlin, "Wegener's hypothesis in general is of the footloose type, in that it takes

FIGURE 13.3 **A.** Direction of ice movement in the southern supercontinent called Gondwanaland by the founders of the continental drift concept. **B.** Glacial striations in the bedrock of Hallet Cove, South Australia, indicate direction of ice movement. (Photo by W. B. Hamilton, U.S. Geological Survey)

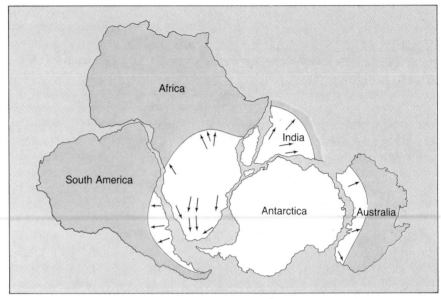

A.

B.

considerable liberty with our globe, and is less bound by restrictions or tied down by awkward, ugly facts than most of its rival theories. Its appeal seems to lie in the fact that it plays a game in which there are few restrictive rules and no sharply drawn code of conduct.''

One of the main objections to Wegener's hypothesis stemmed from his inability to provide a mechanism for continental drift. Wegener proposed two possible energy sources for drift. One of these, the tidal influence of the moon, was presumed by Wegener to be strong enough to give the continents a westward motion. However, the prominent physicist Harold Jeffreys quickly countered with the argument that tidal friction of the magnitude needed to displace the continents would bring the earth's rotation to a halt in a matter of a few years. Further, Wegener proposed that the larger and sturdier continents broke through the oceanic crust, much like ice breakers cut through ice. However, no evidence existed to suggest that the ocean floor was weak enough to permit passage of the continents without themselves being appreciably deformed in the process. By 1929 criticisms of Wegener's ideas were pouring in from all areas of the scientific community. Despite these attacks, Wegener wrote the fourth and final edition of his book, maintaining his basic hypothesis and adding supporting evidence.

Although most of Wegener's contemporaries opposed his views, even to the point of openly ridiculing them, a few considered his ideas plausible. For these few geologists who continued the search, the concept of continents in motion evidently provided enough excitement to hold their interest. Others undoubtedly viewed continental drift as a solution to previously unexplainable observations.

PLATE TECTONICS: A MODERN VERSION OF AN OLD IDEA

During the years that followed Wegener's proposal, great strides in technology permitted mapping of the ocean floor, and extensive data on seismic activity and the earth's magnetic field became available. By 1968 these developments led to the unfolding of a far more encompassing theory than continental drift, known as **plate tectonics**. The implications of plate tectonics are so far-reaching that this theory can be considered the framework from which to view most other geologic processes. Since this concept is relatively new, it most surely will be modified as additional information becomes available; however, the main tenets appear to be sound and are presented here in their current state of refinement.

The theory of plate tectonics holds that the outer, rigid lithosphere consists of about twenty rigid segments called *plates* (see Figure 13.4). Of these, the largest is the Pacific plate, which is located mostly within the ocean proper, except for a small sliver of North America that includes southwestern California

and the Baja Peninsula. Notice from Figure 13.4 that all of the other large plates contain both continental and oceanic crust—a major departure from the continental drift theory, which proposed that the continents moved through, not with, the ocean floor. Most of the smaller plates, on the other hand, consist exclusively of oceanic lithosphere, as for example, the Nazca plate located off the west coast of South America. Although not clearly defined in Figure 13.4, one small plate that roughly coincides with Turkey is located exclusively within a continent.

The lithosphere overlies a zone of much weaker and hotter material known as the asthenosphere. Hence, the lithospheric plates form a rigid outer shell supported from below by the more "plastic" material of the asthenosphere. A relationship appears to exist between the thickness of the lithospheric plates and the nature of the crustal material that caps them. Plates are thinnest in the oceans, where their thickness varies from 80 to 100 kilometers. By contrast, continental blocks are 100 kilometers or more thick, and in some regions may approach 400 kilometers thick.

One of the main tenets of the plate tectonics theory is that each plate moves as a distinct unit in relation to other plates. The mobile behavior of the rock within the asthenosphere is believed to allow this motion in the earth's rigid outer shell. As the plates move, the distance between two cities on the same plate, New York and Denver, for example, remains constant, while the distance between New York and London, which are located on different plates, is continually changing. Since each plate moves as a distinct unit, all major interactions between plates occur along plate boundaries. Thus, most of the earth's seismic activity, volcanism, and mountain building occur along these dynamic margins.

PLATE BOUNDARIES

For some time now, tectonic activity has been known to be restricted to narrow zones, such as the so-called *Ring of Fire* that encircles the Pacific. Thus, the first approximations of plate margins relied on the distribution of earthquake and volcanic activity. Later work indicated that three distinct types of plate boundaries exist, each differentiated by the movement it exhibits (Figure 13.5, page 240). These are:

1 **Divergent boundaries**—where plates move apart, resulting in upwelling of material from the mantle to create new sea floor.

2 **Convergent boundaries**—where plates move together, causing one of the slabs of lithosphere to be consumed into the mantle as it descends beneath an overriding plate.

3 **Transform fault boundaries**—where plates slide past each other without creating or destroying lithosphere.

Each plate is bounded by a combination of these zones (Figure 13.4). Movement along any boundary requires that adjustments be made at the others.

Divergent Boundaries

Divergent boundaries, where plate spreading occurs, are situated at the crests of oceanic ridges. Here, as the plates move away from the ridge axis, the gaps created are immediately filled with molten rock that oozes up from the hot asthenosphere. This material cools slowly to produce new slivers of sea floor. In a continuous manner successive separations and injections of magma add new oceanic crust (lithosphere) between the diverging plates. This mechanism, which has produced the floor of the Atlantic Ocean during the past 165 million years, is called **sea-floor spreading**. The typical rate of spreading at these ridges is estimated to be between 2 and 10 centimeters per year, and averages about 6 centimeters (2 inches) per year. Since new rock is added equally to the trailing edges of both diverging plates, the rate of ocean floor growth is twice the value of the spreading rate. That fact notwithstanding, these seemingly slow rates are rapid enough to have opened and reclosed the Atlantic Ocean more than ten times during the nearly 5-billion-year history of our planet, although this probably did not happen.

Not all spreading centers are as old as the Mid-Atlantic Ridge and not all are found in the middle of large oceans. The Red Sea is believed to be the site of a recently formed divergent boundary. Here the Arabian Peninsula separated from Africa and began to move toward the northeast. Consequently, the Red Sea is providing oceanographers with a view of

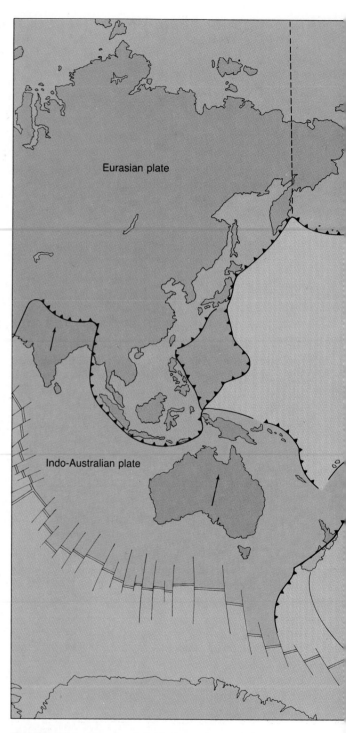

FIGURE 13.4 Mosaic of rigid plates that constitute the earth's outer shell.

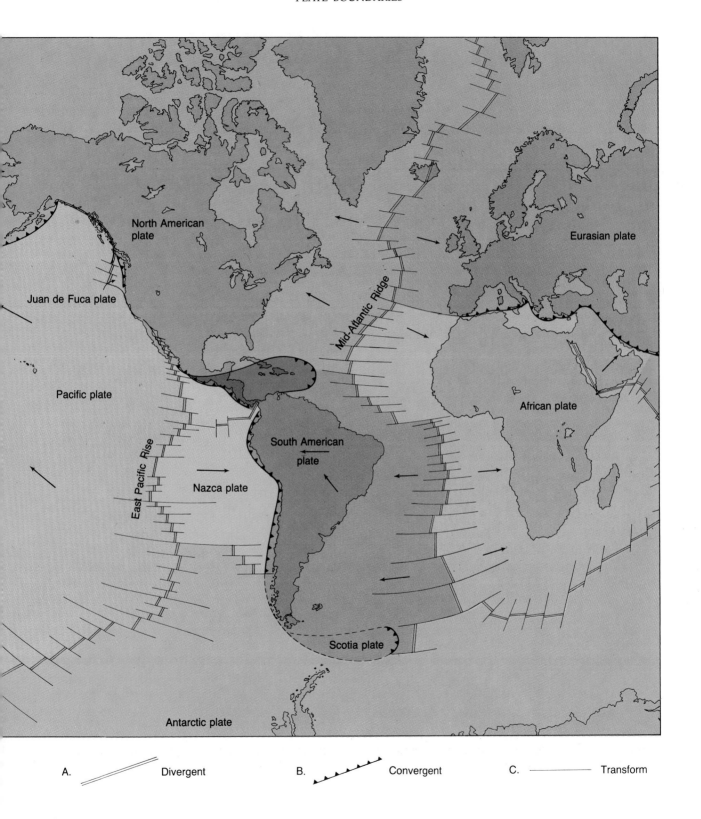

A. Divergent

B. Convergent

C. Transform

FIGURE 13.5 Schematic of plate boundaries. **A.** Divergent boundary. **B.** Convergent boundary. **C.** Transform fault boundary.

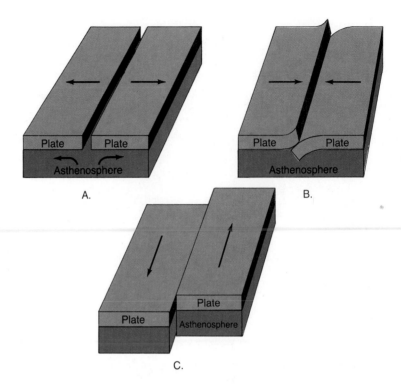

A.

B.

C.

how the Atlantic Ocean may have looked in its infancy. Another result of sea-floor spreading in the recent geologic past has been the formation of the Gulf of California.

When a spreading center develops within a continent, the landmass may split into smaller segments as Wegener had proposed for the breakup of Pangaea. The fragmentation of a continent is thought to be initiated by an upward movement of hot rock from below. The effect of this activity is to upwarp the crust directly above the hot rising plume. Crustal stretching associated with the doming generates numerous tensional cracks as shown in Figure 13.6A. Then, as the hot plume spreads laterally from the region of upwelling, the broken lithosphere is pulled apart. Gradually the broken slabs slide downward into the gaps created by the diverging plates (Figure 13.6B). The large downfaulted valleys generated by this process are called **rifts**, or **rift valleys**. The Great Rift Valley of East Africa is an excellent example of such a feature. If the spreading process continues in East Africa, the rift valley will lengthen and deepen, eventually extending out into the ocean. At this point the valley will become a narrow linear sea with an outlet to the ocean similar to the Red Sea today (Figure 13.6C). The zone of rifting will

remain the site of igneous activity, continually generating new sea floor in an ever-expanding ocean basin (Figure 13.6D).

Our knowledge of oceanic ridge systems, the sites of sea-floor spreading, comes from soundings taken of the ocean floor, core samples obtained from deep-sea drilling, visual inspection using submersibles, and even first-hand inspection of slices of ocean floor which have been shoved up onto dry land. Ocean ridge systems are characterized by an elevated position and numerous volcanic structures which have grown on the newly formed crust. Because of its accessibility, the Mid-Atlantic Ridge has been studied more thoroughly than other ridge systems. The Mid-Atlantic Ridge is a gigantic submerged mountain range standing 2500–3000 meters (8200–10,000 feet) above the adjacent deep-ocean basins. It extends southward from the Arctic Ocean to beyond the southern tip of Africa. In a few places, such as Iceland, the Mid-Atlantic Ridge has actually grown above sea level. Throughout most of its length, however, this divergent boundary lies 2500 meters below sea level.

Although volcanic structures do contribute to the height of a ridge, the warm, buoyant nature of the intruding magma is the primary reason for its ele-

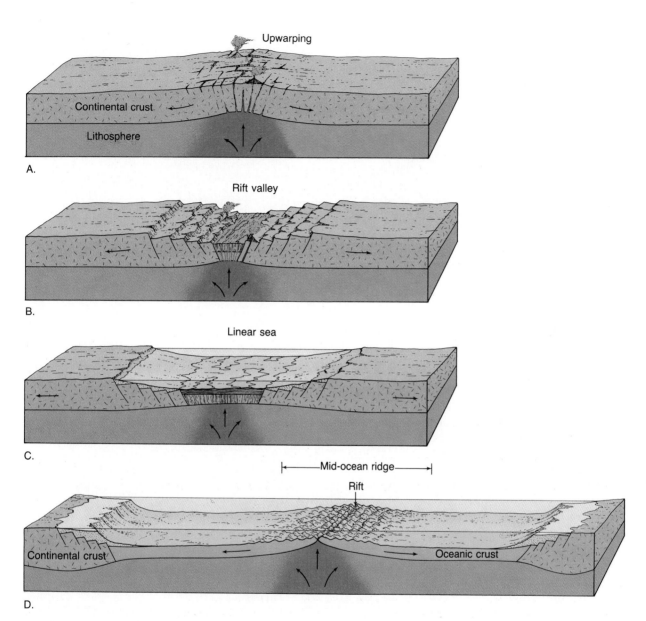

FIGURE 13.6 **A.** Rising magma upwarps the crust, causing numerous cracks in the rigid lithosphere. **B.** As the crust is pulled apart, large slabs of rock sink, generating a rift zone. **C.** Further spreading generates a narrow sea. **D.** Eventually, an expansive ocean basin and ridge system are created.

vated position. As the newly formed lithosphere travels away from the spreading center, it gradually cools and contracts. This thermal contraction accounts in part for the greater ocean depths that exist away from the ridge. Almost 100 million years must pass before cooling and contraction cease completely. By this time rock that was once a part of the majestic ocean mountain system becomes part of the deep-ocean basin.

Convergent Boundaries

At spreading centers new lithosphere is continually being generated; however, since the total surface area of the earth remains constant, lithosphere must also be destroyed. The zones of plate convergence are the sites of this destruction. When two plates collide, the leading edge of one is bent downward, allowing it to descend beneath the other. Upon

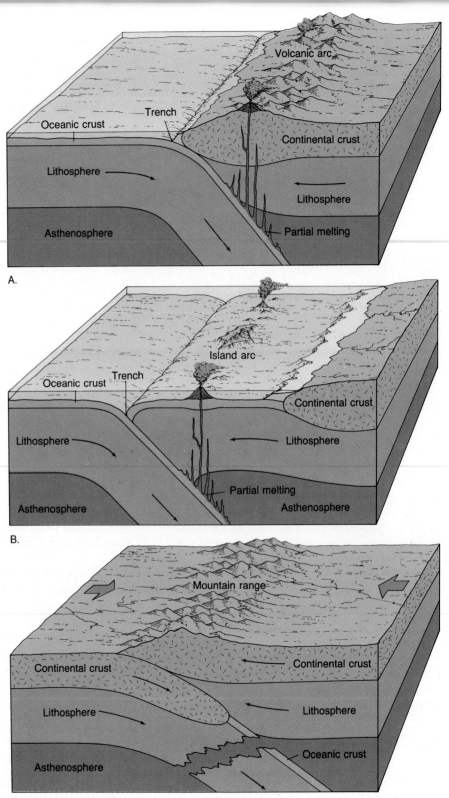

A.

B.

C.

FIGURE 13.7 Zones of plate convergence. **A.** Oceanic-continental. **B.** Oceanic-oceanic. **C.** Continental-continental.

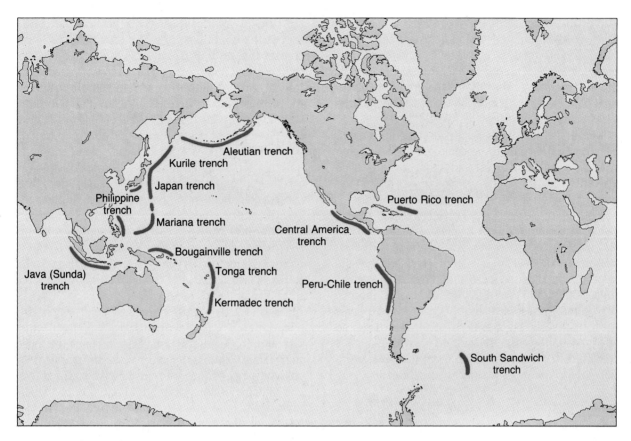

FIGURE 13.8 Distribution of the world's major oceanic trenches.

entering the hot asthenosphere, the plunging plate begins to warm and loses its rigidity. Generally the descending plate is relatively cold and approaches 100 kilometers in thickness. Thus, depending upon its angle of descent, it may reach a depth of 700 kilometers before its leading edge is completely assimilated within the material of the upper mantle.

Although all convergent zones are basically similar, the nature of plate collisions is influenced by the type of crustal material involved. Collisions can occur between two oceanic plates, one oceanic and one continental plate, or two continental plates, as shown in Figure 13.7. Whenever the leading edge of a plate capped with continental crust converges with oceanic crust, the less dense continental material apparently remains "floating," while the more dense oceanic slab sinks into the asthenosphere. The region where an oceanic plate descends into the asthenosphere because of convergence is called a **subduction zone**. As the oceanic plate slides beneath the overriding plate, the oceanic plate bends,

thereby producing a **deep-ocean trench** adjacent to the zone of subduction (Figure 13.7A). Trenches formed in this manner may be thousands of kilometers long and 8 to 11 kilometers deep (Figure 13.8).

Oceanic-Continental Convergence. During a collision between an oceanic slab and a continental block, the oceanic crust is bent, permitting its descent into the asthenosphere (Figure 13.7A). The angle at which the plate descends is usually 45 degrees, or greater. Upon entering the hot asthenosphere, the downward moving plate and the water-soaked sediments carried upon it begin to melt. The newly formed magma created in this manner is less dense than the surrounding mantle rocks and, consequently, when sufficient quantities have gathered, the molten rock will slowly rise. Most of the rising magma will be emplaced in the overlying continental crust, where it will cool and crystallize at a depth of several kilometers. The remaining magma will eventually migrate to the surface where it can give rise to numerous and occasionally explosive volcanic eruptions. The volcanic Andes Mountains are believed to have been produced by such activity when the Nazca plate melted as it plunged beneath the continent of South America (Figure 13.4). The frequent earthquakes that occur within the Andes testify to the activity beyond our view.

Mountains such as the Andes that are believed to be produced in part by volcanic activity associated with the subduction of oceanic lithosphere are called **volcanic arcs**. Two of these volcanic arcs exist in the western United States. One, the Cascade Range, is composed of several well-known volcanic mountains, including Mounts Rainier, Shasta, and St. Helens. The second is the Sierra Nevada, in which Yosemite National Park is located. The Sierra Nevada system is the older of the two and has been inactive for several million years as evidenced by the absence of volcanic cones. Here erosion has stripped away most of the obvious traces of volcanic activity and left exposed the large, crystallized magma chambers that once fed lofty volcanoes. As the recent eruptions of Mount St. Helens testify, the Cascade Range is still quite active. The magma here arises from the melting of a small remaining segment of the Juan de Fuca plate.

Oceanic-Oceanic Convergence. When two oceanic slabs converge, one descends beneath the other, initiating volcanic activity in a manner similar to that which occurs at an oceanic-continental convergent boundary. However, in this case, the volcanoes form on the ocean floor rather than on the continents (Figure 13.7B). If this volcanic activity is sustained, dry land will eventually emerge from the ocean depths. In the early stages of development, this newly formed land consists of a chain of small volcanic islands called an **island arc**. The Aleutian, Mariana, and Tonga islands exemplify such features. Island arcs such as these are generally located a few hundred kilometers from an ocean trench where active subduction of the lithosphere is occurring. Adjacent to the island arcs just mentioned are the Aleutian trench, Mariana trench, and the Tonga trench, respectively.

Over an extended period, numerous episodes of volcanic activity build large volcanic piles on the ocean floor. This volcanic activity, plus the buoyancy of the intrusive igneous rock emplaced within the crust below, gradually increase the size and elevation of the developing arc. This growth, in turn, increases the amount of eroded sediments added to the sea floor. Some of these sediments reach the trench and are deformed and metamorphosed by the compressional forces exerted by the two converging plates. The result of these diverse activities is the development of a mature island arc composed of a complex system of volcanic rocks, folded and metamorphosed sedimentary rocks, and intrusive igneous rocks. Examples of mature island arc systems are the Alaskan Peninsula, the Philippines, and Japan.

Continental-Continental Convergence. When two plates carrying continental crust collide, a continental collision occurs (Figure 13.7C). This is thought to take place because of the light composition, and thus buoyant nature, of continental rocks. Such a collision is believed to have happened when the once-separated continent of India "rammed" into Asia and produced the Himalayas, perhaps the most spectacular mountain range on earth. During this collision, the continental crust buckled, fractured, and was generally shortened. In addition to the Himalayas, several other complex mountain systems, including the Alps, Appalachians, and Urals, are thought to have formed in this manner.

Prior to a continental collision, the landmasses involved are separated by the oceanic crust formed during an earlier episode of sea-floor spreading. As the continental blocks converge, the intervening sea floor is subducted beneath one of the plates. The partial melting of the descending oceanic slab and the sediments carried with it generate a volcanic arc. Depending on the location of the subduction zone, the volcanic arc could develop on either of the converging landmasses, or if the subduction zone developed at an appreciable distance into the ocean, an island arc would form. In any case, erosion of the newly formed volcanic arc would add large quantities of sediment to the already sediment-laden continental margins. Eventually, as the intervening sea floor was consumed, these continental masses would collide, thereby squeezing, folding, and generally deforming the sediments as if they were placed in a gigantic vise. The result would be the formation of a new mountain range composed of deformed sedimentary rocks and fragments of the volcanic arc.

Transform Faults

The third type of plate boundary is the transform fault, where plates slide past one another without the production of crust, as occurs along oceanic ridges, or without the destruction of crust, as occurs at subduction zones. Transform faults roughly parallel the direction of plate movement and were first identified where they join segments of the oceanic ridge system (see Figure 13.4).

The nature of transform faults was proposed in 1965 by J. Tuzo Wilson of the University of Toronto. Wilson suggested that these large fractures connected convergent and divergent plate boundaries into a continuous network that divides the earth's outer shell into several rigid plates. In this role, transform faults provide the means by which the oceanic crust created at the ridge crests can be transported to its site of destruction: the deep-ocean trenches. Figure 13.9 illustrates this activity. Notice that the Juan de Fuca plate moves in a southeasterly direction, eventually being subducted beneath the western margin of the United States. The southern end of this relatively small plate is bounded by the Mendocino escarpment. This transform fault boundary connects an active spreading center to a subduction zone. Therefore, the fault facilitates the movement of the crustal material created at the ridge crest to its destination beneath the North American continent.

Wilson called these special faults *transform faults* because the relative motion of the plates can be changed, or transformed, along them. As we saw in the preceding example, divergence occurring at a spreading center can be transformed into convergence at a subduction zone. Since transform faults connect convergent and divergent boundaries in various combinations, other changes in relative plate motion are possible along transform faults.

Most transform faults are located in oceanic crust, but a few, including the San Andreas fault in California, are situated within continents (Figure 13.9). Along the San Andreas fault, the Pacific plate is moving toward the northwest, past the North American plate. If this movement continues for millions of years, that part of California west of the fault zone, including the Baja Peninsula, will become an island off the west coast of the United States and Canada, and could eventually reach Alaska. However, the

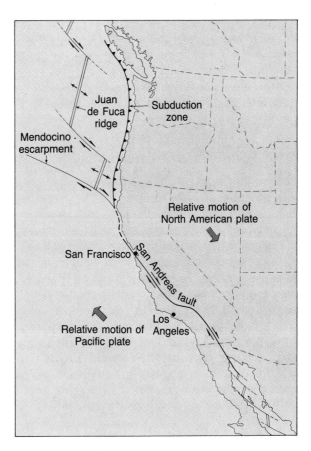

FIGURE 13.9 The role of transform faults in permitting relative motion between adjacent plates. The Mendocino escarpment permits sea floor generated at the Juan de Fuca ridge to move southeastward past the Pacific plate.

more immediate concerns are the earthquakes triggered by movements along this fault system.

TESTING THE MODEL

As the plate tectonics theory was being developed, an avalanche of data was compiled to test this revolutionary idea. The model just presented will surely be modified to fit this wealth of data; however, the basic premises appear able to withstand the test of time.

Although most geologists have accepted this theory energetically, there remains an ever-diminishing number who reject it in part or in total. Some of the evidence supporting continental drift and sea-floor spreading has already been presented in this chapter. In addition, some of the evidence which was instrumental in solidifying the support for this new concept follows. It should be pointed out that some of this evidence was not new, rather it was a new interpretation of old data that swayed the tide of opinion.

Plate Tectonics and Paleomagnetism

Probably the most persuasive evidence to the geologic community for the acceptance of the plate tectonics theory comes from the study of the earth's magnetic field. Anyone who has used a compass to find direction knows that the earth's magnetic field has a north pole and a south pole. These magnetic poles align closely, but not exactly, with the respective geographic poles. In many respects the earth's magnetic field is very much like that produced by a simple bar magnet. Invisible lines of force pass through the earth and extend from one pole to the other. A compass needle, itself a small magnet free to move about, becomes aligned with these lines of force and thus points toward the magnetic poles.

The technique used to study ancient magnetic fields relies on the fact that certain rocks contain minerals which serve as fossil compasses. These iron-rich minerals such as magnetite are abundant, for example, in lava flows of basaltic composition. When heated above a certain temperature called the **Curie point**, these magnetic minerals lose their magnetism. However, when these iron-rich grains cool below their Curie point (about 580°C) they become magnetized in the direction parallel to the existing magnetic field. Once the minerals solidify, the magnetism they possess will remain ''frozen'' in this position. In this regard, they behave much like a compass needle inasmuch as they ''point'' toward the existing magnetic poles. If the rock is moved or the magnetic pole changes position, the rock magnetism will, in most instances, retain its original alignment. Rocks formed thousands or millions of years ago thus ''remember'' the location of the magnetic poles at the time of their formation and are said to possess fossil magnetism, or **paleomagnetism**.

Polar Wandering. A study of lava flows conducted in Europe in the 1950s led to an interesting discov-

ery. The magnetic alignment in the iron-rich minerals in lava flows of different ages was found to vary widely. A plot of the position of the magnetic north pole through time revealed that during the past 500 million years the position of the pole had gradually wandered from a location near Hawaii northward through eastern Siberia and finally to its present location (Figure 13.10). This was clear evidence that either the magnetic poles had migrated through time, an idea known as **polar wandering**, or the continents had drifted.

Although the magnetic poles are known to move, studies of the magnetic field indicate that the average positions of the magnetic poles correspond closely to the positions of the geographic poles. This is consistent with our knowledge of the earth's magnetic field, which is generated in part by the rotation of the earth about its axis. If the geographic poles do not wander appreciably, which we believe is true, neither can the magnetic poles. Therefore, a more acceptable explanation for the apparent polar wandering is provided by the plate tectonics theory. If the magnetic poles remain stationary, their apparent movement can be produced by moving the continents.

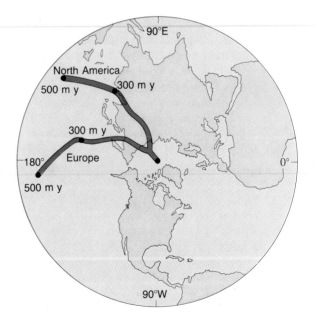

FIGURE 13.10 Simplified apparent polar wandering paths for North America and Europe. If these landmasses are brought together to close the North Atlantic, the paths roughly coincide.

Further evidence for plate tectonics came a few years later when a polar wandering curve was constructed for North America (Figure 13.10). To nearly everyone's surprise the curves for North America and Europe had similar paths, except that they were separated by about 30 degrees of longitude. When these rocks solidified, could there have been two magnetic north poles which migrated parallel to each other? This is very unlikely. The differences in these migration paths, however, can be reconciled if the two presently separated continents are placed next to one another, as we now believe they were prior to opening of the Atlantic Ocean.

Magnetic Reversals. Another revelation in the field of paleomagnetism came when geophysicists learned that the earth's magnetic field periodically reverses polarity; that is, the north magnetic pole becomes the south magnetic pole and vice versa. The cause of these reversals is apparently linked to the fact that the earth's magnetic field changes intensity. Recent calculations, for example, indicate that the magnetic field has weakened about 5 percent over the past century. If this trend continues for the next thousand years or so, we might expect the earth's magnetic field to become very weak or nonexistent. During periods when the earth's magnetic field is very weak, some external influence such as sunspot activity could possibly contribute to a reversal of polarity. After a reversal has taken place, the field would rebuild itself with opposite polarity. A rock solidifying during one of the periods of reverse polarity will be magnetized with the opposite polarity of those rocks being formed today. When rocks exhibit the same magnetism as the present magnetic field, they are said to possess **normal polarity**,

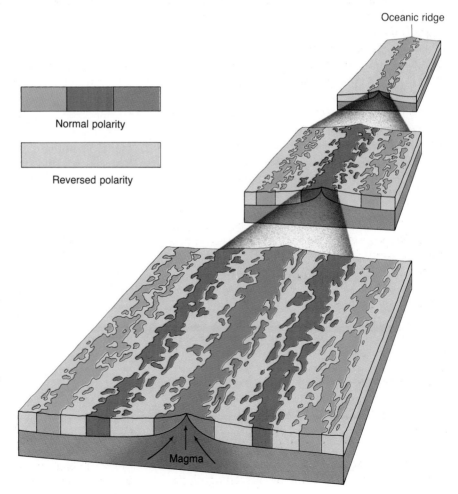

Normal polarity

Reversed polarity

Oceanic ridge

Magma

FIGURE 13.11 New sea floor records the polarity of the magnetic field at the time it formed. Hence it behaves much like a tape recorder, as it continually keeps count of changes in the earth's magnetic field.

while those rocks exhibiting the opposite magnetism are said to have **reverse polarity**. Using the potassium-argon method of radiometric dating, the polarity of the earth's magnetic field has been reconstructed for a period of several million years.

A significant relationship between magnetic reversals and the sea-floor spreading hypothesis was developed from data obtained when very sensitive instruments called *magnetometers* were towed by research vessels across a segment of the ocean floor located off the west coast of the United States (Figure 13.11). Here workers from the Scripps Institution of Oceanography discovered alternating strips of high- and low-intensity magnetism which trended in roughly a north-south direction. This relatively simple pattern of magnetic variation defied explanation until 1963, when Fred Vine and D. H. Matthews tied the discovery of the high- and low-intensity strips to the concept of sea-floor spreading. Vine and Matthews suggested that the strips of high-intensity magnetism are regions where the paleomagnetism of the ocean crust is of the normal type. Consequently, these positively magnetized rocks enhance the existing magnetic field. Conversely, the low-intensity strips represent regions where the ocean crust is polarized in the reverse direction and, therefore, weaken the existing magnetic field. But how do parallel strips of normally and reversely magnetized rock become distributed across the ocean floor?

Vine and Matthews reasoned that as new basalt was added to the ocean floor at the oceanic ridges, it would be magnetized according to the existing magnetic field. Since new rock is added in approximately equal amounts to the trailing edges of both plates, we should expect strips of equal size and polarity to parallel both sides of the ocean ridges as shown in Figure 13.11. This explanation of the alternating strips of normal and reverse polarity, which lay as mirror images across the ocean ridges, was the strongest evidence so far presented in support of the concept of sea-floor spreading.

Plate Tectonics and Earthquakes

By 1968, the basic outline of plate tectonics was firmly established. In this same year, three Lamont-Doherty Observatory seismologists, B. Isacks, J. Oliver, and L. R. Sykes, published papers demonstrating how much more successful the new plate tectonics model was than older models in accounting for the global distribution of earthquakes (Figure 13.12). In particular these seismologists were able to explain the distribution of earthquakes that are associated with deep-ocean trenches. For example, examine the distribution of earthquakes in the vicinity of the Japan trench just to the east of Japan (Figure 13.12). Here most shallow-focus earthquakes occur within, or adjacent to, the trench, whereas intermediate- and deep-focus earthquakes occur toward the mainland. A similar distribution pattern exists along the western margin of South America where the Nazca plate is being subducted beneath the continent.

In the plate tectonics model, deep-ocean trenches are produced where slabs of oceanic lithosphere plunge into the mantle (Figure 13.13). Very shallow focus earthquakes are generated at the outer edge of the trenches where the plate is bent to permit its descent. Shallow-focus earthquakes are also produced as the descending plate scrapes beneath the overriding slab. As the slab continues to descend into the asthenosphere, deeper-focus earthquakes are generated. The deep-focus earthquakes appear to originate within the descending plate and are caused by resistance to the plate's downward movement. Since the earthquakes occur within the rigid lithosphere rather than within the mobile asthenosphere, they provide a method for tracking the plate's descent into the mantle. Very few earthquakes have been recorded below 700 kilometers (435 miles), possibly because the lithosphere has been completely assimilated into the mantle by the time it reaches this depth.

The plate tectonics model also explains why deep-focus earthquakes are confined to areas adjacent to oceanic trenches (subduction zones), while earthquakes which occur along divergent boundaries and transform fault boundaries have only shallow foci. Recall that earthquakes result from the rapid release of strain that can only build up in the rigid material of the lithosphere. Since subduction zones are the only regions where rigid material is forced to great depths, these should be the only sites of deep-focus earthquakes. Indeed, the absence of deep-focus earthquakes along divergent boundaries and transform faults also substantiates the plate tectonics theory.

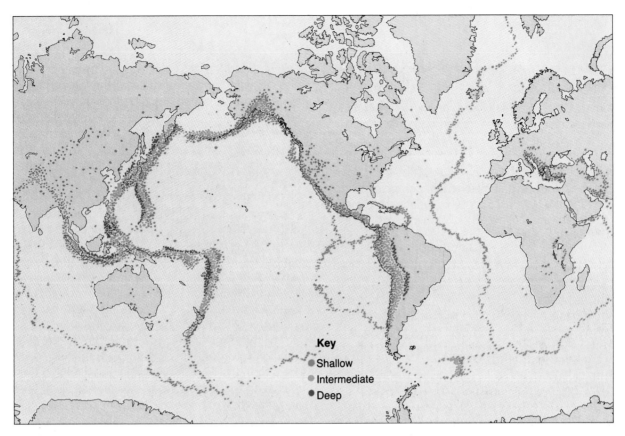

FIGURE 13.12 Distribution of shallow- , intermediate- , and deep-focus earthquakes. (Data from NOAA)

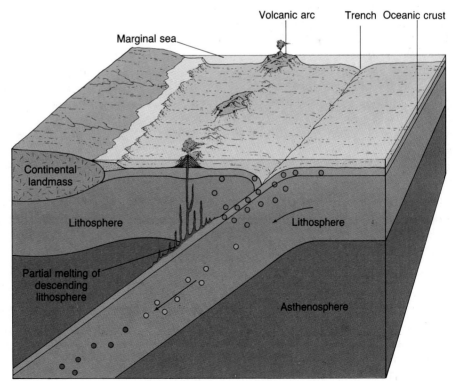

FIGURE 13.13 Relationship between the descending plate and depth of earthquake foci.

249

Evidence from the Deep Sea Drilling Project

Some of the most convincing evidence confirming plate tectonics has come from drilling into the sediment on the ocean floor. The source of these important data was the Deep Sea Drilling Project, a program begun in the late 1960s under the joint sponsorship of several major oceanographic institutions and the National Science Foundation.* The primary goal was (and indeed still is) to gather firsthand information about the age and processes of ocean basin formation. To accomplish this, a new drilling ship, the *Glomar Challenger*, was built (Figure 13.14). The *Glomar Challenger* represented a significant technological breakthrough, because this ship was capable of lowering drill pipe thousands of meters to the ocean floor and then drilling hundreds of meters into the sediments and underlying basaltic crust.

At several sites, holes were drilled through the entire thickness of sediments to the basaltic rock below. An important objective was to gather samples of sediment from just above the igneous crust as a means of dating the sea floor at each site.* Since sedimentation begins immediately after the oceanic crust forms, fossils found in the oldest sediments (that is, those resting directly above the basalt) can be used to date the ocean floor at that site. When the oldest sediment from each drill site was plotted against its distance from the ridge crest, it was revealed that the age of the sediment increased with increasing distance from the ridge.

The data from the Deep Sea Drilling Project also reinforced the idea that the ocean basins are geologically youthful. To date, no sediment with an age in excess of 160 million years has been found. By comparison, some continental crust has been dated at more than 3.8 billion years.

The thickness of ocean-floor sediments provided additional verification of sea-floor spreading. Drill cores from the *Glomar Challenger* revealed that sediments on the ridge crest are almost entirely absent and that the sediment thickens with increasing dis-

*In 1984 the National Science Foundation began a new program to replace the Deep Sea Drilling Project. The new ten-year project is called the Ocean Drilling Program and will use a new drill ship with capabilities beyond those of the *Glomar Challenger*.

*Radiometric dates of the ocean crust itself are unreliable because of the alteration of basalt by seawater.

A.

B.

FIGURE 13.14 A. *Glomar Challenger*, the drilling ship of the Deep Sea Drilling Project. This pioneering vessel was a significant technological breakthrough because it was capable of lowering drill pipe thousands of meters to the ocean floor and then drilling hundreds of meters into the sediments and underlying crust. **B.** Amidships is the towering 42-meter- (140-foot-) high derrick. This is a view looking straight down from near the top of the derrick. (Photos courtesy of Victor S. Sotelo, Deep Sea Drilling Project)

tance from the ridge. Since the ridge crest is younger than the areas farther away from it, this pattern of sediment distribution is exactly what was predicted by the theory of plate tectonics.

Hot Spots

Another line of evidence was developed when mapping of seamounts in the Pacific revealed a chain of volcanic structures extending from the Hawaiian Islands to Midway Island and then continuing north-

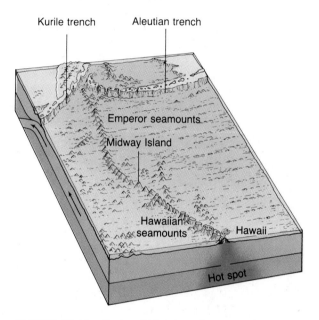

FIGURE 13.15 Chain of islands and seamounts extending from Hawaii to the Aleutian trench.

ward toward the Aleutian trench. Potassium-argon dating of 27 volcanoes in this chain revealed an increase in age with an increase in distance from Hawaii. Suiko Seamount, which is located near the Aleutian trench, is 65 million years old, Midway Island is 27 million years old, and the island of Hawaii rose from the sea less than 1 million years ago (Figure 13.15).

Researchers have proposed that a **hot spot** exists within the mantle and emits magma onto the overlying sea floor. Presumably, as the Pacific plate moved over the hot spot, successive volcanic structures emerged. The age of each volcano indicates the time when it was situated over the relatively stationary hot spot. Kauai is the oldest of the large islands in the Hawaiian chain. Five million years ago, when it was positioned over the hot spot, Kauai was the only Hawaiian Island in existence (Figure 13.16). Evidence of the age of Kauai can be seen by examining the extinct volcanoes which have been eroded into jagged peaks and vast canyons. By contrast, the south slopes of the island of Hawaii consist of fresh lava flows and two of Hawaii's volcanoes, Mauna Loa and Kilauea, remain active. Recent evidence indicates that a new volcanic pile is forming on the ocean floor just off the coast of Hawaii. Geologically speaking, it should not be long before another tropical island will be added to the Hawaiian chain.

PANGAEA: BEFORE AND AFTER

Robert Dietz and John Holden have rather precisely projected the gross details of the migrations of

FIGURE 13.16 Radiometric dating of the Hawaiian Islands revealed the decreasing age of the volcanic activity toward Hawaii. (Data from I. McDougall)

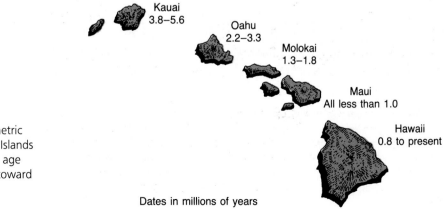

Dates in millions of years

individual continents over the past 500 million years. By extrapolating plate motion back in time using such evidence as the orientation of volcanic structures left behind on moving plates, the distribution and movements of transform faults, and paleomagnetism, Dietz and Holden were able to reconstruct Pangaea (see Figure 13.1). The use of radiometric dating helped them establish the time frame for the formation and eventual breakup of Pangaea, and the relatively stationary positions of hot spots through time helped to fix the locations of the continents.

Breakup of Pangaea

The fragmentation of Pangaea began about 200 million years ago. Figure 13.17 illustrates the breakup and subsequent paths taken by the landmasses involved. As we can readily see in Figure 13.17A, two major rifts initiated the breakup. The rift zone between North America and Africa generated numerous outpourings of Triassic-age basalts which are presently visible along the eastern seaboard of the United States. Radiometric dating of these basalts indicates that rifting occurred between 200 and 165 million years ago. This date can be used as the birth date of this section of the North Atlantic. The rift that formed in the southern landmass of Gondwanaland developed a ''Y''-shaped fracture which sent India on a northward journey and simultaneously separated South America-Africa from Australia-Antarctica.

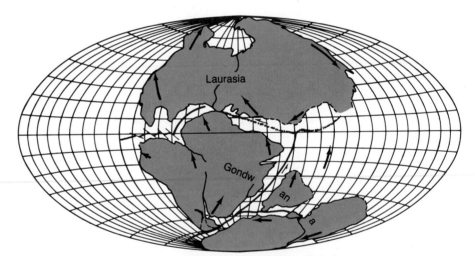

A. 180 Million Years Ago (Triassic Period)

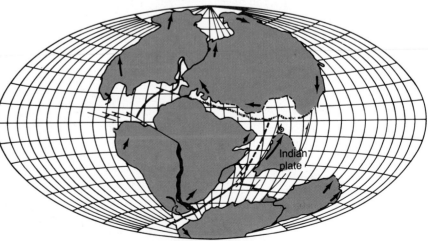

FIGURE 13.17 Several views of the breakup of Pangaea over a period of 200 million years according to Dietz and Holden. (Robert S. Dietz and John C. Holden. *Journal of Geophysical Research* 75: 4939–56, 1970. Copyright © by American Geophysical Union)

B. 135 Million Years Ago (Jurassic Period)

Figure 13.17B illustrates the position of the continents 135 million years ago, about the time Africa and South America began splitting apart to form the South Atlantic. India can be seen halfway into its journey to Asia, and the southern portion of the North Atlantic has widened considerably. By the end of the Cretaceous period, about 65 million years ago, Madagascar had separated from Africa, and the South Atlantic had emerged as a full-fledged ocean (Figure 13.17C).

The current map (Figure 13.17D) shows India in contact with Asia, an event that occurred about 45 million years ago and created the highest mountains on earth, the Himalayas, along with the Tibetan Highlands. By comparing Figures 13.17C and 13.17D, we can see that the separation of Greenland from Eurasia was a recent event in geologic history. Also notice the recent formation of the Baja Peninsula and the Gulf of California. This event is thought to have occurred less than 10 million years ago.

Before Pangaea

Prior to the formation of Pangaea, the landmasses had probably gone through several episodes of fragmentation similar to what we see happening today. Also like today, these ancient continents moved away from each other only to collide again at some other location. During the period between 500 and

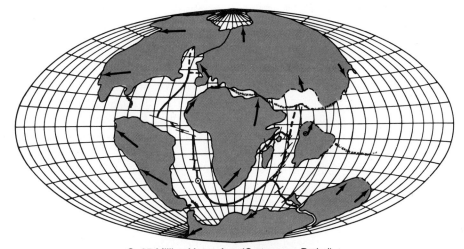

C. 65 Million Years Ago (Cretaceous Period)

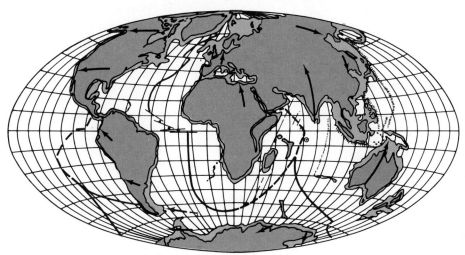

D. Present

225 million years ago, the fragments of an earlier dispersal began collecting to form the continent of Pangaea. Evidence of these earlier continental collisions include the Ural Mountains of the Soviet Union and the Appalachians of North America.

A Look into the Future

After Dietz and Holden drew together the events that over the past 500 million years resulted in the current configuration of the continents, they went one step further and extrapolated plate motion into the future. Figure 13.18 illustrates what they envision the earth's landmasses will look like 50 million years from now. Important changes are seen in Africa where a new sea emerges as East Africa parts company with the mainland. In North America we see that the Baja Peninsula and the portion of southern California that lies west of the San Andreas fault have slid past the North American plate. If this northward migration takes place as predicted, Los Angeles and San Francisco will pass each other.

These projections into the future, although interesting, must be viewed with caution since many assumptions must be correct for these events to unfold as just described. Nevertheless, similar types of changes in the shapes and positions of the continents will undoubtedly occur for many millions of years to come.

THE DRIVING MECHANISM

The plate tectonics theory describes plate motion and the effects of this motion. Therefore, acceptance does not rely on a knowledge of the force or forces moving the plates. This is fortunate, since none of the driving mechanisms yet proposed can account for all of the major facets of plate motion. Nevertheless, the unequal distribution of heat within the earth is accepted by most as the underlying cause of plate movement. The unequal distribution of heat, in turn, is thought by many geologists to generate large convection cells within the mantle (Figure 13.19A,B). The warm, less dense material of the lower mantle rises very slowly in the regions of oceanic ridges. As the material spreads laterally it cools, becomes more dense, and begins to sink back into the mantle, only to be reheated. Note that rocks need not be molten to flow. Just as a hot, solid metal can be pounded into different shapes, so too can rock move when subjected to heat and stress over an extended period. Measurements indicate a higher rate of heat flow at the oceanic ridges than in other oceanic regions—a good indication that some type of thermal convection cells might exist. However, many of the details of their motions remain unclear. How many cells exist? At what depth do they originate? What is the structure of these thermal cells?

Although unequal distribution of heat is generally accepted as the underlying driving force of plates,

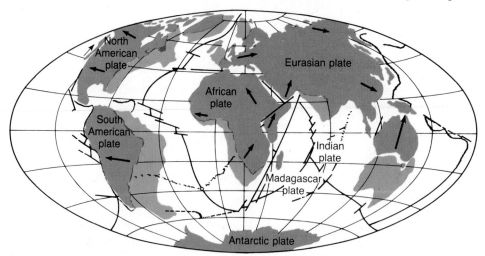

FIGURE 13.18   The world as it may look 50 million years from now. (From "The Breakup of Pangaea," Robert S. Dietz and John C. Holden. Copyright © 1970 by Scientific American, Inc. All rights reserved)

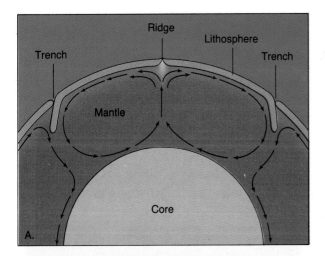
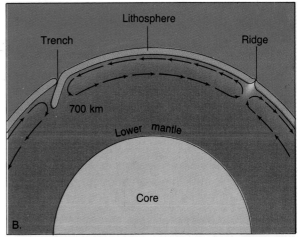
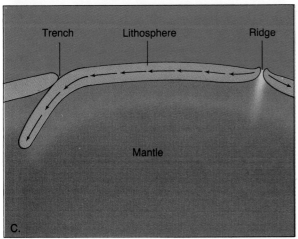
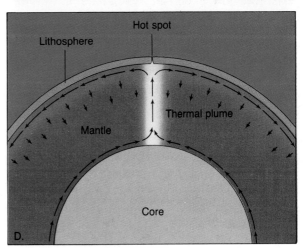

FIGURE 13.19 Proposed models of the driving force for plate tectonics. **A.** Large convection cells in the mantle carry the lithosphere in a conveyor belt fashion. **B.** Convection cells confined to the mobile upper mantle are responsible for plate motion. **C.** Push-pull models are also a type of convection. Here the cold, sinking oceanic slab pulls the trailing sea floor along, while gravity-sliding of material down the elevated ridge crest pushes the slab. **D.** The hot plume model suggests that all upward movement is confined to a few narrow plumes, while downward flow occurs slowly throughout the remaining mantle. (Parts A, B, and D after "The Earth's Mantle," by Peter J. Wyllie, © 1975, by Scientific American, Inc. All rights reserved)

many geologists are not convinced that large convection cells can exist in the mantle. Thus, many other mechanisms have been suggested. One mechanism relies on the fact that the cold oceanic slab has a greater density than the asthenosphere supporting it from below. This being the case, it has been proposed that as a plate begins to descend, the heavy, sinking slab might pull the trailing lithosphere along. This hypothesis is similar to another model which suggests that the elevated position of an ocean ridge could cause the lithosphere to slide under the influence of gravity (Figure 13.19C). These push-pull models are themselves a type of convection current. As the sinking plate enters the mantle, material is forced aside which then migrates toward the ridge systems. The cell is completed as molten rock moves up from the asthenosphere to fill the gap in the diverging oceanic plates.

Some oceans, notably the Atlantic, lack subduction zones; thus, the slab-pull mechanism cannot adequately explain the spreading occurring at these ridges. Other ridge systems are rather subdued, which would reduce the effectiveness of the alternate slab-push model. Perhaps the slab-push and slab-pull phenomena are active in different ridge systems and occasionally even work in tandem.

Another version of the thermal convection model suggests that relatively narrow, hot plumes (hot spots) of mantle rock generate plate motion (Figure 13.19D). These hot plumes are thought to originate near the mantle-core boundary. Upon reaching the lithosphere, these plumes would spread laterally and carry the plates away from the zone of upwelling. The hot plumes usually reveal themselves as volcanic structures growing up from the ocean floor in such places as Iceland. About twenty hot spots have been identified along ridge systems, where they may contribute to plate divergence. Recall, however, that some hot plumes, for example, the one which generated the Hawaiian Islands, are not located in ridge areas. Therefore, we must conclude that this model is not without its shortcomings. Perhaps a combination of all these phenomena generates the plate motion observed.

REVIEW QUESTIONS

1 What was probably the first bit of evidence which led scientists to believe that the continents were once connected?

2 What was Pangaea?

3 List the evidence that Wegener and his associates gathered to support the continental drift hypothesis.

4 Early in this century, what was the prevailing view of how land animals migrated across vast expanses of ocean?

5 On what basis were plate boundaries first established?

6 Where and how is new lithosphere being formed? Destroyed? Why must the production and destruction of lithosphere be going on at approximately the same rate?

7 Why is the oceanic portion of a lithospheric plate destroyed and the continental portion is not?

8 Relate the formation of the Andes Mountains to the movement of plates.

9 Discuss the formation of an island arc.

10 In what ways may the origin of the Japanese Islands be considered similar to the formation of the Andes Mountains? How do they differ?

11 Differentiate between transform faults and the other two types of plate boundaries.

12 Some predict that California will sink into the ocean. Is this idea consistent with the theory of plate tectonics?

13 Describe the distribution of earthquake epicenters and foci depths as they relate to oceanic trench systems.

14 If the "hot spot" concept proves correct, what direction was the Pacific plate moving while the Emperor seamounts were being produced (see Figure 13.15)? While the Hawaiian seamounts were being produced?

15 With what type of plate boundary are the following places or features associated (be as specific as possible): Himalayas, Aleutian Islands, Red Sea, Andes Mountains, San Andreas fault, Iceland, Japan?

CHAPTER FOURTEEN

MOUNTAIN BUILDING

Mountains often are spectacular features which rise several hundred meters or more above the surrounding terrain. Some occur as single isolated masses; the volcanic cone Kilimanjaro, for example, stands almost 6000 meters (20,000 feet) above sea level overlooking the grasslands of East Africa. Others make up a portion of an extensive mountainous chain, such as the American Cordillera, which reaches from the tip of South America northward through Alaska. Chains such as the Himalayas are youthful, gigantic mountains that are still rising, while others are very old and nearly worn down, as exemplified by the Appalachian Mountains in the eastern United States.

The name for the processes which collectively produce a mountain system is **orogenesis**, from the Greek *oros* ("mountain") and *genesis* ("to come into being"). Mountain systems show evidence of enormous forces which have folded, faulted, and generally deformed large sections of the earth's crust. Although the processes of folding and faulting have contributed to the majestic appearance of mountains, much of the credit for their beauty must be given to the work of running water and glacial ice, which sculpture these uplifted masses in an unending effort to lower them to sea level.

The first encompassing explanation of orogenesis came a little more than a decade ago as part of the plate tectonics theory. As noted before, the idea of plates colliding has opened many new and exciting avenues to geologists. Before examining mountain building according to the plate tectonics model however, it will be advantageous first to view the processes of crustal uplifting and rock deformation.

Mount Timpanogos in the Wasatch Range, Utah. (Photo by Stephen Trimble)

CRUSTAL UPLIFT

The fossilized shells of marine invertebrates are often found in mountain regions, an indication that the sedimentary rock composing the mountain was once below sea level. This is rather convincing evidence that some dramatic changes occurred between the time these animals died and when their fossilized remains were discovered. Evidence for crustal uplift such as this is common in the geologic record and is even present in the historical record. For example, Figure 14.1 depicts three columns remaining from a Roman temple. The columns have clam borings to a height of about 6 meters (20 feet), indicating that the land upon which the temple was built submerged and was later partially uplifted. These elevated clam borings might also be explained by a recent change in sea level; however, a similar change in sea level is not recorded at any other location for that same time period. Further evidence for crustal uplift can be found along the coastline of the western United States. When a coastal area remains undisturbed for an extended period, wave action cuts a gently sloping bench. In parts of California, ancient wave-cut benches can now be found as terraces, hundreds of meters above sea level (Figure 14.2). Each terrace represents a period when that area was at sea level. Unfortunately, the reasons for uplift are not always as easy to determine as the evidence for the movements.

We know that the force of gravity must play an important role in determining the elevation of the land. In particular, the less dense lithosphere is believed to float on top of the denser and more easily deformed rocks of the asthenosphere. The concept of a floating lithosphere in gravitational balance is called **isostasy**. Perhaps the easiest way to envision isostatic balance is to compare the lithosphere to floating logs. Imagine two logs, one much thicker

FIGURE 14.1 Remaining columns of the ancient Roman temple of Serapis, Pozzuoli, Italy, in 1836. Clam borings 6 meters (20 feet) above sea level indicate former submergence. (After Charles Lyell, *Principles of Geology*, 10th ed. 1867)

than the other, floating in water. The larger log will float higher in the water than the smaller log. In the same manner, mountainous regions are believed to represent unusually thick sections of the earth's crust, whereas areas of low elevation do not have such crustal thickness. Mountains, like thick logs, not only stand high above the surface, but also extend farther into the supporting material below (Figure 14.3). This fact has been confirmed by seismic and gravitational data.

Carrying this idea one step further, the lithosphere beneath the oceans must be thinner than that of the continents because its elevation is lower. Although this is true, oceanic rocks also have a greater density than continental rocks, another factor contributing to their lower position.

If the concept of isostasy is correct, we should expect that when weight is added to the crust, the crust will respond by subsiding, and that when weight is removed there will be uplifting. (Visualize what happens to a ship as cargo is being loaded and unloaded.) Evidence for this type of movement exists, strongly supporting the theory of *isostatic adjustment.* For example, when Hoover Dam was built in the 1930s, the impounded waters of Lake Mead, and to a lesser degree the millions of tons of sediment collected by it, caused regional subsidence and a marked increase in seismic activity. Another classic example is provided by nature. When continental glaciers occupied portions of North America during the Pleistocene epoch, the added weight of the 3-kilometer-thick masses of ice caused downwarping of the earth's crust. In the 8000 to 10,000 years since the last ice sheets melted, uplifting of as much as 330 meters has occurred in the Hudson Bay region, where the thickest ice had accumulated.

As the foregoing examples illustrate, isostatic adjustment can account for considerable crustal movement. Thus, we can now understand why, as erosion lowers the summits of mountains, the crust will rise in response to the reduced load. The processes of uplifting and erosion will continue until the deeply buried portions of the mountains have reached the same height as the surrounding crust (Figure 14.3). In addition, as the mountains wear down, the weight of the eroded sediment deposited on the adjacent continental margin will cause it to subside.

To summarize, mountains are unusually thick portions of the earth's crust that remain elevated above their surroundings because of isostasy. As erosion removes material, isostatic adjustment gradually raises the mountains in response. Eventually the deepest portions of the mountains are brought up to the shallower depths of the surrounding crust. The question still to be answered is, How do these thick sections of the earth's crust come into existence?

ROCK DEFORMATION

When rocks are subjected to stresses greater than their own strength, they begin to deform, usually by folding or fracturing. It is easy to visualize how individual rocks break, but how are large rock units bent into intricate folds without being appreciably broken

FIGURE 14.2 Wave-cut terraces on San Clemente Island south of Los Angeles, California. Once at sea level, the highest terraces are now about 400 meters (1320 feet) above it. (Photo by John S. Shelton)

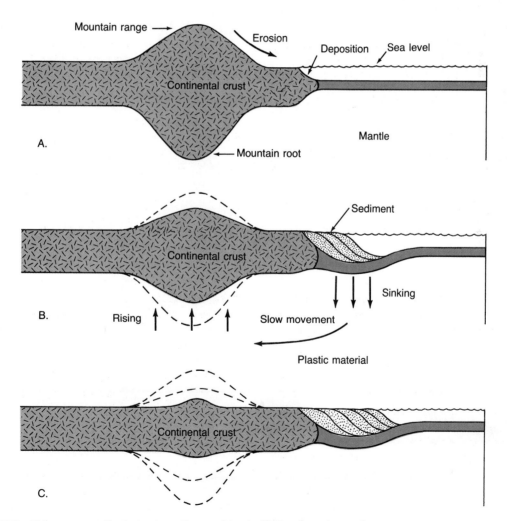

FIGURE 14.3 This sequence illustrates how the combined effects of erosion and isostatic adjustment result in a thinning of the crust in mountainous regions.

during the process? In an attempt to answer this, geologists turned to the laboratory and subjected rocks to stresses while simulating those conditions believed to exist at various depths within the crust. Although all rock types respond somewhat differently to stresses, the general characteristics of rock deformation were determined from these experiments. Geologists discovered that when stress is applied slowly and under low pressure, rocks first respond by deforming elastically. Changes resulting from **elastic deformation** are reversible; that is, like a rubber band, the rock will return to nearly its original size and shape when the stress is removed. However, once the elastic limit is surpassed, rocks either rupture or deform plastically. **Plastic deformation** results in permanent changes; that is, the size and shape of a rock unit are altered through folding and flowing. Laboratory experiments confirmed that at high temperatures and pressures, most rocks deform plastically once their elastic limit is surpassed. Rocks tested under surface conditions also deform elastically, but once they exceed their elastic limit, most behave like a brittle solid and rupture. Recall that the energy for most earthquakes comes from stored elastic energy that is released as rock ruptures and snaps back to its original shape.

One factor that researchers cannot duplicate in the laboratory is geologic time. We know that if stress is applied quickly, as with a hammer, rocks tend to fracture. On the other hand, these same materials may deform plastically if stress is applied over an extended period. For example, marble benches have been known to sag under their own weight over a period of a hundred years or so. In nature, small forces applied over long time periods surely play an important role in the deformation of rock strata.

Faulting

As indicated in our discussion of earthquakes, faults are fractures in the earth's crust along which appreciable movement has taken place. Faults are categorized on the basis of the relative movement between the blocks on both sides of the fault plane. The movement can be horizontal, vertical, or oblique.

Faults having primarily vertical movement are called **dip-slip faults**, since the displacement is in the direction of the inclination, or dip, of the fault plane (Figure 14.4). Dip-slip faults are classified as **normal faults** when the rock above the fault plane moves *down* relative to the rock below (Figure 14.5). **Reverse faults** are created when the rock above the fault plane moves *up* relative to the rock below (Figure 14.6). Reverse faults having a very low angle to the horizontal are also referred to as **thrust faults**. In mountainous regions such as the Alps and the Appalachians, thrust faults have displaced rock as far as 50 kilometers over adjacent strata. Thrust faults of this type result from strong compressional stresses.

Faults in which the dominant displacement is along the trend, or strike, of the fault are called **strike-slip faults**. Many large strike-slip faults are associated with plate boundaries and are called

FIGURE 14.4 Faulting caused the vertical displacement of these sedimentary beds in southern Nevada. (Photo by E. J. Tarbuck)

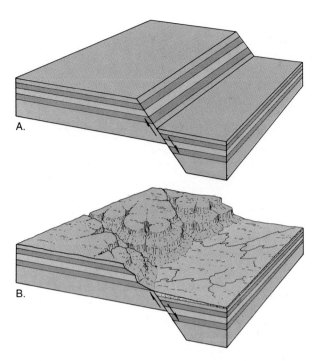

FIGURE 14.5 Block diagrams of a normal fault. **A.** The relative movement of displaced blocks. **B.** How erosion would alter the upfaulted block.

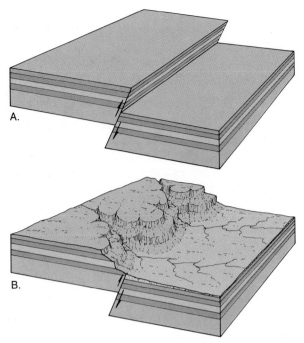

FIGURE 14.6 Block diagrams of a reverse fault. **A.** The relative movement of displaced blocks. **B.** How erosion would alter the upfaulted block.

transform faults. Transform faults have nearly vertical dips and serve to connect large structures such as segments of an oceanic ridge. The San Andreas fault in California is a well-known transform fault in which the displacement has been on the order of several hundred kilometers. When faults have both vertical and horizontal movement, they are called **oblique-slip faults**.

Fault motion provides the geologist with a method of determining the nature of the forces at work within the earth. Normal faults indicate the existence of *tensional stresses* that pull the crust apart. This "pulling apart" can be accomplished either by uplifting that causes the surface to stretch and break, or by horizontal forces that actually rip the crust apart. Normal faulting is known to occur at spreading centers, where plate divergence is prevalent. Here a central block called a **graben** is bounded by normal faults and drops as the plates separate (Figure 14.7). These grabens produce an elongated valley bounded by upfaulted structures called **horsts**. The Great Rift Valley of East Africa consists of several large grabens, above which tilted horsts produce a linear mountainous topography.

This valley, nearly 6000 kilometers (3700 miles) long, contains the excavation sites of some of the earliest human fossils. Other rift valleys include the Rhine Valley in Germany and the valley of the Dead Sea in the Middle East.

Since the blocks involved in reverse and thrust faulting are displaced toward one another, geologists conclude that *compressional forces* are at work. The primary regions of this activity are

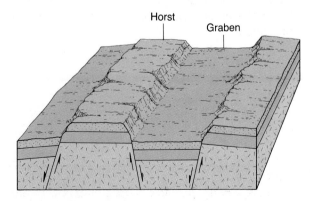

FIGURE 14.7 Diagrammatic sketch of downfaulted block (graben) and upfaulted block (horst).

FIGURE 14.8 Block diagram of principal types of folded strata.

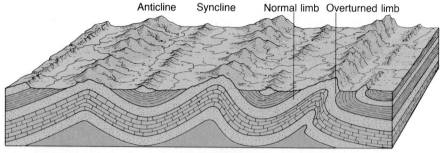

thought to be the convergent zones, where plates are colliding. Compressional forces generally produce folds as well as faults, and result in a general thickening and shortening of the material involved.

Folding

During mountain building, flat-lying sedimentary and volcanic rocks are often bent into a series of broad folds, much like those that would form if you were to hold the ends of a sheet of paper and then push them together. Such folding shortens and thickens the crust. Experimental evidence shows that when sedimentary rocks are buried deep within the crust, where confining pressures are great, strata can be deformed into very tight folds stacked one atop the other, without appreciable fracturing. Figure 14.8 illustrates some common folded structures. Linear upfolded forms are commonly called **anticlines**, whereas downfolded structures are typically referred to as **synclines**. Depending upon their

orientation, anticlines and synclines are said to be *symmetrical, asymmetrical,* or *overturned* if one limb has been tilted beyond the vertical.

Folds do not continue on forever; rather, their ends die out much like the wrinkles in cloth. Some folds are said to be *plunging,* since the axis of the fold is plunging into the ground (Figure 14.9). Figure 14.10 shows some examples of plunging folds and the pattern produced when erosion removes the upper layers of these structures and exposes their interiors. Note that the outcrop pattern of an anticline points in the direction it is plunging, while the opposite is true for synclines. A good example of the kind of topography that results when erosional forces attack folded sedimentary strata is found in the Valley and Ridge Province of the Appalachians. Here resistant sandstone beds remain as imposing ridges separated by valleys cut into more easily eroded shale or limestone beds.

Although most folds are caused by compressional stresses that squeeze and crumble strata, some folds are a consequence of vertical displacement. When

FIGURE 14.9 Sheep Mountain, a doubly plunging anticline. Note that erosion has cut the flanking sedimentary beds into low ridges that make a ''V'' pointing in the direction of plunge. (Photo by John S. Shelton)

FIGURE 14.10 Plunging folds. **A.** Idealized view. **B.** View after extensive erosion.

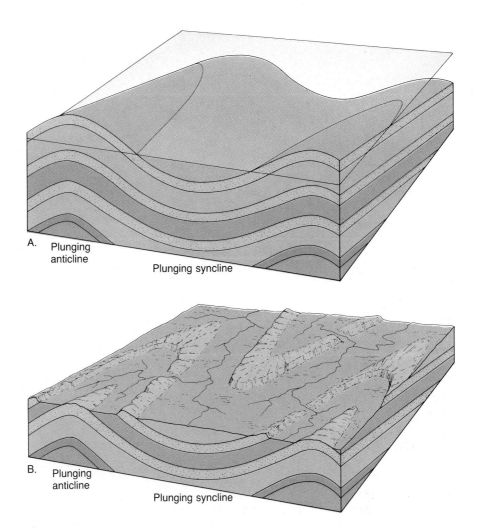

A. Plunging anticline

Plunging syncline

B. Plunging anticline

Plunging syncline

upwarping produces a circular or somewhat elongated structure, the feature is called a **dome**. Downwarped structures having a similar shape are termed **basins**. The Black Hills of western South Dakota is one such domal structure in which erosion has stripped away the upwarped sedimentary beds, exposing older igneous and metamorphic rocks in the center. Several basins also exist in the United States. The basins of Michigan and Illinois have very gently sloping beds similar to saucers. Because large basins contain sedimentary beds sloping at such low angles, they are usually identified by the age of the rocks composing them. The youngest rocks are found near the center and the oldest rocks are at the flanks. This is just the opposite order of a domal structure such as the Black Hills, where the oldest rocks form the core.

MOUNTAIN TYPES

Even though no two mountain ranges are exactly alike, they can be classified according to their most dominant characteristics. Using this approach, four main categories of mountains result: (1) Folded mountains (complex mountains); (2) Volcanic mountains; (3) Fault-block mountains; and (4) Upwarped mountains. Mountain ranges of the same type are commonly found in close proximity forming a mountain system. For example, nearly the entire state of Nevada is composed of numerous elongated fault-block mountains that are separated by faulted basins. Further, within any mountainous belt, such as that portion of the American Cordillera in the western United States, mountain ranges representing each of these groups can be found. In addition to these basic varieties, some regions have mountainous topography that was produced without appreciable crustal deformation. For example, plateaus (areas of high-standing rocks that are essentially horizontal) can be deeply dissected into rugged terrains. Although these highlands have the topographical expression of mountains, they lack the structure associated with orogenesis.

In the following sections we will examine three of the four basic mountain types. Volcanic mountains are treated in detail in Chapter Three.

Folded Mountains

The largest and most complex mountain systems are the so-called **folded mountains**. Although folding is often more conspicuous, faulting, metamorphism, and igneous activity are always present in varying degrees. All major mountain belts, including the Alps, Urals, Himalayas, and Appalachians, are of this type. Since folded mountains represent the world's major mountain systems, the process of mountain building is usually described in terms of their forma-

tion. Thus a separate section on orogenesis is devoted to the evolution of these often majestic and always complex mountain systems.

Fault-Block Mountains

Fault-block mountains are generated when large rock units are displaced, often accompanied by tilting, along high-angle normal faults. Recall that tensional stresses, which stretch crustal material, produce normal faults and the displacement along them.

Excellent examples of fault-block mountains are found in the Basin and Range Province, a region that encompasses Nevada and portions of Utah, New Mexico, Arizona, and California. Here the crust has literally been broken into hundreds of pieces, giving rise to nearly parallel mountain ranges that average about 80 kilometers in length and rise precipitously above the adjacent sediment-laden basins.

The most widely accepted proposal for fault-block mountain formation in the Basin and Range Province

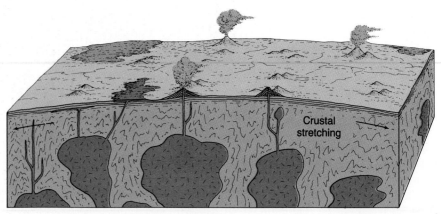

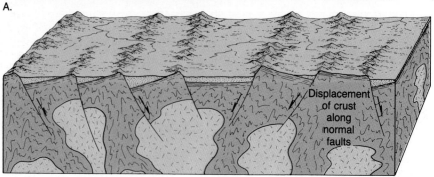

FIGURE 14.11 Formation of the fault-block mountains of the Basin and Range Province. **A.** Tensional forces extended and faulted the rocks, giving rise to a period of volcanism. **B.** Continued extension displaced rock units along numerous high-angle faults.

comes from the plate tectonics theory. According to this model, the nature of the boundary between the Pacific plate and the North American plate changed about 30 million years ago. The relative movement along this boundary changed, and the forces acting upon the region became tensional rather than compressional. The resultant tensional forces extended (stretched) and faulted the Basin and Range rock (Figure 14.11A). This deformation, in turn, gave rise to a period of volcanism. Although the igneous activity has since waned, continued extension has displaced large rock units along high-angle normal faults (Figure 14.11B).

Other fault-block mountains in the United States are the Teton Range of Wyoming and the Sierra Nevada of California. Both are faulted along their eastern flanks, which were uplifted as the blocks tilted downward to the west. Looking west from Jackson Hole, Wyoming, and Owens Valley, California, respectively, the eastern fronts of these ranges rise over 2 kilometers, making them two of the most precipitous mountain fronts in the United States.

Upwarped Mountains

Upwarped mountains are possibly the most diverse mountain type. Some, such as the Black Hills in western South Dakota and the Adirondack Mountains in upstate New York, consist of older igneous and metamorphic bedrock that was once eroded flat and subsequently mantled with sediment. As the regions were upwarped, erosion removed the veneer of sedimentary strata, leaving a core of igneous and metamorphic rocks standing above the surrounding terrain (Figure 14.12).

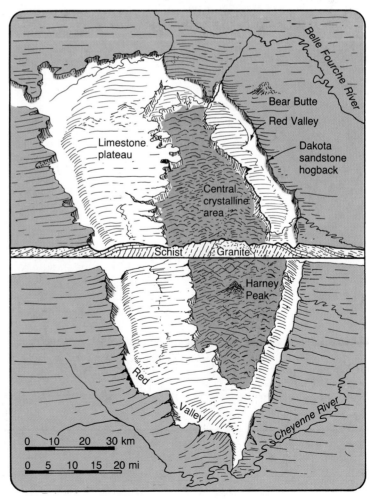

FIGURE 14.12 The Black Hills of South Dakota, an upwarped mountain whose resistant igneous and metamorphic central core has been exposed by erosion. (After Arthur N. Strahler. *Introduction to Physical Geography*, 3rd ed. Copyright © 1973, by John Wiley & Sons, Inc. Reprinted by permission)

Upwarped mountains are also found in the middle and southern Rockies from Montana south through Colorado and into New Mexico. Included in this group are the Front Range of Colorado and the Bighorns of Wyoming. These mountains border the Great Plains and were pushed almost vertically upward as part of a broad crustal upwarping, or in some instances, were displaced upward along high-angle faults. It has been suggested that the upwarping of the eastern Rockies was in response to compressional stresses generated as small crustal fragments were added to the Pacific margin of North America (see discussion entitled Orogenesis and Continental Accretion on page 274).

In general the upwarped portions of the Rockies consist of older basement rocks covered by relatively thin layers of younger sedimentary strata. However, since the time of deformation, much of this sedimentary blanket has been eroded from the highest portions of the uplifted blocks, exposing the igneous and metamorphic core. In many areas, remnants of these sedimentary layers flank the crystalline cores of the mountain ranges. They are often easy to identify because the upturned strata form prominent angular ridges called **hogbacks** (Figure 14.13). Examples of the exposed cores include a number of granitic outcrops that project as steep summits, such as Pikes Peak and Longs Peak in Colorado's Front Range.

MOUNTAIN BUILDING AND PLATE TECTONICS

It has long been recognized that complex mountain systems have many features in common—enough, in fact, that geologists have concluded that they must have comparable orogenic histories. Many young mountains parallel the coasts of continents. They are made up of thick sequences of sedimentary rock, occasionally totaling more than 15,000 meters (50,000 feet), which have been folded, faulted, and intruded by igneous bodies (Figure 14.14). Until the last decade it had generally been accepted that these sediments accumulated in a slowly subsiding trough called a **geosyncline.** After great thicknesses of sediment had built up, horizontal forces from the seaward side of the geosyncline began to squeeze the sediments, shortening and thickening the crust, producing a high-standing mountain system while simultaneously pushing much of the sediment deeper into the earth. It was believed that the melting of these deeply buried sediments generated magma which moved upward, intruding the overlying unmelted sediments. Thus, a complex mountain chain containing folded and faulted sedimentary rocks surrounding a core of igneous intrusions and metamorphic rocks was formed.

Although the geosynclinal concept of mountain building has many merits, the underlying cause of

FIGURE 14.13 A view looking north along the east flank of the Front Range of the Colorado Rockies. The upturned remnants of sedimentary strata (center) once covered the igneous and metamorphic terrain which lies to the west. (Photo by T. S. Lovering, U.S. Geological Survey)

FIGURE 14.14 Intensely folded rock strata provide evidence of the forces altering the earth's crust. (Photo by W. B. Hamilton, U.S. Geological Survey)

orogenesis is not explained. What produced the subsidence in the geosyncline? Why did sediment accumulate, relatively undisturbed for millions of years, and then suddenly go through a period of deformation? Such questions forced geologists to continue to evaluate the complex problem of mountain building.

With the development of the plate tectonics theory, many questions from the geosynclinal theory were answered. This new and widely accepted theory suggests that orogenesis results as large segments of the earth's lithosphere are displaced. According to plate tectonics, mountain building occurs at convergent plate boundaries. Here the colliding plates provide the compressional stress to fold, fault, and metamorphose the thick accumulations of sediments that are deposited along the flanks of landmasses, while melting of the subducted oceanic crust provides a source of magma that intrudes and further deforms these deposits.

Orogenesis at Subduction Zones

The first stage in the development of a complex mountain system is thought to occur prior to the formation of the subduction zone. During this period the continental margin is passive; that is, it is not a plate boundary but a part of the same plate as the adjoining oceanic crust. The east coast of the United States provides an example of a present-day passive continental margin. Here, as at other passive continental margins surrounding the Atlantic, deposition of sediment is producing a thick wedge of sandstones, limestones, and shales (Figure 14.15A).

At some time the continental margin becomes active; a subduction zone forms and deformation begins (Figure 14.15B). A good place to examine an active continental margin is the west coast of South America. Here the Nazca plate is being subducted beneath the South American plate along the Peru-Chile trench. This subduction zone probably formed in conjunction with the breakup of the supercontinent Pangaea. As the South American plate gradually separated from Africa and migrated westward, the oceanic crust adjacent to the west coast of South America was bent and thrust under the continental plate. However, the oceanic crust does not give way without some effect on the overriding plate. In the case of South America, the Nazca plate apparently deformed sediments that flanked the continental margin, producing the original folded and faulted portion of the complex mountain system we now call the eastern Andes.

Subduction and partial melting of the Nazca plate initiated yet another stage—the development of the volcanic arc (Figure 14.15C). In any arc system the volcanic activity is the most noticeable, but far

Continental margin

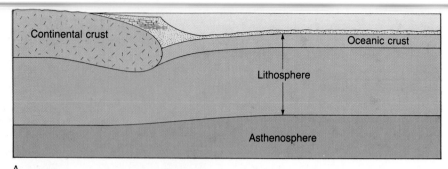

Continental crust

Oceanic crust

Lithosphere

Asthenosphere

A.

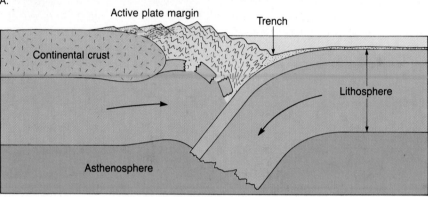

Active plate margin

Trench

Continental crust

Lithosphere

Asthenosphere

B.

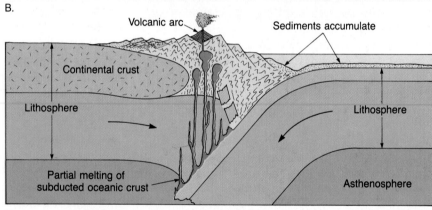

Volcanic arc

Sediments accumulate

Continental crust

Lithosphere

Lithosphere

Partial melting of
subducted oceanic crust

Asthenosphere

C.

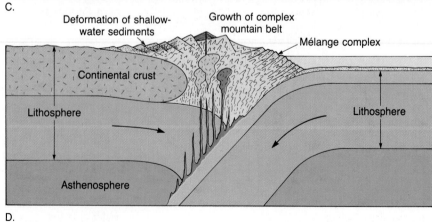

Deformation of shallow-
water sediments

Growth of complex
mountain belt

Mélange complex

Continental crust

Lithosphere

Lithosphere

Asthenosphere

D.

FIGURE 14.15 Subduction-
type orogenesis along an active continental margin.
A. Passive plate margin.
B. Plate convergence generates a subduction zone.
C. Partial melting of subducted plate generates the
volcanic arc. **D.** Continued
growth of the complex mountain system through deformation of the shallow- and
deep-water sediments.

greater quantities of magma are emplaced below the surface, forming numerous batholiths. The effect of this is to further thicken the crust. In response to isostatic adjustment, uplift will follow. As a consequence of crustal thickening, the Andes Mountains rise many kilometers above the adjacent oceanic trench.

During the development of the volcanic arc, sediment derived from the land as well as that scraped from the subducting plate is plastered against the landward side of the trench. This chaotic accumulation of metamorphosed rocks and scraps of oceanic crust is called a **mélange** (Figure 14.15D). The metamorphic rocks of a mélange are formed under great pressure from the converging plates, but at rather low temperatures. Consequently, they can be distin-

guished from the metamorphic rocks that form at much higher temperatures in association with intrusive igneous bodies. When a mélange is found in the interior of a continent, it represents the relic of a former subduction zone. Such a circumstance provides an important clue for the interpretation of the geologic history of that region.

An excellent example of a volcanic arc-mélange complex as just described is found in the western United States and includes the Sierra Nevada and the Coast Ranges of California (Figure 14.16). These parallel mountain belts were produced by subduction of a portion of the Pacific basin under the western edge of the North American plate. The Sierra Nevada batholith is a remnant of a portion of the volcanic arc that was produced by several surges of magma over a period of tens of millions of years. Subsequent uplifting and erosion have removed most evidence of past volcanic activity and exposed a core of crystalline rocks. In the trench region, sediments scraped from the subducting plate and those provided by the eroding volcanic arc were intensely folded and faulted into the complex mélange which presently constitutes the Franciscan Formation of California's Coast Ranges. Uplifting of the Coast Ranges took place quite recently, as evidenced by the unconsolidated sediments still mantling portions of these highlands. In this region the subduction of the oceanic plate ceased about 30 million years ago when the North American plate collided with a spreading ridge in the Pacific. At that time the San Andreas fault became the boundary between the plates, changing a convergent plate boundary into a transform fault boundary.

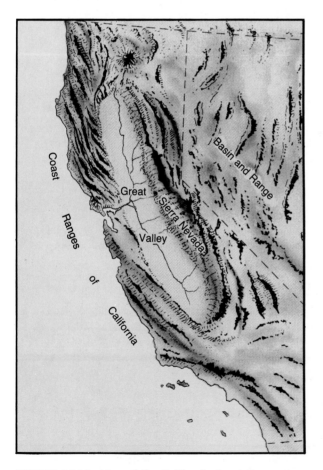

FIGURE 14.16 Map of the California Coast Ranges and the Sierra Nevada.

Continental Collisions

Up to this point we have discussed the formation of orogenic belts where the leading edge of only one of the two converging plates contains continental crust. However, both of the colliding plates may be carrying continental crust. Because continental rocks are evidently too buoyant to undergo any appreciable subduction, a collision between the continental fragments eventually results (Figure 14.17). An excellent example of such a collision occurred about 45 million years ago when India collided with Asia.

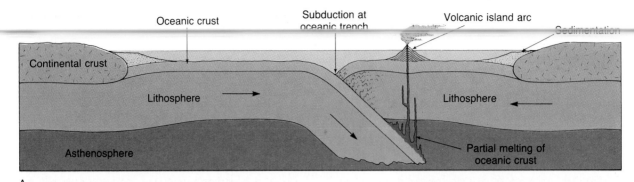

Oceanic crust

Subduction at
oceanic trench

Volcanic island arc

Sedimentation

Continental crust

Lithosphere

Lithosphere

Asthenosphere

Partial melting of
oceanic crust

A.

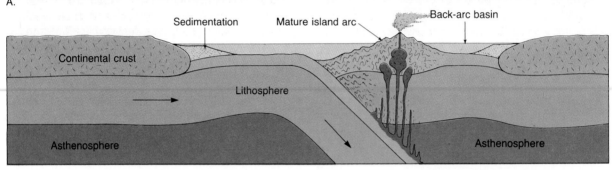

Sedimentation

Mature island arc

Back-arc basin

Continental crust

Lithosphere

Asthenosphere

Asthenosphere

B.

Deformation of marginal
sediments and the volcanic arc

Continental crust

Continental crust

Lithosphere

Deformed oceanic crust

Asthenosphere

C.

FIGURE 14.17 Orogenesis
and continental collisions.
A. Converging plates generate
a subduction zone and initiate
island arc volcanism.
B. Sediments scraped from
the subducting plate and igneous activity add to the size
of the volcanic arc. **C.** Closing
of the back-arc basin deforms
the entrapped marginal sediments and volcanic arc. **D.** A
continental collision closes the
ocean basin, resulting in further deformation and metamorphism of the sediments
and volcanic rocks.

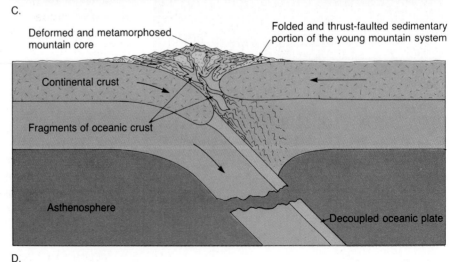

Deformed and metamorphosed
mountain core

Folded and thrust-faulted sedimentary
portion of the young mountain system

Continental crust

Fragments of oceanic crust

Asthenosphere

Decoupled oceanic plate

D.

272

India, thought to have been previously attached to Antarctica, was rafted nearly 5000 kilometers (3100 miles) due north before the collision occurred. The result was the formation of the spectacular Himalaya Mountains and the Tibetan Highlands. Although most oceanic crust separating these landmasses prior to the collision was subducted, some was caught up in the squeeze along with sediment that lay offshore and can now be found elevated high above sea level. After such a collision the subducted oceanic plate is believed to decouple from the rigid continental plate and continue its downward path.

Other mountain ranges displaying evidence of continental collisions are the Alps, Urals, and Appalachians. The Appalachians are thought to have resulted from a collision between North America, Europe, and northern Africa. Although they have since separated, these landmasses were juxtaposed as part of the supercontinent Pangaea less than 200 million years ago. Detailed studies in the southern Appalachians indicate that the formation of this mountain belt was more complex than once thought. Rather than forming during a single continental collision, the Appalachians resulted from several distinct episodes of mountain building that occurred over a period of nearly 250 million years. As the two continental blocks began to converge, a subduction zone formed seaward of the ancient coastline of North America. The igneous activity associated with the subducting plate gave rise to a vol-

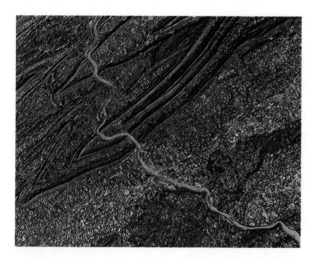

FIGURE 14.18 Valley and Ridge Province. (Courtesy of NASA)

canic island arc, perhaps similar to those volcanic arcs presently rimming the western Pacific (Figure 14.17A). Then, about 380 million years ago, the ocean basin behind the volcanic arc began to close.

The final orogeny occurred about 250–300 million years ago when Africa and Europe collided with North America. This event is thought to have displaced the earlier-accreted volcanic arc farther inland along low-angle thrust faults. At some locations the total displacement may have exceeded 250 kilometers (155 miles). This landward displacement further deformed the shallow-water sediments which had flanked North America. Today these folded and faulted sandstones, limestones, and shales compose the essentially unmetamorphosed rocks of the Valley and Ridge Province (Figure 14.18).

In summary, the orogenesis of a complex mountain chain, as typified by the Appalachians, is thought to occur as follows:

1 After the breakup of a continental landmass, a thick wedge of sediments is deposited along passive continental margins, thereby increasing the size of the newly formed continent.

2 For reasons not yet understood, the ocean basin then begins to close and the continents start to converge.

3 Plate convergence results in subduction of the intervening oceanic slab and initiates an extended period of igneous activity. This activity results in the formation of a volcanic arc often located a few hundred kilometers seaward of the ancient coastline (Figure 14.17A).

4 Debris eroded from the volcanic arc and the mainland, plus sediment scraped from the descending plate, add to the wedge of sediment along the continental margin (Figure 14.17B).

5 Further convergence causes the narrow sea behind the volcanic arc to close. This orogenic event deforms and metamorphoses the back-arc sediments and associated volcanic debris as well as the volcanic arc itself (Figure 14.17C).

6 Eventually the continents collide. This event and the associated igneous activity further deform and metamorphose the entrapped sediments and volcanic arc to produce the crystalline core of the young mountain belt. As this deformed terrain is thrust landward, the shallow-water deposits that once formed the continental shelf

are folded and displaced inland along low-angle thrust faults (Figure 14.17D).

7 Finally, a change in the plate boundary ends mountain belt growth. Only then does erosion become the dominant process in altering the landscape.

This sequence of events is thought to have been duplicated many times in the geologic past. However, the rate of deformation and the geologic and climatic settings varied in each instance. Thus, the formation of each mountain chain must be regarded as a unique event.

Orogenesis and Continental Accretion

When originally formulated, the plate tectonics theory suggested two mechanisms for orogenesis. First, continental collisions were proposed to explain the formation of such mountainous regions as the Alps, Himalayas, and Appalachians. Second, as typified by the Andes, orogenesis associated with the subduction of oceanic lithosphere was thought to be the underlying tectonic process for many circum-Pacific mountain chains. Recent investigations, however, indicate yet another mechanism for orogenesis. This new proposal suggests that relatively small crustal fragments collide and merge with continental margins and that through this process of collision and accretion, many of the mountainous regions rimming the Pacific have been generated.

What is the nature of the small crustal fragments and how did they originate? Researchers believe that prior to their accretion (attachment) to a continental block, some fragments may have been small continental blocks similar in nature to the present-day island of Madagascar. Many others may have been located below sea level and are represented today by submerged platforms rising high above the floor of the western Pacific. Over one hundred of these so-called oceanic plateaus are known to exist. Such plateaus are believed to be submerged continental fragments, extinct volcanic arcs, or submerged volcanic chains produced by hot spot activity.

The widely accepted view today is that as oceanic plates move, they carry the embedded oceanic plateaus or continental fragments to a subduction zone. Here the upper portions of these crustal frag-

ments are peeled from the descending plate and thrust in relatively thin sheets upon the adjacent continental block. This newly added material, called **terrane**, increases the width of the continent and may later be overridden and displaced farther inland by colliding with other fragments.

The idea that mountain building occurs in association with the accretion of small crustal fragments to a continental mass arose chiefly from studies of the northern portion of the North American Cordillera. Here it was learned that some terranes, principally those in the mountainous belts of Alaska and British Columbia, contain fossil and magnetic evidence to indicate these rocks originated nearer the equator. It is now believed that these terranes were once scattered throughout the eastern Pacific, much as we find oceanic plateaus distributed in the western Pacific today. Within the last 200 million years, these fragments migrated toward and collided with the west coast of North America. Apparently, this activity has resulted in the piecemeal addition of fragments to the entire Pacific Coast from the Baja Peninsula to northern Alaska. If this is true, such collisions are responsible for the orogenesis of much of the North American Cordillera. In a like manner, many of the present-day ocean plateaus will eventually be accreted to active continental margins, thus resulting in the formation of new orogenic belts.

THE ORIGIN AND EVOLUTION OF CONTINENTAL CRUST

In the preceding section, we learned that the theory of plate tectonics provides a model from which to examine the formation of complex mountainous belts. But what roles have plate tectonics and mountain building played in events leading to the origin and evolution of continents? At this time no single answer has met with overwhelming acceptance. The lack of agreement among geologists can in part be attributed to the complex nature and antiquity of most continental material, which makes deciphering its history very difficult. Nevertheless, during the last two decades, great strides have been made in unraveling the secrets held by rocks composing stable continental interiors.

One view, which has gained support in recent years, contends that the continents have grown larger through geologic time by the gradual accre-

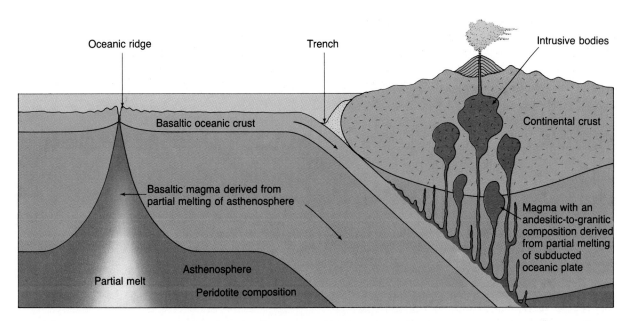

FIGURE 14.19 The two-stage process for transforming material from the asthenosphere into continental crust. Once continental crust is generated, its low density apparently keeps it afloat indefinitely.

tion of material derived from the upper mantle. A main tenet of this hypothesis is that the primitive crust was of an oceanic type and the continents were small or possibly nonexistent. Further, this proposal suggests that the formation of continental material takes place in two distinct phases as shown in Figure 14.19. The first step occurs in the upper mantle directly beneath the oceanic ridges. Here partial melting of the rock peridotite yields basaltic magma that rises to form oceanic crust. The ocean floor rocks are higher in silica, potassium, and sodium, and lower in iron and magnesium than the rocks of the upper mantle from which they were derived. As new ocean floor is generated at the ridge crests, older oceanic crust is being destroyed at the oceanic trenches. In trench regions, the subducted oceanic crust is heated sufficiently to cause partial melting. This gives rise to relatively light, silica-rich rocks which are then emplaced in volcanic arcs. The subducted oceanic crust, depleted of its lighter constituents, continues to sink and is no longer involved in the process of generating crustal rocks.

According to this view, the earliest continental rocks came into existence at a few isolated island arcs. Once formed, these island arcs coalesced to form larger continental masses, while deforming the

volcanic and sedimentary rocks which were deposited in the intervening oceans. Eventually this process generated masses of continental crust having the size and thickness of modern continents. In this way, the process of mountain building not only restructures continental rocks, but also generates new continental materials.

If the continents do in fact grow by accretion of material at their flanks, then the continents have grown larger at the expense of oceanic crust. This view assumes the buoyancy and indestructibility of continental crust. Even sediment eroded from a continent and carried to great depths on an adjacent subducting plate eventually melts and returns to the continents. Although crustal rock apparently remains afloat indefinitely, some continents are occasionally fragmented and carried along in a conveyor belt fashion until they collide with other landmasses. Australia, which separated from Antarctica, is presently being rafted northward and will probably join Asia in much the same manner as India did about 45 million years ago. According to this view, fragmentation and the formation of new crustal rocks that accompanied the reshuffling of these fragments are responsible for the present volume, structure, and configuration of continents.

A word of caution is in order. The views set forth in this section to explain the origin and evolution of the continents are still somewhat speculative. Plate tectonics appears to be the major force in crustal evolution over the last 600 million years. However, during the earth's early history, heat released by the decay of uranium, thorium, and potassium must have been at least twice as great as it is today. Was plate tectonics active early in the earth's history, only at a different rate, or were there much different processes in operation? Was the primitive crust composed primarily of continental rocks, or was it of the oceanic type? These are some of the questions that still require definitive answers.

REVIEW QUESTIONS

1 State the three lines of evidence which support the concept of crustal uplift. Can you think of another?

2 What happens to a floating object when weight is added? Subtracted? How do these principles apply to changes in the elevations of mountains? What term is applied to the adjustment that causes crustal uplift of this type?

3 List two lines of evidence which support the idea of a floating crust.

4 What conditions favor rock deformation by folding? By faulting?

5 Using the concept of elastic rebound, explain why earthquakes cannot occur at great depths.

6 Compare the movement of normal and reverse faults. What type of force produces each?

7 At which of the three types of plate boundaries does normal faulting predominate? Reverse faulting? Strike-slip faulting?

8 Describe a horst and a graben. Explain how a graben valley forms and name one.

9 Compare and contrast anticlines and synclines. Domes and basins. Anticlines and domes.

10 Although we classify many mountains as folded, why might this description be misleading?

11 Name the site where sediments are deposited and have a good chance of being squeezed into a mountain range.

12 Which type of plate boundary is most directly associated with mountain building?

13 Describe a mélange and explain its formation.

14 Would the discovery of a sliver of oceanic crust in the interior of a continent tend to support or refute the theory of plate tectonics? Why?

15 Why might it have been difficult for geologists to conclude that the Appalachian Mountains were formed by plate collision if examples like the Himalayas did not exist?

16 How does the plate tectonics theory explain the existence of fossil marine life on top of the Ural Mountains?

17 In your own words, briefly enumerate the steps involved in the formation of a mountain system according to the plate tectonics model.

CHAPTER FIFTEEN

GEOLOGIC TIME

In 1869 John Wesley Powell, who was later to head the U.S. Geological Survey, led a pioneering expedition down the Colorado River and through the Grand Canyon (Figure 15.1). Writing about the strata that were exposed by the downcutting of the river, Powell said that, " . . . the canyons of this region would be a Book of Revelations in the rock-leaved Bible of geology." Powell was undoubtedly impressed with the countless millions of years of earth history exposed along the walls of the Grand Canyon. Interpreting earth history is a prime goal of the science of geology. Like a modern-day sleuth, the geologist must interpret the clues found preserved in the rocks. By studying rocks, especially sedimentary rocks, and the features they contain, geologists can often unravel the complexities of the past.

Events by themselves, however, have little meaning until they are put into a time perspective. Studying history, whether it be the Civil War or the Age of Dinosaurs, requires a calendar. Among the major contributions that geology has made to our knowledge is the geologic calendar and the concept that earth history is exceedingly long. Over many years geologists have devised a time scale of earth history— a calendar where geologic events can be put in their proper place. Geologists, recognizing that earth history has spanned an immense amount of time, worked at finding out just how old the earth is.

EARLY METHODS OF DATING THE EARTH

Current methods of radiometric dating put the age of the earth between 4.6 and 4.8 billion years. How-

ever, this great age for the earth is a relatively recent discovery. Although James Hutton and others who accepted the principle of uniformitarianism believed the earth was very old, they had no way of knowing its exact age. Solutions to this dating problem were sought, and several methods were subsequently devised.

One method involved the rate at which sediment is deposited. Some geologists reasoned that if they could determine the rate that sediment accumulates, and could further ascertain the total thickness of sedimentary rock that had been deposited during earth history, they could accurately estimate the length of geologic time. All that was necessary was to divide the rate of sediment accumulation into the total thickness of sedimentary rock. Unfortunately this method was riddled with difficulties, some of which are as follows:

1 Different sediments accumulate at different rates under varying conditions. Thus, determining an overall rate of sediment accumulation is extremely difficult. Further, if such a rate is determined, it does not necessarily mean that the same rate can be applied to the past.
2 Since no single locality has a complete geologic column, estimates of the total thickness of sedimentary rocks had to be complied by adding together the maximum known thickness of rocks of each age. These estimates had to be revised each time a thicker section was discovered.
3 Sediment compacts when it is lithified; thus, a correction for compaction had to be made.

Needless to say, estimates of the earth's age varied considerably as different scientists attempted this method. The figure representing the maximum thickness of sedimentary rock ranged from 9600

The strata exposed in the Grand Canyon contain clues to millions of years of earth history. (Photo by E. J. Tarbuck)

A.

B.

FIGURE 15.1 A. Start of the expedition from Green River station. A drawing from Powell's 1875 book. **B.** Major John Wesley Powell, pioneering geologist and the second director of the U.S. Geological Survey. (Courtesy of the U.S. Geological Survey)

meters (32,000 feet) to over 100,500 meters (330,000 feet). The amount of time for 0.3 meter (1 foot) of sediment to accumulate varied from 100 years to over 8600 years. The age of the earth as calculated by this method therefore ranged from 3 million to 1.5 billion years!

Another method for dating the earth involved the salinity of the oceans, which were assumed to originally have been fresh water. Scientists felt that if they could accurately estimate the quantity of salt being carried to the ocean each year by rivers and the total amount of salt currently in the oceans, they could determine the length of geologic time by dividing the latter figure by the former. Near the turn of the twentieth century John Joly calculated the age of the earth at about 90 million years using this method. Joly, however, had no accurate notion of the amount of salt lost from the oceans because of deposition and winds blowing salt inland. It is also probable that the rate of salt accumulation has not always been constant. Thus, Joly's estimate for the age of the earth was not accurate. However, both of the methods for dating the earth that have just been described indicated that the earth was considerably older than the 6000 years given it by Archbishop Ussher.*

Perhaps the most influential estimates of the age of the earth were compiled by the well-known and highly respected physicist Lord Kelvin in the latter part of the nineteenth century. Since Kelvin's estimates required few assumptions and were based on precise measurements, they were widely accepted for a time. One of Kelvin's methods was founded on the widely held assumption that the earth had originally been molten and had cooled to its present condition. Although his data and calculations were limited, Kelvin still made it quite obvious that the earth could not be more than 100 million years old, and likely much less. The second of Kelvin's estimates was based on the fact that the source of the sun's tremendous output of energy was of a conventional nature (nuclear fusion and radioactivity had not yet been discovered). His calculations indicated that the sun could only have illuminated the earth for a few tens of millions of years. Furthermore, he said that in the past it had been much hotter and in the future it

---

*See the section entitled *Catastrophism* in the Introduction.

would become much cooler. He believed the earth was inhabitable for organisms for a period of only 20–40 million years. Kelvin's apparently irrefutable estimates had a rather profound impact:

> Evolutionists found it virtually impossible to accept these figures, but all they had were educated guesses in the face of Kelvin's potent mathematics. Darwin and others compromised their original theories in their later years in an effort to reconcile evolution and uniformitarianism with the physicists' estimates. Eventually, however, they were vindicated.*

RADIOACTIVITY AND RADIOMETRIC DATING

Most atoms are stable and do not change. However, some are unstable, constantly releasing heat as their nuclei break apart or decay. This is the heat that helps maintain the high temperatures in the earth's interior and is the source of the heat which Kelvin was measuring when he thought he was measuring the "cooling" earth.

In Chapter One we learned that an atom is composed of electrons, protons, and neutrons. *Electrons* have a negative charge and *protons* have a positive charge. Since a *neutron* is actually a proton and electron combined, it has no charge. Protons and neutrons are found in the center, or *nucleus*, of the atom, and electrons spin around the nucleus in definite paths, or orbits. Practically all (99.9 percent) of the mass of an atom is found in the nucleus, indicating that electrons have practically no mass at all. By adding together the number of protons and neutrons in the nucleus, the *mass number* of the atom is determined. The *atomic number* (the atom's identifying number) is equal to the number of protons. Each one of the more than 100 known elements has a different number of protons in the nucleus, and thus a different atomic number. Atoms of the same element may have different numbers of neutrons in the nucleus. Such atoms, called *isotopes*, have different mass numbers but the same atomic number.

The forces that bind protons and neutrons together in the nucleus are very strong; however, the nature of these forces is still poorly understood. Some isotopes have unstable nuclei; that is, the forces that bind the protons and neutrons together are not sufficiently strong. As a result, the nuclei spontaneously break apart, or decay, a process called **radioactivity**. What happens when unstable nuclei break apart? Two common types of radioactive decay are illustrated in Figure 15.2 and are summarized as follows:

1 *Alpha particles* (α particles) may be emitted from the nucleus. An alpha particle is composed of 2 protons and 2 neutrons. Thus, the emission of an alpha particle means that the mass number of the isotope is reduced by 4 and the atomic number is lowered by 2.
2 When a *beta particle* (β particle), or electron, is given off from a nucleus, the mass number remains unchanged, because electrons have practically no mass. However, since the electron must have come from a neutron (remember, a neutron is a combination of a proton and an electron), the nucleus contains one more proton than before. Therefore, the atomic number increases by 1.

The radioactive isotope is often referred to as the **parent**, and the isotopes resulting from the decay of the parent are termed the **daughter products**. Figure 15.3 provides an example of radioactive decay. Here it may be seen that when the radioactive parent, uranium-238 (atomic number 92, mass number 238) decays, it emits 8 alpha particles and 6 beta particles before becoming the stable daughter product lead-206 (atomic number 82, mass number 206).

Certainly among the most important results of the discovery of radioactivity is that it provided a reliable means of calculating the ages of rocks and minerals which contain radioactive isotopes, a procedure referred to as **radiometric dating**. Why is radiometric dating reliable? The answer lies in the fact that the rate at which radioactive isotopes decay is constant and unaffected by any chemical or physical agents.

The amount of time required for one-half of the nuclei in a sample to decay, called its **half-life**, is a

---

*Leigh W. Mintz. *Historical Geology: The Science of a Dynamic Earth,* 2nd ed. (Columbus, Ohio: Merrill, 1977), pp. 84–85.

FIGURE 15.2 Common types of radioactive decay. Notice that in each case the number of protons (atomic number) in the nucleus changes, thus yielding a different element.

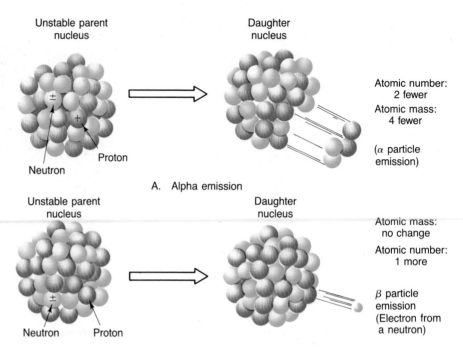

Unstable parent nucleus — Daughter nucleus

Atomic number: 2 fewer
Atomic mass: 4 fewer

(α particle emission)

Neutron — Proton

A. Alpha emission

Unstable parent nucleus — Daughter nucleus

Atomic mass: no change
Atomic number: 1 more

β particle emission (Electron from a neutron)

Neutron — Proton

B. Beta emission

common way of expressing the rate of radioactive disintegration. If we began with a pound of radioactive material, half a pound would decay after one half-life, half the remaining amount would break down after another half-life, and so on.

Figure 15.4 illustrates the principle of radiometric dating using a hypothetical radioactive parent that decays directly into the stable daughter product. Its half-life is 1 million years. By calculating the percentages of radioactive parent and stable daughter product, the age of the specimen can be determined. In this example, when the quantities of parent and daughter are equal (ratio 1:1), we know that one half-life has transpired and that the specimen is 1 million years old. When the ratio of parent to daughter reaches 1:15, we know the sample is 4 million years old.

Of the many radioactive isotopes that exist in nature, five have proven significant in providing radiometric ages for ancient rocks. Table 15.1 summarizes these most frequently used isotopes. Others are either very rare or have half-lives that are too short or much too long to be useful. Rubidium-87 and the two isotopes of uranium are used only for dating rocks that are millions of years old, but potassium-40 is more versatile. Although the half-life of potassium-40 is 1.3 billion years, recent analytic techniques have made it possible to detect the tiny amounts of its stable daughter product, argon-40, in rocks as young as 50,000 years.

To date more recent events, carbon-14 (also called **radiocarbon**), the radioactive isotope of carbon, is used. Since it has a half-life of only 5730 years, it can be used for dating events from the his-

TABLE 15.1 Frequently used radioactive isotopes.

Radioactive Parent	Stable Daughter Product	Currently Accepted Half-life Values
Uranium-238	Lead-206	4.5 billion years
Uranium-235	Lead-207	713 million years
Thorium-232	Lead-208	14.1 billion years
Rubidium-87	Strontium-87	47.0 billion years
Potassium-40	Argon-40	1.3 billion years

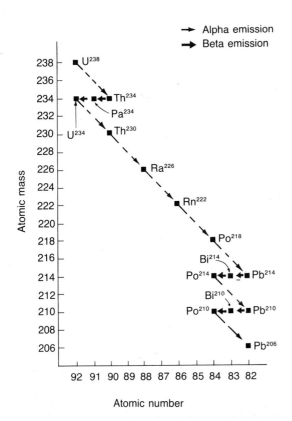

FIGURE 15.3 The most common isotope of uranium (U-238) is an example of a radioactive decay series. Before the stable end product (Pb-206) is reached, many different isotopes are produced as intermediate steps.

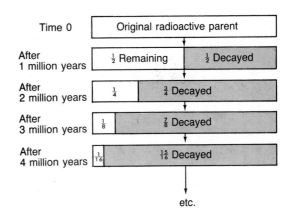

FIGURE 15.4 Decay of a hypothetical radioactive isotope with a half-life of 1 million years.

toric past, as well as those from recent geologic history. Until the late 1970s radiocarbon was useful in dating events only as far back as 40,000–50,000 years. However, as was the case with potassium-40, the development of more sophisticated analytical techniques has increased the usefulness of this "clock." Carbon-14 can now be used to date events as far back as 75,000 years. This is a significant accomplishment because it means that geologists can now date many ice-age phenomena that previously could not be dated accurately.

Carbon-14 is continuously produced in the upper atmosphere as a consequence of cosmic ray bombardment, in which cosmic rays (high-energy nuclear particles) shatter the nuclei of gases to release neutrons. The neutrons are absorbed by nitrogen (atomic number 7, mass number 14), causing its nucleus to emit a proton. Thus, the atomic number drops by 1 (to 6), and a new element, carbon-14, is

created (Figure 15.5A). This isotope of carbon is quickly incorporated into carbon dioxide, circulates in the atmosphere, and is absorbed by living matter. As a result, all organisms contain a small amount of carbon-14.

While an organism is alive, the decaying radiocarbon is continually replaced. As a result, the ratio of carbon-14 to carbon-12 (the most common isotope of carbon) remains constant. However, when the plant or animal dies, the amount of carbon-14 gradually decreases as it decays to nitrogen-14 by beta emission (Figure 15.5B). Therefore, by comparing the proportions of carbon-14 and carbon-12 in a sample, radiocarbon dates can be determined. Although carbon-14 is only useful in dating the last small fraction of geologic time, it has become a very valuable tool for anthropologists, archeologists, and historians, as well as for geologists who study very recent earth history. In fact, the development of radiocarbon dating was considered so important that the chemist who discovered this application, Willard F. Libby, received a Nobel Prize.

Bear in mind that although the basic principle of radiometric dating is relatively simple, the actual procedure is quite complex, for the chemical analysis which determines the quantities of parent and daughter that are present must be painstakingly precise. In addition, some radioactive materials do not decay directly into the stable daughter product as was the case with our hypothetical example, a fact which may further complicate the analysis. In the case of uranium-238, there are thirteen intermediate unstable daughter products formed before the

FIGURE 15.5 A. Production and **B.** Decay of carbon-14.

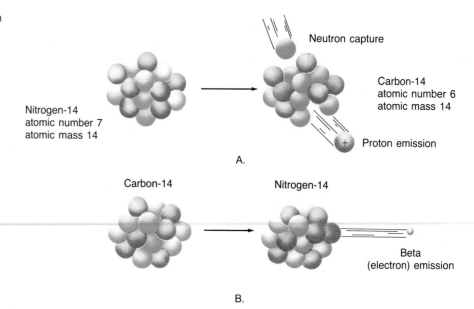

fourteenth, and last daughter product, the stable isotope lead-206, is produced (Figure 15.3).

Radiometric dating methods have produced literally thousands of dates for events in earth history. Rocks from several localities have been dated at more than 3 billion years, and geologists realize that still older rocks exist. For example, a granite from South Africa which has been dated at 3.2 billion years contains inclusions of quartzite. Quartzite is a metamorphic rock which originally was the sedimentary rock sandstone. Since sandstone is the product of the lithification of sediments produced by the weathering of pre-existing rocks, we have a positive indication that older rocks existed.

Radiometric dating has vindicated the ideas of Hutton, Darwin, and others who over 150 years ago assumed that geologic time must be immense. Indeed, it has proven that there has been enough time for the slow processes we observe to have accomplished tremendous tasks.

THE MAGNITUDE OF GEOLOGIC TIME

The magnitude of geologic time is difficult to grasp, because we must learn to think in spans of time that far exceed our common experience. Earth features, which seem to be everlasting and unchanging to us and in fact to generations of people, are indeed slowly changing. Over millions of years, mountains rise and are eroded to hills, and rivers excavate deep canyons. How long is 5 billion years? If you were to begin counting to 5 billion at the rate of one number per second and continued 24 hours a day, 7 days a week, and never stopped, it would take about two lifetimes (150 years) to reach 5 billion! Don L. Eicher gives us another basis for comparison:

Compress for example, the entire 4.5 billion years of geologic time into a single year. On that scale, the oldest rocks we know date from about mid-March. Living things first appeared in the sea in May. Land plants and animals emerged in late November and the widespread swamps that formed the Pennsylvanian coal deposits flourished for about four days in early December. Dinosaurs became dominant in mid-December, but disappeared on the 26th, at about the time the Rocky Mountains were first uplifted. Manlike creatures appeared sometime during the evening of December 31st, and the most recent continental ice sheets began to recede from the Great Lakes area and from northern Europe about 1 minute and 15 seconds before midnight on the 31st. Rome ruled the Western world for 5 seconds from 11:59:45 to 11:59:50. Columbus discovered America 3 seconds before midnight, and the science of geol-

FIGURE 15.6 Applying the law of superposition to this cross section, the shale bed is oldest and the sandstone is youngest.

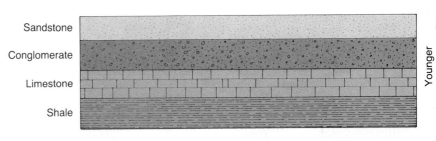

Sandstone

Conglomerate

Limestone

Shale

Younger

ogy was born with the writings of James Hutton just slightly more than one second before the end of our eventful year of years.*

RELATIVE DATING

Radiometric dating results in specific dates for rock units which represent various events in the earth's distant past. We can now state with some confidence that particular geologic events took place a certain number of years ago. Such dates are referred to as **absolute dates,** for they pinpoint the time in history when something took place. Prior to the discovery of radioactivity and the development of the technology of radiometric dating, geologists had no precise method of absolute dating and had to rely solely on relative dating. **Relative dating** means that rocks are placed in their proper sequence or order. Relative dating will not tell us how long ago something took place, only that it followed one event and preceded another. The relative dating techniques which were developed are still widely used. Absolute dating methods did not replace these techniques; they simply supplemented them. To establish a relative time scale, a few simple principles or rules had to be discovered and applied. Although they may seem rather obvious to us today, their discovery was a very important scientific achievement.

Nicolaus Steno, a physician in Florence, Italy, is credited with being the first to recognize a sequence of historical events in an outcrop of sedimentary rock layers. Working in the mountains of western Italy, Steno applied a very simple rule that has come to be the most basic principle of relative dating—the **law of superposition**. The law simply states that in an undeformed sequence of sedimentary rocks, each bed is older than the one above it and younger than the one below. Although it may seem obvious

that a layer could not be deposited with nothing beneath it for support, it was not until 1669 that Steno clearly stated the principle. This rule also applies to other surface-deposited materials such as lava flows and beds of ash from volcanic eruptions. Applying the law of superposition to the beds shown in Figure 15.6, we can easily place the layers in their proper order. The sandstone is youngest and the shale is oldest.

Steno is also credited with recognizing the importance of another basic principle, called the **principle of original horizontality**. Simply stated, it means that layers of sediment are generally deposited in a nearly horizontal position. Thus, if we observe rock layers that are inclined at a steep angle, they must have been moved into that position by crustal disturbances sometime after their deposition.

When igneous intrusions or faults cut through other rocks, they are assumed to be younger than the rocks they cut. For example, when two dikes

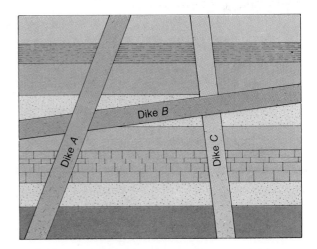

FIGURE 15.7 Cross-cutting relationships. All of the dikes are younger than the rock into which they were intruded. Since dike *B* cuts through dike *C* and dike *A* cuts through dike *B*, the order of intrusion, from oldest to youngest, is dike *C*, dike *B*, dike *A*.

*Don L. Eicher. *Geologic Time* (Englewood Cliffs, New Jersey: Prentice-Hall, 1968), p. 19.

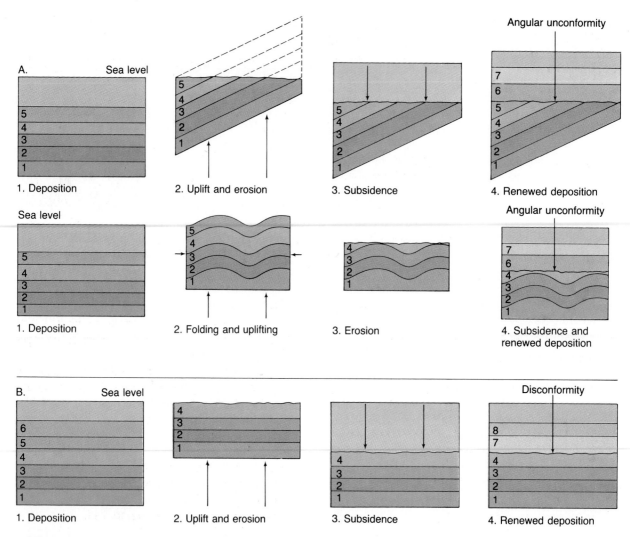

FIGURE 15.8 Development of unconformities. **A.** Two examples of angular unconformity. **B.** Erosion resulting in a disconformity. In each case a gap in the rock record has been created.

intersect, the older one must have been opened up in order to allow the younger one to cut through it. The younger dike would be continuous, while the older dike would be interrupted at the point of their intersection. Figure 15.7 on the previous page illustrates this principle of **cross-cutting**.

Layers of rock are said to be **conformable** when they are found to have been deposited without interruption. However, there is no place on earth that contains a complete set of conformable strata. Even for a particular span of time, many locations do not have a complete sequence of rocks representing the entire period. All such breaks in the rock record are termed **unconformities**. Figure 15.8 illustrates

some of the ways in which unconformities may develop. Perhaps the most easily recognized type of unconformity consists of tilted or folded sedimentary rocks that are overlain by other, more flat-lying strata. These are called **angular unconformities** and indicate that the period of deformation (folding or tilting) and erosion is not represented by sedimentary rocks (Figure 15.9). **Disconformities** may be more difficult to recognize because the strata on either side of these unconformities are essentially parallel. Disconformities may represent either a period of nondeposition or a period of erosion.

By applying the principles of relative dating to the hypothetical geologic cross section shown in Figure

FIGURE 15.9 Angular un-conformity as seen from Desert View, Grand Canyon National Park. Here tilted and eroded Precambrian rocks are overlain by younger horizontal strata of Paleozoic age. (Photo by E. J. Tarbuck)

15.10, the rocks and the events in earth history they represent may be placed in their proper sequence. The following statements summarize the logic used to interpret the cross section:

1 Applying the law of superposition, beds A, B, C, and E were deposited, in that order. Since bed D is a sill (a concordant igneous intrusion), it is younger than the rocks that were intruded. Further evidence that the sill is younger than beds C and E are the inclusions in the sill of fragments from these beds. If the igneous mass contains pieces of surrounding rock, the surrounding rock must have been there first.

2 Following the intrusion of the sill (D), the intrusion of the dike (F) occurred. Since the dike cuts through beds A through E, it must be younger than all of them.

3 Next, the rocks were tilted and then eroded. We know the tilting happened first because the up-turned ends of the strata have been eroded. The tilting and erosion, followed by further deposition, produced an angular unconformity.

4 Beds G, H, I, J, and K were deposited in that order, again using the law of superposition. Although the lava flow (bed H) is not a sedimentary rock layer, it is a surface-deposited layer, and thus superposition may be applied.

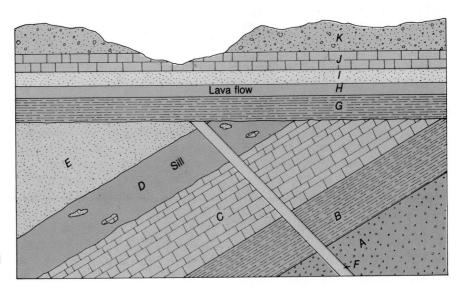

FIGURE 15.10 Geologic cross section of a hypothetical region.

FIGURE 15.11 Correlation of strata within a small area. (After U.S. Geological Survey)

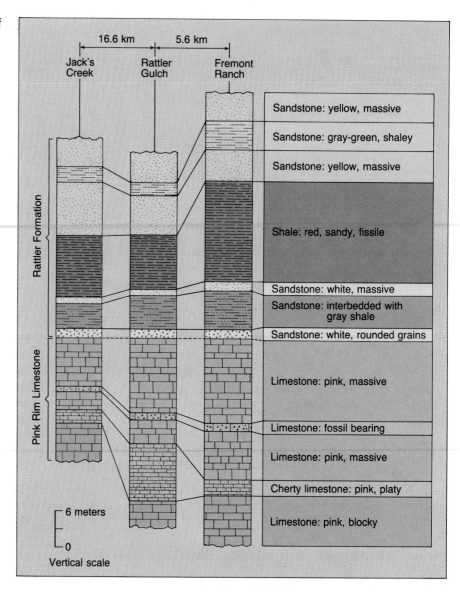

5   Finally, the irregular surface and the stream valley indicate that another gap in the rock record is being produced by erosion.

In the foregoing example our goal was to establish a relative time scale for the rocks and events in the area of the cross section. Remember, we do not have any idea how many years of earth history are represented, nor do we know how the ages of the strata in this area compare to any other area.

CORRELATION

In order to develop a geologic calendar that is applicable to the whole earth, rocks of similar age in dif-

ferent regions must be matched up. Such a task is referred to as **correlation**. Within a limited area there are several methods of correlating the rocks of one locality with those of another. A bed or series of beds may be traced simply by walking along the outcropping edges. This may not be possible, however, when the continuity of the bed is interrupted. Correlation over short distances is often achieved by noting the place of a bed in a sequence of strata, or a bed may be identified in another location if it is composed of very distinctive or uncommon minerals (Figure 15.11). By correlating the rocks from one place to another, a more comprehensive view of the

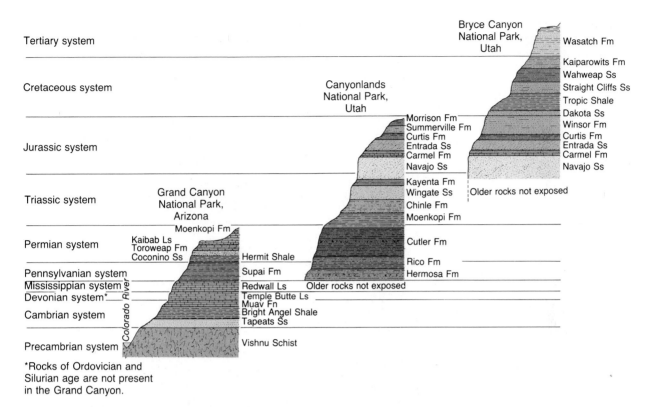

Tertiary system

Cretaceous system

Jurassic system

Triassic system

Permian system

Pennsylvanian system

Mississippian system

Devonian system*

Cambrian system

Precambrian system

Bryce Canyon National Park, Utah — Wasatch Fm
Kaiparowits Fm
Wahweap Ss
Straight Cliffs Ss
Tropic Shale
Dakota Ss
Winsor Fm
Curtis Fm
Entrada Ss
Carmel Fm
Navajo Ss

Canyonlands National Park, Utah

Morrison Fm
Summerville Fm
Curtis Fm
Entrada Ss
Carmel Fm
Navajo Ss

Older rocks not exposed

Kayenta Fm
Wingate Ss
Chinle Fm
Moenkopi Fm

Cutler Fm

Rico Fm

Hermosa Fm

Older rocks not exposed

Grand Canyon National Park, Arizona

Moenkopi Fm
Kaibab Ls
Toroweap Fm
Coconino Ss
Hermit Shale
Supai Fm
Redwall Ls
Temple Butte Ls
Muav Fn
Bright Angel Shale
Tapeats Ss
Vishnu Schist

Colorado River

*Rocks of Ordovician and Silurian age are not present in the Grand Canyon.

FIGURE 15.12 Correlation of strata at three locations on the Colorado Plateau reveals the total extent of sedimentary rocks in the region. (After U.S. Geological Survey)

geologic history of a region is possible (Figure 15.12).

Most geologic studies involve rather small areas. Although they are usually important in their own right, their full value is realized only when the rocks are correlated with those of other regions. Although the methods just described may be sufficient to trace a rock formation over relatively short distances, they are not adequate for matching rocks separated by great distances. When correlation between widely separated areas or between continents is the objective, the geologist must rely upon fossils.

FOSSILS

Paleontology, the branch of geology devoted to the study of ancient life, is based on the study of fossils. Originally the term **fossil** referred to any curious object dug from the ground, but it is now used to mean the remains or traces of organisms pre-

served from the geologic past. Although geologists studying past life must have a firm background in the biological sciences, they are often quick to point out the differences between the two fields: "If the remains stink they belong to zoology, but if not, to paleontology."

Fossils are of many types. The remains of relatively recent organisms may not have been altered at all. Such objects as teeth, bones, and shells are common examples. Far less common are entire animals, flesh included, that have been preserved because of rather unusual circumstances. Remains of prehistoric elephants called mammoths that were frozen in the Arctic tundra of Siberia and Alaska are examples, as are the mummified remains of sloths preserved in a dry cave in Nevada.

Given enough time, the remains of an organism are likely to be modified. Often fossils become *petrified* (literally, "turned into stone") meaning that the original substance, such as wood or bone, has been replaced by mineral matter from circulating solutions

FIGURE 15.13 Petrified wood in Petrified Forest National Park, Arizona. (Photo by E. J. Tarbuck)

FIGURE 15.14 Natural casts of shelled invertebrates. (Photo by E. J. Tarbuck)

or that open spaces have been filled by minerals (Figure 15.13). Although many minerals may act as petrifying agents, silica (SiO_2) is among the more common materials composing petrified remains.

Molds and casts constitute another common class of fossils. When a shell or other structure is buried in sediment and then dissolved by underground water, a *mold* is created. The mold faithfully reflects only the shape and surface markings of the organism, but does not reveal any information concerning its internal structure. If these hollow spaces are subsequently filled with mineral matter, *casts* are created (Figure 15.14).

A type of fossilization called *carbonization* is particularly effective in preserving leaves and delicate animal forms. It occurs when fine sediment encases the remains of an organism. As time passes, pressure squeezes out the liquid and gaseous components and leaves behind a thin residue of carbon (Figure 15.15). Black shales deposited as organic-rich mud in oxygen-poor environments often contain abundant carbonized remains. If the film of carbon is lost from a fossil preserved in fine-grained sediment, a

replica of the surface, called an *impression,* may still show considerable detail (Figure 15.16).

Delicate organisms, such as insects, are difficult to preserve and consequently are quite rare in the fossil record. Not only must they be protected from decay, they must not be subjected to any pressure that would crush them. One way in which some insects have been preserved is in *amber,* the hardened resin of ancient trees. The fly in Figure 15.17 was preserved after being trapped in a drop of sticky resin. Resin sealed off the insect from the atmosphere and protected the remains from damage by water and

FIGURE 15.15 A fossil bee preserved as a thin carbon film. (Photo courtesy of the National Park Service)

FIGURE 15.16 Impressions are common fossils and often show considerable detail. (Photo by E. J. Tarbuck)

air. As the resin hardened, a protective, pressure-resistant case was formed.

In addition to the fossils already mentioned, there are numerous other types, many of them only traces of prehistoric life. Examples of such indirect evidence include:

1   Tracks—footprints made by animals in soft sediment that was later lithified.
2   Burrows—tubes in sediment, wood, or rock made by an animal. These holes may later become filled with mineral matter and preserved. Some of the oldest-known fossils are believed to be worm burrows.
3   Coprolites—fossil dung and stomach contents that can provide useful information pertaining to food habits of organisms.
4   Gastroliths—highly polished stomach stones that were used in the grinding of food by some extinct reptiles.

Sometimes objects which people think are fossils are merely rocks with an accidental resemblance. George Gaylord Simpson, a noted paleontologist, related this experience:

> Only the other day I was offered for sale at a large price "the petrified leg of a woman." I was called a liar and a cheat when I explained that it was only a piece of volcanic rock with an accidental (and very slight) resemblance to the vision in the mind of the owner.

Conditions Favoring Preservation

Only a tiny fraction of the organisms that have lived during the geologic past have been preserved as fossils. Normally the remains of an animal or plant are totally destroyed. Under what circumstances are they preserved? Two special conditions appear to be necessary: rapid burial and the possession of hard parts.

Usually when an organism perishes, its soft parts are quickly eaten by scavengers or decomposed by bacteria. The remaining hard parts are then weathered, eventually crumbling into dust. Occasionally, however, the remains are buried by sediment. In this situation, scavengers and weathering cannot disturb them because they have been removed from the environment where these destructive processes operate. Rapid burial following death therefore is an important condition favoring preservation.

In addition, animals and plants have a much better chance of being preserved as part of the fossil record if they have hard parts. Although traces and imprints of soft-bodied animals such as jellyfish, worms, and insects exist, they are rare. Flesh usually decays so rapidly that preservation is exceedingly

FIGURE 15.17 Insect in amber. (Courtesy of Ward's Natural Science Establishment, Inc., Rochester, N.Y.)

remote. Hard parts like shells, bones, and teeth predominate in the record of past life.

Because preservation is contingent on special conditions, the record of life in the geologic past is slanted. While the record of those organisms with hard parts that lived in areas of sedimentation is quite complete, we only get an occasional glimpse at the rest.

Fossils and Correlation

The existence of fossils had been known for centuries, yet it was not until the late 1700s and early 1800s that their significance as geologic tools was made evident. During this period an English engineer and canal builder, William Smith, discovered that each rock formation in the canals contained fossils unlike those in the beds either above or below. Further, he noted that sedimentary strata in widely separated areas could be identified by their distinctive fossil content. Based upon Smith's classic observations and the findings of many geologists who followed, one of the most important and basic principles in historical geology was formulated: Fossil organisms succeed one another in a definite and determinable order, and therefore any time period can be recognized by its fossil content. This has come to be known as the **principle of faunal succession**. In other words, when fossils are arranged according to their age by using the law of superposition on the rocks in which they are found, they do not present a random or haphazard picture. To the contrary, fossils show progressive changes from simple to complex and reveal the advancement of life through time. For example, an Age of Trilobites is recognized quite early in the fossil record. Then, in succession, paleontologists recognize an Age of Fishes, an Age of Coal Swamps, an Age of Reptiles, and an Age of Mammals. These "ages" pertain to groups that were especially plentiful and characteristic during particular time periods. Within each of the "ages" there are many subdivisions based, for example, on certain species of trilobites, and certain types of fish, reptiles, and so on. This same succession of dominant organisms, never out of order, is found on every major landmass.

Since fossils were found to be time indicators, they became the most useful means of correlating rocks of similar age in different regions. Geologists pay particular attention to certain fossils called **index fossils**. Due to the fact that these fossils are widespread geographically and are limited to a short span of geologic time, their presence provides an important method of matching rocks of the same age. Rock formations, however, do not always contain a specific index fossil. In such situations, groups of fossils are used to establish the age of the bed. Figure 15.18 illustrates how a group of fossils can be used to date rocks more precisely than could be accomplished by the use of only one of the fossils.

In addition to being important and often essential tools for correlation, fossils are important environmental indicators. Although much can be deduced about past environments by studying the nature and characteristics of sedimentary rocks, a close examination of the fossils present can usually provide a great deal more information. For example, when the remains of certain clam shells are found in lime-

FIGURE 15.18 Overlapping ranges of fossils help date rocks more exactly than using a single fossil.

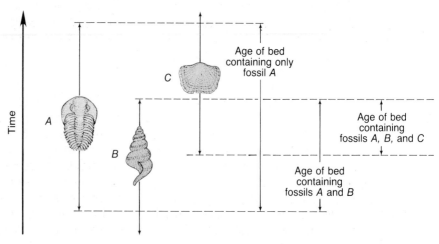

stone, the geologist feels it is quite reasonable to assume that the region was once covered by a shallow sea. Also, by using what we know of living organisms, we can conclude that fossil animals with thick shells capable of withstanding pounding and surging waves inhabited shorelines. On the other hand, animals with thin, delicate shells probably indicate deep, calm offshore waters. Hence by taking a closer look at the types of fossils, the approximate shoreline may be identified. Further, fossils can be used to indicate the former temperature of the water. Certain kinds of present-day corals must live in warm and shallow tropical seas like those around Florida and the Bahamas. When similar types of coral are found in ancient limestones, they give a good estimate of the marine environment that must have existed when they were alive. The preceding are just a few brief examples of how fossils can help unravel the complex story of earth history.

THE GEOLOGIC CALENDAR

The whole of geologic history has been subdivided into units of varying magnitude which together comprise the calendar of earth history (Figure 15.19). The major units of the calendar were delineated during the nineteenth century, principally by workers in western Europe and Great Britain. Since absolute dating was not a reality during this time, the entire calendar was created using methods of relative dating. Absolute dates have only recently been added to the calendar.

By examining Figure 15.19, you can see that the largest of the subdivisions of the geologic calendar are called **eras**. Three eras are currently recognized: the **Paleozoic** ("ancient life"), the **Mesozoic** ("middle life"), and the **Cenozoic** ("recent life"). As the names imply, the eras are bounded by quite profound worldwide changes in life forms. Each era is subdivided into time units known as **periods**. The Paleozoic has seven, the Mesozoic three, and the Cenozoic two. Since we are currently living in the Cenozoic era, there may be more periods yet to come. Each period is characterized by a somewhat less profound change in life forms as compared with the eras. The major divisions, with brief explanations of each, are shown in Table 15.2 on page 295. Finally, each of the twelve periods is divided into still smaller units called **epochs**. Except for the seven

epochs which have been named for the periods of the Cenozoic era, those of other periods are not commonly referred to by specific names. Rather, the terms *early, middle,* and *late* are generally applied to the epochs of these earlier periods.

Notice that the detail of the geologic calendar does not begin until about 600 million years ago, the date for the beginning of the first period of the Paleozoic era: the Cambrian period. The more than 4 billion years prior to the Cambrian is simply referred to as the **Precambrian**. In terms of life forms, the Precambrian is often given another name, the *Cryptozoic eon.* The term *Cryptozoic* is derived from the Greek words for *hidden* and *life,* and refers to the relatively obscure fossil record of this time span. By contrast, the time span beginning with the Cambrian period and extending to the present is referred to as the *Phanerozoic eon.* This term is also derived from the Greek and means *visible life.* Thus, the Phanerozoic eon represents the age of abundant life.

Why is the huge expanse of Precambrian time not divided into numerous eras, periods, and epochs? The reason is that Precambrian history is not known in great enough detail. The quantity of information geologists have deciphered about the earth's past is somewhat analogous to the detail of human history. The farther back we go, the less that is known. Certainly more data and information exist about the past ten years than for the first decade of the twentieth century; the events of the nineteenth century have been documented much better than the events of the first century A.D.; and so on. So it is with earth history. The more recent past has the freshest, least disturbed, and most observable record. The farther back in time the geologist goes, the more fragmented the record and clues become.

DIFFICULTIES IN DATING THE GEOLOGIC CALENDAR

Although reasonably accurate absolute dates have been worked out for the periods of the geologic calendar (see Figure 15.19), the task is not without its difficulties. The primary difficulty in assigning absolute dates to units of time lies in the fact that radioactive elements are typically restricted to igneous rocks. Even if a detrital rock included sediment which contained a radioactive mineral, the rock

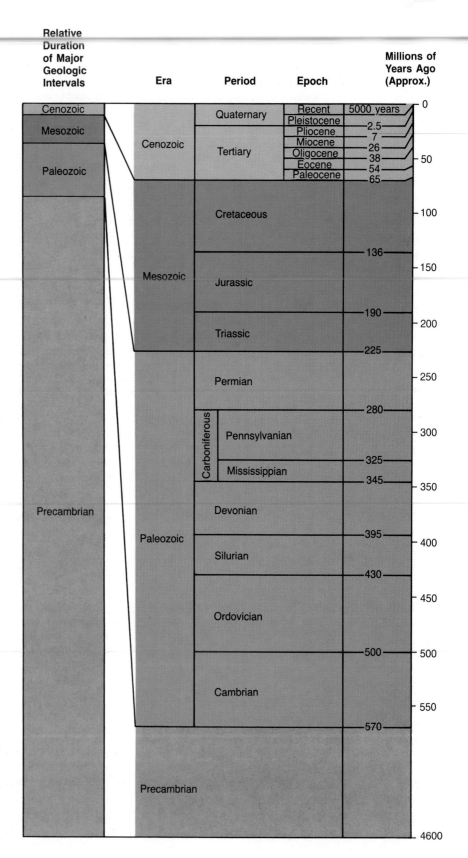

Relative Duration of Major Geologic Intervals	Era	Period	Epoch		Millions of Years Ago (Approx.)
Cenozoic	Cenozoic	Quaternary	Recent	5000 years	0
Mesozoic			Pleistocene		2.5
		Tertiary	Pliocene		7
Paleozoic			Miocene		26
			Oligocene		38
			Eocene		54
			Paleocene		65
	Mesozoic	Cretaceous			100
					136
		Jurassic			150
					190
		Triassic			200
					225
	Paleozoic	Permian			250
					280
		Carboniferous Pennsylvanian			300
					325
		Carboniferous Mississippian			345
		Devonian			350
					395
		Silurian			400
					430
		Ordovician			450
					500
Precambrian		Cambrian			550
					570
		Precambrian			4600

FIGURE 15.19 The geologic calendar. Absolute dates were added quite recently, long after the calendar had been established using relative dating techniques. The Precambrian accounts for more than 85 percent of geologic time.

TABLE 15.2 Major divisions of geologic time.

Cenozoic Era **(Age of Recent Life)**	Quaternary period	The several geologic eras were originally named Primary, Secondary, Tertiary, and Quaternary. The first two names are no longer used; Tertiary and Quaternary have been retained but used as period designations.
	Tertiary period	
Mesozoic Era **(Age of Middle Life)**	Cretaceous period	Derived from Latin word for chalk (creta) and first applied to extensive deposits that form white cliffs along the English Channel.
	Jurassic period	Named for the Jura Mountains, located between France and Switzerland, where rocks of this age were first studied.
	Triassic period	Taken from word "trias" in recognition of the threefold character of these rocks in Europe.
Paleozoic Era **(Age of Ancient Life)**	Permian period	Named after the province of Perm, U.S.S.R., where these rocks were first studied.
	Pennsylvaniar period	Named for the state of Pennsylvania where these rocks have produced much coal.
	Mississippian period	Named for the Mississippi River valley where these rocks are well exposed.
	Devonian period	Named after Devonshire County, England, where these rocks were first studied.
	Silurian period	Named after Celtic tribes, the Silures and the Ordovices, that lived in Wales during the Roman Conquest.
	Ordovician period	
	Cambrian period	Taken from Roman name for Wales (Cambria) where rocks containing the earliest evidence of complex forms of life were first studied.
Precambrian		The time between the birth of the planet and the appearance of complex forms of life. More than 85 percent of the earth's estimated 4.6–4.8 billion years falls into this span.

Source: U.S. Geological Survey.

would not be dated, for the grains in a detrital sedimentary rock are not the same age as the rock in which they occur. The sediments composing such a rock may have been weathered from rocks of diverse ages. Thus, the age of a mineral in a sedimentary rock only tells us that the rock can be no older.

On the other hand, in igneous rocks the minerals and rock form simultaneously; the age of the mineral containing a radioactive isotope is the same age as the rock. Therefore, in order to date sedimentary strata, the geologist must relate them to igneous masses, as in Figure 15.20. In this example, the age

FIGURE 15.20 Absolute dates for sedimentary layers are usually determined by examining their relationship to igneous rocks. (After U.S. Geological Survey)

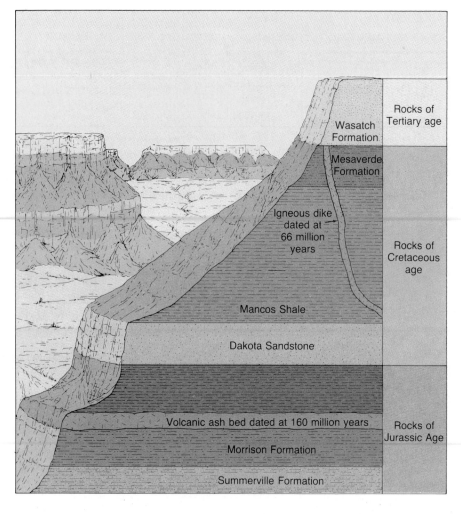

Rocks of Tertiary age

Wasatch Formation

Mesaverde Formation

Igneous dike dated at 66 million years

Rocks of Cretaceous age

Mancos Shale

Dakota Sandstone

Volcanic ash bed dated at 160 million years

Rocks of Jurassic Age

Morrison Formation

Summerville Formation

of the volcanic ash bed within the Morrison Formation and the dike cutting the Mancos Shale and Mesaverde Formation are known. The sedimentary beds below the ash are obviously older than the ash, and all the layers above the ash are younger. The dike is younger than the Mancos Shale and the Mesaverde Formation but older than the Wasatch Formation because the dike does not intrude the Tertiary rocks. From this kind of evidence, geologists estimate that a part of the Morrison Formation was

deposited about 160 million years ago as indicated by the ash bed. Further, they conclude that the Tertiary period began after the intrusion of the dike, 66 million years ago. This is one example of the literally thousands that illustrates how dated materials are used to bracket the various episodes in the history of the earth within specific time periods and illustrates the necessity of combining laboratory dating methods with field observations of rocks.

REVIEW QUESTIONS

1 Describe two early methods for dating the earth. How old was the earth thought to be according to these estimates? List some weaknesses of each method.

2 If a radioactive isotope of thorium (atomic number 90, mass number 232) emits 6 alpha particles and 4 beta particles during the course of radioactive decay, what is the atomic number and mass number of the stable daughter product?

3 Why is radiometric dating the most reliable method of dating the geologic past?

4 A hypothetical radioactive isotope has a half-life of 10,000 years. If the ratio of radioactive parent to stable daughter product is 1:3, how old is the rock containing the radioactive material?

5 Assume that the age of the earth is 5 billion years.
  a What fraction of geologic time is represented by recorded history (assume 5000 years for the length of recorded history)?
  b The first abundant fossil evidence does not appear until the beginning of the Cambrian period (600 million years ago). What percent of geologic time is represented by abundant fossil evidence?

6 Distinguish between absolute and relative dating.

7 What is the law of superposition? How are cross-cutting relationships used in relative dating?

8 When you observe an outcrop of steeply inclined sedimentary layers, what principle allows you to assume that the beds were tilted after they were deposited?

9 What is meant by the term *correlation?*

10 Describe several types of fossils. What organisms have the best chance of being preserved as fossils?

11 Describe William Smith's important contribution to the science of geology.

12 Why are fossils such useful tools in correlation?

13 In addition to being important aids in dating and correlating rocks, how else are fossils helpful in geologic investigations?

14 What subdivisions make up the geologic calendar? What is the primary basis for differentiating the eras?

15 Briefly describe the difficulties in assigning absolute dates to layers of sedimentary rock.

CHAPTER SIXTEEN

EARTH HISTORY:
A BRIEF SUMMARY

ORIGIN OF THE EARTH

The earth is one of nine planets that along with several dozen moons and numerous smaller bodies revolve around the sun. The orderly nature of our solar system led most astronomers to conclude that its members formed at essentially the same time and from the same primordial material. This proposal, known as the **nebular hypothesis**, suggests that the bodies of our solar system formed from an enormous cloud composed mostly of hydrogen and helium with only a small percentage of all the other heavier elements.

About 5 billion years ago, and for reasons that are not yet fully understood, this huge cloud of minute rocky fragments and gases began to contract under its own gravitational influence (Figure 16.1). The contracting material is assumed to have had some component of rotational motion, which, like a spinning ice skater pulling in her arms, rotated faster and faster as it contracted. This rotation in turn caused the nebular cloud to assume a flattened disklike shape. Within the rotating disk, relatively small eddylike contractions formed the nuclei from which the planets would eventually develop. However, the greatest concentration of material was pulled toward the center of this rotating mass and gravitationally heated, forming the hot *protosun*.

In a relatively short time after the formation of the protosun, the temperature within the rotating disk dropped significantly. This decrease in temperature caused substances with high melting points to condense into small particles, perhaps the size of sand grains. Materials such as iron and nickel solidified first. Next to condense were the elements of which rocky substances are composed. As these fragments collided, they joined into larger objects that in a few

tens of millions of years accreted into the planets. In the same manner, but on a lesser scale, the processes of condensation and accretion acted to form the moons and other small bodies of the solar system.

As the *protoplanets* (planets in the making) accumulated more and more debris, the solar system began to clear. The removal of debris allowed sunlight to heat the surfaces of the newly-formed planets. The resulting high surface temperatures of the inner planets, coupled with the fact that these bodies possessed comparatively weak gravitational fields, meant that the earth and its neighbors, Mercury, Venus, and Mars, were unable to retain appreciable amounts of the lighter components of the primordial cloud. These materials, which included hydrogen, helium, ammonia, methane, and water, vaporized from their surfaces and were eventually wisked from the inner solar system by the solar winds. At distances beyond Mars temperatures are quite low. Consequently the large outer planets, Jupiter, Saturn, Uranus, and Neptune, accumulated huge amounts of hydrogen and other light materials from the primordial cloud. The accumulation of these gaseous substances is thought to account for the comparatively large sizes and low densities of the outer planets.

Shortly after the earth formed, the decay of radioactive elements, coupled with heat released by colliding particles, produced at least some melting of the interior. Melting, in turn, is thought to have allowed the heavier elements, principally iron and nickel, to sink, while the lighter rocky components floated upward. This segregation of material, which began early in the earth's history, is believed to be occurring still, but on a much smaller scale. As a result of this chemical differentiation, the earth's interior is not homogeneous. Rather, it consists of shells or spheres composed of materials having different properties.

A coiled cephalopod of Cretaceous age. (Photo by E. J. Tarbuck)

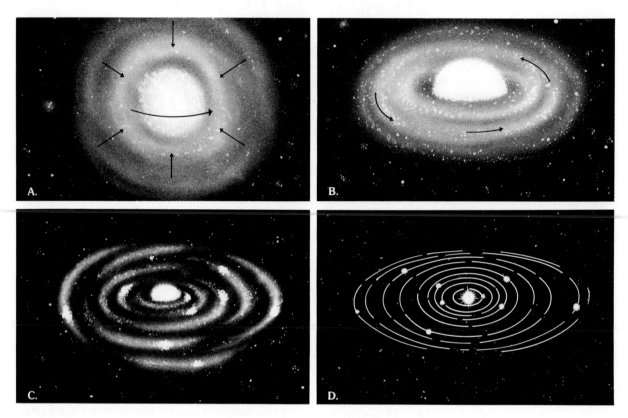

FIGURE 16.1 Nebular hypothesis. **A.** A huge cloud of dust and gases begins to contract. **B.** Due to its rotational motion the cloud forms into a flattened disk. **C.** The planets then begin to accrete along the disk while most of the material is gravitationally swept toward the center, producing the sun. **D.** In time most of the remaining debris is collected into the nine planets and their moons.

An important consequence of the period of chemical differentiation is that gaseous materials were allowed to escape from the earth's interior, similar to what happens today during volcanic eruptions. By this process an atmosphere composed chiefly of gases expelled from within the planet gradually evolved. It is on this planet, with this atmosphere that life as we know it came into existence.

ORIGIN OF THE ATMOSPHERE

The atmosphere has not always consisted of the same relatively stable mixture of gases that we breathe today. On the contrary, the present mixture of gases that makes up our atmosphere is the result

of very gradual change, a slow evolutionary process that began soon after the earth formed.

Scientists believe that the earth's earliest atmosphere was swept away by solar winds—vast streams of particles emitted by the sun. As the earth cooled, a solid crust formed and the gases that had been dissolved in the molten rock were gradually released. This process is termed **outgassing**. Thus, an atmosphere believed to be made up of gases similar to those released during volcanic eruptions came into being. The principal components of this atmosphere were probably water vapor, carbon dioxide, and nitrogen.

As the earth continued to cool, clouds formed and great rains commenced. At first, the water evaporated before reaching the surface or was quickly boiled away. This helped speed the cooling of the earth's surface. When the earth had cooled sufficiently, torrential rains filled the ocean basins. This event not only diminished the amount of water vapor in the air, but it carried away much of the carbon dioxide as well.

We are now faced with an interesting paradox. If the earth's primitive atmosphere resulted from volcanic outgassing, it could not have contained free oxygen, because free oxygen is not emitted during this process. How then did our present oxygen-rich atmosphere come into existence? Scientists have proposed two probable sources of the free oxygen in the atmosphere. It is known that water vapor that is carried into the upper atmosphere is dissociated into hydrogen and oxygen by the action of the sun's ultraviolent radiation. Hydrogen, a very light gas, escapes the atmosphere, whereas the heavier oxygen atoms remain and combine to form molecular oxygen (O_2). Although there is little doubt that some of the atmosphere's free oxygen was created in this manner, this very slow process does not adequately account for the present percentage of oxygen in our atmosphere.

A second, more important, source of oxygen is believed to have been (and, indeed, continues to be) green plants. By the process of photosynthesis, plant life utilizes sunlight to generate oxygen by changing water and atmospheric carbon dioxide into organic matter. This method of oxygen production obviously implies the need for life on earth prior to the time that free oxygen was present in the atmosphere. Scientists believe that the first life forms, probably bacteria, carried out their metabolic processes without oxygen. Even today many of these anaerobic bacteria still exist. Next, primitive green plants evolved, which, in turn, supplied most of the free oxygen to support higher life forms. Slowly the amount of oxygen in the atmosphere increased. Evidence from the Precambrian rock record suggests that the first free oxygen combined with certain substances dissolved in water, especially iron. Then, once these mineral oxidation needs were met, substantial quantities of free oxygen began to accumulate in the atmosphere. By the beginning of the Paleozoic era, the fossil record reveals that organisms that require oxygen were abundant in the sea. Hence, the makeup of the atmosphere is directly related to the life forms on earth, and its composition evolved through time

from an oxygen-free environment to one that contained significant amounts of free oxygen.

THE PRECAMBRIAN

The Precambrian encompasses an immense amount of geologic time, from the earth's distant beginnings more than 4.6 billion years ago until the start of the Cambrian period about 4 billion years later. A glance at the portion of Figure 15.19 which illustrates the relative duration of major geologic intervals reveals that the Precambrian represents most of geologic time. In fact, it accounts for about seven-eighths of earth history.

Unraveling the Precambrian

Over large expanses of the continents outcrops of Precambrian rocks are rare because of an overlying blanket of younger strata. In such regions exposures are confined to the cores of extensively eroded mountains such as the Rockies and to the bottoms of deep canyons such as the Grand Canyon (Figure 16.2). There are, however, extensive regions known as **shields**, where Precambrian rocks are dominant. These geologically complex areas consist largely of deformed metamorphic rocks and were termed shields because of their gently convex profiles, similar to a warrior's shield in cross section. Every continent has one or more of these core regions of ancient Precambrian rocks. In North America, the Canadian shield encompasses 7.2 million square kilometers (2.8 million square miles), when the nearby Greenland shield is included.

The Precambrian is the least understood span of earth history. As a consequence it has not been divided into numerous time units that have gained worldwide acceptance. Indeed, the history of this vast segment of geologic time consists mainly of scattered episodes that are more sketchy and speculative than the geologic history of more recent times. One important reason for this fact is that shields and other Precambrian exposures consist of complex metamorphic rocks that in many cases have been intruded by large igneous bodies. Some rocks have been folded, faulted, intruded, and metamorphosed many times. Each event has tended to obscure

FIGURE 16.2 Precambrian Vishnu Schist exposed at the bottom of the Grand Canyon. (Photo by E. J. Tarbuck)

earlier events, making Precambrian history difficult to decipher.

An additional factor that complicates the study of Precambrian history is the scarcity of fossils. Unfortunately for those attempting to study the Precambrian, abundant fossil evidence does not appear in the geological record until the beginning of the Paleozoic era. Life existed during the Precambrian, but the forms were simple and soft bodied. For that reason, there is only a meager Precambrian fossil record. Although many Precambrian exposures have been studied in great detail, correlation is exceedingly difficult without fossils. This fact helps explain why no widely accepted subdivisions are established for Precambrian time. With the development and application of radiometric dating, a solution to the troublesome task of dating and correlating Precambrian rocks now exists. Untangling the complex

Precambrian record, however, is still far from being accomplished. In summary, understanding Precambrian history is difficult because the meager fossil content hinders dating and correlation, the rocks of this age are often highly metamorphosed and deformed, and the record is fragmentary owing to extensive erosion and the presence of younger overlying strata.

Before concluding this discussion of Precambrian rocks, mention should be made of the many Precambrian regions that contain important ore deposits. In fact, the search for these economically significant minerals yields the information that forms the basis of our understanding of Precambrian geology. Precambrian rocks are important sources of almost every kind of mineral resource except fossil fuels. Iron, nickel, gold, silver, copper, chromium, uranium, and diamonds are mined in important quantities in Precambrian regions. Among the more noteworthy are extensive iron ore deposits. Rocks from the middle Precambrian (1.2–2.5 billion years ago) contain most of the world's large iron deposits, mainly in the form of hematite (Fe_2O_3). These iron-rich sedimentary rocks appear to represent a time when oxygen became sufficiently abundant to react with the dissolved iron in shallow lakes and seas. Later, after much of the iron was removed and deposited, the formation of these rich iron deposits declined, and the oxygen levels in the ocean and the atmosphere began to rise. Since most of the earth's free oxygen results from photosynthesis, the presence of abundant oxygen and consequently the formation of the extensive Precambrian iron ore deposits is linked to life in the sea.

Precambrian Fossils

A century ago the earliest fossils known dated from the beginning of the Cambrian period. A major unanswered problem facing science at that time was the abrupt appearance in the geologic record of complex organisms. Today our knowledge of Precambrian life, although far from complete, is nevertheless quite extensive. Precambrian fossils are not nearly as plentiful as in younger strata, but they do exist.

Perhaps the most common Precambrian fossils consist of distinctly layered mounds or columns of calcium carbonate (Figure 16.3). These structures,

FIGURE 16.3 Stromatolites are layered structures composed of calcium carbonate that is deposited by algae. They are among the most common Precambrian fossils. (Geological Survey of Canada photo no. 155607)

called **stromatolites**, are not the remains of actual organisms, but rather are material deposited by algae. The origin of stromatolites is clear because of their close resemblance to deposits made by modern algae as well as the presence of scattered algal bodies within them. Stromatolites from the early Precambrian are known; however, they do not become common until the middle Precambrian. Although stromatolites are often quite large structures, most other Precambrian fossils are microscopically small. Well-preserved remains of many minute organisms have been discovered that extend the record of life back beyond 3.1 billion years ago. Many of these most ancient fossils are preserved in the sedimentary rock called chert.* This material commonly precipitates from water as nodules or as thin beds. Because it is very durable and resists compression, chert is

excellent for preserving simple organisms such as bacteria and algae. When a sample is believed to contain fossils of these tiny microorganisms, the rock must be cut into very thin slices and studied carefully under strong microscopes.

Microfossils have been discovered at several localities around the world. Ancient Precambrian rocks in southern Africa, dated at more than 3.1 billion years, were found to contain bacteria and blue-green algae. In North America another well-known and important group of Precambrian fossils (approximately 1.7 billion years old) comes from the Gunflint Chert on the north shore of Lake Superior in Canada. Here, as in southern Africa, bacteria and blue-green algae were found. The fossils discovered at these two localities are of the most primitive organisms, termed *prokaryotes*. The cells of such organisms lack organized nuclei and reproduce asexually. Examples of more advanced organisms (termed *eukaryotes* because their cells contain nuclei) include

*Recall from Chapter Four that chert is a very hard, dense chemical sedimentary rock made of microcrystalline silica (SiO_2).

fossils discovered at Bitter Springs in northern Australia. Here the approximately one-billion-year-old remains included green algae. Unlike prokaryotes, eukaryotes reproduce sexually, which means genetic materials are exchanged. Since this mode of reproduction allows for greatly increased genetic variation, the rate of evolutionary change may have increased dramatically after this point.

The fossil record of animals dates from the late Precambrian and includes distinctive trails and burrows believed to have been created by elongate, wormlike animals. In addition, impressions showing complete shapes have also been discovered. Areas in southern Australia and Newfoundland have yielded many hundreds of these specimens. Although most, if not all, Precambrian animals were soft bodied and lacked shells, the fossil record is sufficient to demonstrate the existence of a diverse and complex group of multicelled organisms as the Precambrian came to a close. Thus, the stage was ready for the appearance of even more complex organisms with preservable hard parts as the Cambrian period opened.

THE PALEOZOIC ERA

The geological calendar (Figure 15.19) shows that the last 570 million years of earth history are divided into three eras: Paleozoic, Mesozoic, and Cenozoic. The Paleozoic era, which encompasses more than 345 million years, has by far the longest duration of the three. Seven periods make up the Paleozoic era. They are, in order from most distant to most recent, the Cambrian, Ordovician, Silurian, Devonian, Mississippian, Pennsylvanian, and Permian.*

The beginning of the Paleozoic is marked by the appearance of life forms that possessed hard parts. The most abundant of these early organisms were the trilobites (Figure 16.4). Hard parts greatly enhanced an organism's chance of being preserved as part of the fossil record. Therefore, our knowledge of the diversification of life improves greatly from the Paleozoic onward. This abundant fossil evidence has allowed geologists to construct a far more detailed calendar for the last one-eighth of geologic time than for the preceding seven-eighths. More-

*Outside of North America, the Mississippian and Pennsylvanian periods are combined into the Carboniferous period.

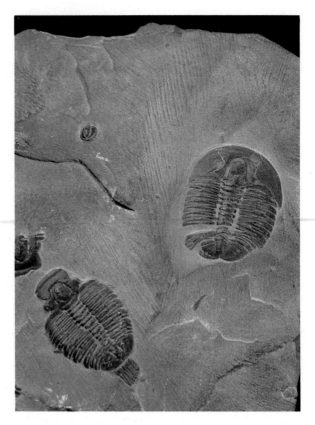

FIGURE 16.4 Fossil trilobites embedded in shale. (Photo by E. J. Tarbuck)

over, since every organism is associated with a particular environment, the greatly improved fossil record provided vast and invaluable information for deciphering past geologic environments.

Early Paleozoic History

For the purpose of this discussion, the early Paleozoic consists of the Cambrian, Ordovician, and Silurian periods. Beginning with the early Paleozoic, geologists have been able to determine approximate positions of the major landmasses (Figure 16.5). At this time, the vast southern continent of Gondwanaland was essentially intact and consisted of present-day South America, Africa, Australia, Antarctica, India, and perhaps China. Evidence of an extensive continental glaciation in western Africa places it near the South Pole. The landmasses that were not part of Gondwanaland existed as three separate units,

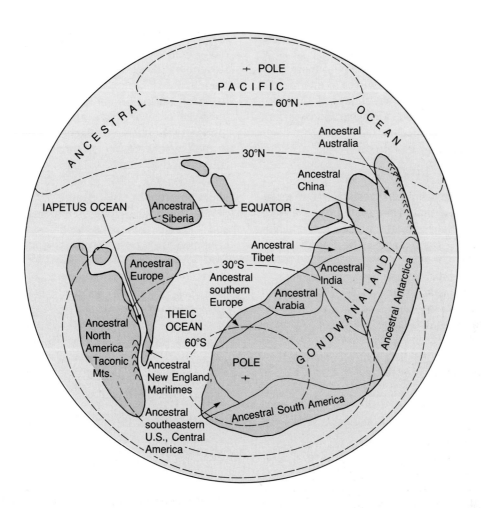

FIGURE 16.5 Reconstruction of the earth as it may have existed in early Paleozoic times. (From Burchfiel et al. *Physical Geology.* Columbus, Ohio: Merrill, 1982)

plus several scattered fragments. Although the exact positions of these "northern" continents are not known with certainty, ancestral North America and Europe are thought to have been located near the equator and separated by a narrow sea.

As the Paleozoic opened, North America was a land with no living things, either plant or animal. There were no Appalachian or Rocky mountains; rather, the continent was largely a barren lowland. Several times during the Cambrian and Ordovician periods shallow seas moved inland and then receded from the interior of the continent. Deposits of clean sandstones, used today to make glass, mark the edge of these shallow seas in the midcontinent. As

the Ordovician drew to a close, a mountain-building event occurred that affected eastern North America from the present-day central Appalachians to Newfoundland. The mountains produced during this event, which is often called the Taconic orogeny, have since been worn away, leaving behind deformed strata and a large volume of detrital sedimentary rocks that were derived from the weathering of these mountains.

During the Silurian period much of North America was once again inundated by shallow seas. This time, however, many large barrier reefs were present and several shallow marine basins experienced restricted circulation with the open ocean. Climatic conditions were such that evaporation from these basins was great and triggered deposition of large quantities of rock salt and gypsum. Today these thick evaporite beds are important sources of raw materials for chemical, rubber, plasterboard, and photographic industries in Ohio, Michigan, and western New York state.

Early Paleozoic Life

Life in early Paleozoic time was restricted to the seas and consisted of several invertebrate groups (Figure 16.6). Included were trilobites, graptolites, brachiopods, bryozoans, and mollusks. The Cambrian pe-

riod was the golden age of trilobites. There were more than 600 genera of these mud-burrowing scavengers. By Ordovician times, a group of small marine invertebrates called brachiopods outnumbered the trilobites. Brachiopods are among the most common and abundant Paleozoic fossils. Although the adults lived attached to the sea floor, the young larvae were free-swimming. This mobility accounts for the group's wide geographic distribution. The Ordovician also marked the appearance of abundant cephalopods (Figure 16.6). These highly developed mollusks took on a more mobile life style and became the major predators of their time. Cephalopods, whose descendants include the mod-

FIGURE 16.6 The shallow waters of an Ordovician inland sea contained an abundance of marine invertebrates. Shown here are straight-shelled cephalopods, trilobites, brachiopods, snails, and corals. (Courtesy of the Field Museum of Natural History, Chicago)

ern squid, octopus, and nautilus, were the first truly large organisms on earth. Whereas the largest trilobites seldom exceeded 30 centimeters (12 inches) in length and the biggest brachiopods were no more than about 20 centimeters (8 inches) across, one species of cephalopod reached a length of nearly 10 meters (30 feet).

The beginning of the Cambrian period marks an important event in animal evolution. For the first time organisms existed that had the capability of secreting hard parts. Why several diverse life forms began to develop hard parts at about the same time remains unanswered. One proposal suggested that Precambrian oceans were deficient in calcium, a major constituent in the shells of many marine invertebrates. However, conspicuous late Precambrian limestone (calcium carbonate) deposits are now known, indicating that plenty of calcium was available. Another proposal suggested that since an external skeleton would have provided protection from predators, it would have had great survival value. Yet, the sequence of events revealed in the fossil record does not seem to support this hypothesis. Organisms with hard parts were plentiful in the Cambrian period, whereas predators such as the cephalopods were not abundant until the Ordovician period. Whatever the answer, hard parts clearly served many useful purposes and aided adaptations to new ways of life. Sponges, for example, developed a network of fine interwoven silica spicules that allowed them to grow larger and more erect, capable of extending above the surface in search of food. Mollusks (clams and snails) secreted external shells of calcium carbonate that provided protection as well as a covering that allowed body organs to function in a more controlled environment. The successful trilobites developed an exoskeleton composed of a protein called chitin, which permitted them to burrow through soft sediment in search of food.

Late Paleozoic History

The late Paleozoic consists of four periods known as the Devonian, Mississippian, Pennsylvanian, and Permian. This span of earth history is marked by a major reorganization of the earth's landmasses, which culminated during Permian times in the formation of the supercontinent Pangaea (Figure 16.7).

The first major event in the formation of Pangaea occurred as ancestrial North America collided with Europe. Beginning at the close of the Silurian and extending through the Devonian period, the narrow sea that separated these landmasses began to close slowly (see Figure 16.5). The strong compressional forces associated with this continental collision deformed the rocks that flanked these continents and produced the original northern Appalachian Mountains of eastern North America and the Caledonian Mountains of northwestern Europe. The fusion of these two landmasses left the continents arranged in three major groups: Gondwanaland (the southern supercontinent), Euramerica (Europe and North America), and Siberia. By Permian times, the newly formed continent of Euramerica had collided with the Siberian landmass along the line of the Ural

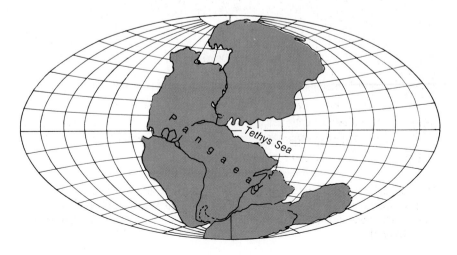

FIGURE 16.7 Reconstruction of Pangaea as it is thought to have appeared 200 million years ago. (After Robert S. Dietz and John C. Holden. *Journal of Geophysical Research* 75: 4943. Copyright © by American Geophysical Union)

Mountains. Through this union the northern continent of Laurasia, which included North America, Europe, and Siberia, was born.

During the late Paleozoic, as Laurasia was being formed, Gondwanaland was migrating northward. By the Pennsylvanian period, Gondwanaland collided with Laurasia to form a mountainous belt through central Europe. Simultaneously, a collision occurred between the African fragment of Gondwanaland and the southeastern edge of North America to produce the southern Appalachians.

By the close of the Paleozoic, all the continents had collided and fused together into the supercontinent of Pangaea. With only a single vast continent, the world's climate became very seasonal, having extremes far greater than those presently experienced on earth. As we shall see, these altered climatic conditions resulted in a period of extinction that represents one of the most dramatic biological declines in all of earth history.

Late Paleozoic Life

During most of the late Paleozoic, numerous groups of organisms underwent a period of great diversification. As the Silurian gave way to the Devonian, some 400 million years ago, plants that had adapted to survival at the water's edge began to move inland. These earliest land plants were leafless, vertical spikes about the size of a person's index finger. However, by the end of the Devonian, the fossil record indicates the existence of forests with trees tens of meters high.

In the oceans, armor-plated fishes that had evolved during the Ordovician made major adaptive radiations. Their armor plates became thin, lightweight scales which increased their speed and mobility. Other fishes evolved during the Devonian, including primitive sharks and boney fishes, the groups to which virtually all modern fishes belong. Because the Devonian period was a time when fishes evolved rapidly, it is often called the "age of fishes."

By late Devonian time, two groups of boney fishes, the lung fish and the lobe-finned fish, became adapted to land environments. Not unlike their modern relatives, these fishes had primitive lungs that supplemented the exchange of gases that took place through the gills. It is believed that the lobe-finned fish occupied tidal flats or small ponds. More-

over, it is thought that in times of drought they may have been able to use their boney fins to "walk" from dried-up pools in search of other ponds. Through time, the lobe-finned fish began to rely more on their lungs and less on their gills. By late Devonian time, true air-breathing amphibians with fish-like heads and tails began to invade the land.

Although modern amphibians, which include frogs, toads, and salamanders, are small in size and occupy rather limited biological niches, the conditions during the remainder of the Paleozoic were ideal for these newcomers to the land. Plants and insects, which were their main diet, were very abundant and large. Having only minimal competition from other land dwellers, the amphibians rapidly diversified. Some groups took on roles and forms that were more similar to modern reptiles, such as crocodiles, than to modern amphibians.

By the Pennsylvanian period, large tropical swamps extended across North America, Europe, and Siberia (Figure 16.8). Trees grew to heights approaching 30 meters (100 feet) with trunks over one meter across. Some common trees of the time included the scale trees *Lepidodendron* and *Sigillaria,* the seed ferns, and the scouring rushes called *Calamites.* The coal deposits that fueled the industrial revolution originated in these vast swamps. Further, it was in the lush coal swamp environment of the late Paleozoic that the amphibians radiated quickly into a variety of species.

The Paleozoic ended with the Permian period, a time when the earth's major landmasses joined to form the supercontinent Pangaea. The redistribution of land and water that resulted from the creation of Pangaea, as well as the changes in elevations of the landmasses, brought about pronounced changes in world climates. Much of the northern continent was elevated above sea level and the climate trended toward a period of greater aridity. This climatic change caused the great scale trees of the coal swamps to become nearly extinct. By the close of the Permian, 75 percent of the amphibian families disappeared, and plants declined in numbers and varieties. Although the amphibians declined, their descendants became the most successful and advanced animals on earth. Marine life was not spared during this time of mass extinctions. Many marine invertebrates that had been dominant during the Paleozoic, including all of the remaining trilo-

FIGURE 16.8 Restoration of a Pennsylvanian coal swamp. Shown are scale trees (left), seed ferns (lower left), and scouring rushes (right). Also note the large dragonfly. (Courtesy of the Field Museum of Natural History, Chicago)

bites, as well as some types of corals and brachiopods, failed to adapt to the widespread environmental changes at the close of the Permian. These animals, like many others, became extinct.

THE MESOZOIC ERA

Spanning nearly 160 million years, the Mesozoic era is divided into three periods: the Triassic, Jurassic, and Cretaceous. This episode of geologic history records the final days of the supercontinent Pangaea and its initial breakup. Moreover, it was a time when organisms that had survived the great Permian extinction began to diversify in spectacular ways. On land, dinosaurs became dominant and remained unchallenged for over 100 million years. Because of this the Mesozoic era is often referred to as the "age of dinosaurs."

Early geologists recognized a profound difference between the kinds of fossils in Permian strata and those discovered in younger Triassic rocks. Clearly one-half of the fossil groups that occurred in late Paleozoic rocks were missing in Mesozoic age rocks. On this basis, it was decided to separate the Paleozoic and Mesozoic at the Permian-Triassic boundary.

Mesozoic History

The Mesozoic era begins with the Triassic period, a time when much of the world's land was above sea level. In fact, in North America no period exhibits a more meager marine sedimentary record than the Triassic. Of the exposed Triassic strata, most are red sandstones and mudstones that lack fossils and contain features indicating a terrestrial environment.

As the Jurassic opened, a major invasion of the sea occurred in western North America. Adjacent to this shallow Jurassic sea, in what is presently the Colorado Plateau region, extensive continental sediments were deposited. The thickest and most prominent of these terrestrial deposits is the Navajo Sandstone, a windblown, white quartz sandstone that in places approaches a thickness of 300 meters (1000 feet). The extent of these massive dunes indicates that a major desert occupied much of the American Southwest during early Jurassic times.

Among the best-known Jurassic deposits is the Morrison Formation, which consists of shales, siltstones, and conglomerates that were deposited on an expansive floodplain. Preserved within the Morrison strata is the world's richest-known storehouse of dinosaur fossils. Included are fossilized bones of huge dinosaurs such as *Brontosaurus, Brachiosaurus,* and *Stegosaurus.*

As the Jurassic period gave way to the Cretaceous, shallow seas once again invaded and retreated from western North America as well as the Atlantic and Gulf coastal regions. The submergence of North America during the Cretaceous was extensive. This is evidenced by the fact that late Cretaceous marine deposits are more widespread than those of any period since the Ordovician. In the west, fossiliferous marine strata are interbedded with nonmarine deposits, such as the well-known Mesa Verde Formation. Further, coal deposits of Cretaceous age are very important economically in both the western United States and Canada. For example, on the Crow Indian reservation in Montana, coal deposits of Cretaceous age contain nearly 20 billion tons of high quality coal.

A major event of the Mesozoic era was the breakup of Pangaea. Throughout most of the Triassic, this great supercontinent existed much as it had during Permian times. However, evidence found in late Triassic rocks indicates that a rift developed between the eastern United States and western Africa, which marked the opening of the Atlantic. The formation of the incipient North Atlantic Ocean represents the beginning of the breakup of Pangaea, a process that continued throughout the Mesozoic and into the Cenozoic.*

As Pangaea fragmented, the westward-moving North American plate overrode the adjacent portion of the Pacific plate. This tectonic event was the beginning of a continuous wave of deformation that moved inland along the entire western margin of the continent. By Jurassic times, subduction of the Pacific plate had begun to produce the chaotic mixture of rocks that are today represented in the Coast Ranges of California. Further inland igneous activity was widespread. During this episode of igneous activity, which lasted for nearly 60 million years, huge quantities of magma were emplaced a few miles below the surface. The remnants of this intrusive activity include the granitic rocks of the Sierra Nevada as well as the Idaho batholith and the Coast Range batholith of British Columbia.

---

*Additional discussion of this topic is found in the section entitled "Pangaea: Before and After" in Chapter Thirteen.

The tectonic activity that began in the Jurassic continued throughout the Cretaceous, ultimately forming the vast Cordilleran of western North America. Compressional forces associated with the converging plates moved huge rock units in a shingle-like fashion toward the east. Throughout much of the western margin of North America, older rocks were thrust 150 kilometers or more eastward over younger strata. Today, these deformed strata are best represented in the Northern Rockies of Idaho, western Montana, and British Columbia.

A later tectonic event, named the Laramide orogeny, occurred in the late Cretaceous and continued into the early Tertiary. The Laramide orogeny produced several scattered highlands, as large blocks of Precambrian basement rocks were nearly vertically uplifted. The mountains produced by this period of tectonism include the ancestrial middle and southern Rockies of Colorado and Wyoming. The mountain building in the North American Cordilleran did not end with the Mesozoic, but rather continued throughout much of the Cenozoic period.

Mesozoic Life

The life forms at the dawn of the Mesozoic era were those that had survived the great Permian extinctions. These survivors began to diversify in many new ways to fill the voids created by the biological vacuum at the close of the Paleozoic. On land, conditions favored those groups that were best able to adapt to drier climates. Among the members of the plant kingdom, the gymnosperms were one such group. Unlike the first plants to invade the land, the seed-bearing gymnosperms were not dependent on free-standing water for fertilization. Consequently, these plants were not restricted to a life near the water's edge.

The gymnosperms quickly replaced the scale trees as the dominant trees of the Mesozoic. Ancient gymnosperms included the cycads, the conifers, and the ginkgoes. The cycads had large, palmlike leaves and resembled a large pineapple plant, whereas the ginkgoes had fan-shaped leaves much like their modern relatives. The largest plants in the Mesozoic forests were the conifers, whose modern descendants include the pines, firs, and junipers. The best-known fossil occurrence of these ancient trees is found in northern Arizona's Petrified Forest National

FIGURE 16.9 A composite Mesozoic landscape showing large carnivorous and herbivorous dinosaurs. (Courtesy of the Peabody Museum of Natural History, Yale University)

Park. Here, huge pertrified logs lay exposed at the surface, having been weathered from rocks of the Triassic Chinle Formation (see Figure 15.13).

Another group that readily adapted to the drier Mesozoic environment was the reptiles, the first true terrestrial life forms. Unlike the amphibians, reptiles have shell-covered eggs that can be laid on land. The elimination of a water-dwelling stage, as for example the tadpole stage in the development of frogs, was an important evolutionary step. Of interest is the fact that the watery fluid within the reptilian egg closely resembles seawater in chemical composition. Because the reptile embryo develops in this watery environment, the shelled egg has been characterized as a "private aquarium" in which the embryos of these land vertebrates spend their water-dwelling stage of life.

With the perfection of the shelled egg, reptiles quickly became the dominant land animals. They continued this dominance for more than 160 million years. The largest, most awe-inspiring of the Mesozoic reptiles were the dinosaurs (Figure 16.9). Some of the huge dinosaurs such as *Tyrannosaurus* were carnivorous whereas others such as the ponderous *Brontosaurus* were herbivorous. The extremely long neck of *Brontosaurus* was thought to be an adaptation for feeding on tall conifer trees. However, not all dinosaurs were large. In fact, certain forms closely resembled modern fleet-footed lizards. Further, evidence indicates that some dinosaurs, unlike their present-day relatives, were probably warm blooded.

The reptiles made one of the most spectacular adaptive radiations in all of earth history (Figure 16.10). One group, the pterosaurs, took to the air. These "dragons of the sky" possessed huge membranous wings that allowed them to exhibit rudimentary flight. Another group of reptiles, exemplified by the fossil *Archaeopteryx*, led to more successful flyers: the birds. Whereas some reptiles took to the skies, others returned to the sea. Included in this latter group were the fish-eating reptiles: the plesiosaurs and ichthyosaurs. Although these reptiles became proficient swimmers, they retained their reptilian teeth and breathed by means of lungs.

At the close of the Mesozoic many reptile groups became extinct, as had other dominant life forms before them. Of the large number of Mesozoic reptiles, only a few survived to recent times. Among these are the turtles, snakes, crocodiles, and lizards. The huge land-dwelling dinosaurs, the marine plesiosaurs, and the flying pterosaurs all became extinct (Figure 16.10). The demise of the great reptiles is generally attributed to this group's inability to adapt to some radical change in environmental conditions. What event could have triggered the sudden extinction of the dinosaurs—the most successful group of land animals ever to have lived?

One modern view proposes that about 65 million years ago a large asteroid collided with the earth. The impact of such a body would have produced a cloud of dust thousands of times greater than that released during the 1980 eruption of Mount St. Helens. For many months the dust-laden atmosphere would have greatly restricted the amount of sunlight that penetrated to the earth's surface. Without sunlight for photosynthesis, delicate food chains would collapse. It is further hypothesized that large dinosaurs would be affected more adversely by this chain of events than would smaller life forms.

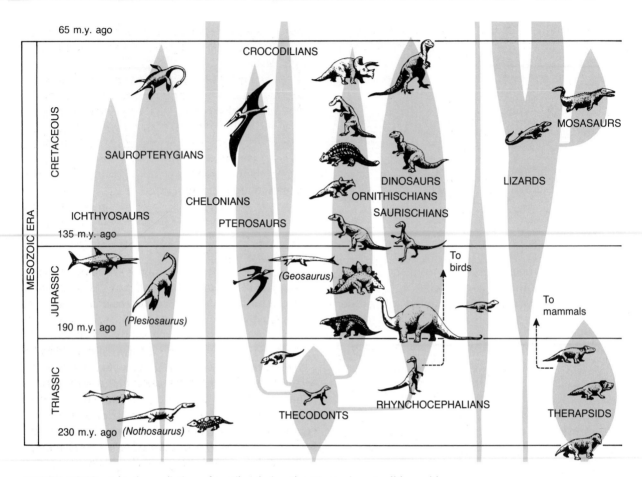

FIGURE 16.10 Adaptive radiation of reptiles during the Mesozoic era—"the golden age of reptiles." Dinosaurs, pterosaurs (flying reptiles), and large marine reptiles all became extinct by the end of the Mesozoic. (Drawing by Robert W. Tope. From Burchfiel et al. *Physical Geology*. Columbus, Ohio: Merrill, 1982)

What evidence points to this catastrophic collision 65 million years ago? First, a thin layer of sediment, nearly one centimeter thick has been discovered at the Mesozoic-Cenozoic boundary. This particular sediment is found worldwide and contains a high level of the element iridium, which is also found in similar proportions in meteorites. Could this layer represent the scattered remains of an asteroid that was responsible for the environmental changes which led to the demise of many reptile groups? Secondly, this period of mass extinction appears to have affected all land animals larger than dogs. Supporters of this catastrophic-event scenario suggest that small rat-like mammals could survive a break-

down of food chains lasting perhaps several months. Large animals, they contend, could not survive such an event.

Other scientists disagree with the foregoing hypothesis. They claim that what appears to be a mass extinction over a short period of time did, in fact, occur over millions of years. Based upon an examination of the fossil record at the Mesozoic-Cenozoic boundary, these geologists conclude that the decline of the dinosaurs was gradual.

Whichever group is correct, the fact remains that one of the more successful groups ever to inhabit the earth died out at the close of the Mesozoic. The decline of the reptiles provided vacancies in habitats for the mammals. Although small and inconspicuous during the Mesozoic, the mammals rose to dominance during the Cenozoic.

THE CENOZOIC ERA

The Cenozoic era, or "era of recent life," encompasses the past 65 million years of earth history. It is during this span that the physical landscapes and life forms of our modern world came into being. By examining the dates on the geologic calendar (see Figure 15.19), it is clear that the Cenozoic era represents a much smaller fraction of geologic time than either the Paleozoic era or the Mesozoic era. Although the length of the Cenozoic is relatively short, it nevertheless possesses a rich history, because the completeness of the geologic record improves as time approaches the present. The rock formations of this time span are more widespread and less disturbed than those of any preceding era. Consequently, the available geological data about this era is much more abundant than for any other.

The Cenozoic era is divided into seven epochs which, in turn, are grouped into two periods having very unequal durations. The Tertiary period includes five epochs and embraces about 63 million years, practically all of the Cenozoic era.* The Quaternary period consists of two epochs that represent only the last two million years of geologic time.

North America in the Cenozoic

At the beginning of the Cenozoic era most of North America was above sea level, a situation that would remain throughout the era. During this span, the eastern and western margins of the continent experienced markedly contrasting events due to their different relationships with plate boundaries. The Atlantic and Gulf coastal regions, being far removed from an active plate boundary, were tectonically stable. Western North America, on the other hand, was the leading edge of the North American plate.

---

*An alternative means of dividing the Cenozoic era does not recognize a Tertiary period. Rather, this time span is divided into two periods. The *Paleogene* period includes the Paleocene, Eocene, and Oligocene epochs, whereas the Miocene and Pliocene epochs make up the *Neogene* period. The use of Paleogene and Neogene instead of Tertiary is most common in Europe.

As a result, plate interactions during the Cenozoic gave rise to many orogenic and volcanic events in the West.

The continental margin of eastern North America was tectonically stable during the Cenozoic era; for that reason, it was the site of abundant marine sedimentation. The most extensive region of sediment deposition surrounded the Gulf of Mexico, from the Yucatán peninsula to Florida. Here the great buildup of sediment caused the crust to downwarp and produced numerous faults that in many instances created traps in which petroleum accumulated. Today these and other petroleum traps are the source of the most economically important resources derived from the Cenozoic strata of the Gulf Coast.

By early Cenozoic time most of the once-majestic Appalachians had been eroded to a low plain. Then, by the mid-Cenozoic, isostatic adjustments raised the region once again, rejuvenating its rivers. The new base level caused the streams to erode with renewed vigor and sculpture the surface into its present day topography (Figure 16.11). The sediments derived from the erosion of the Appalachians were deposited along the eastern margin of the continent where they attained a thickness of many kilometers. Today portions of the strata deposited during the Cenozoic are exposed and constitute the gently sloping Atlantic and Gulf coastal plains.

In the West, the Laramide orogeny that built the Rocky Mountains came to an end in the Tertiary period. At this time erosion lowered the mountains, and the basins between uplifted ranges filled with detritus. Meanwhile a great wedge of sediment derived from the eroding Rockies was building eastward, creating the Great Plains.

Beginning in the Miocene epoch, a broad region extending from northern Nevada southeastward into Mexico experienced crustal movements that produced a system of north-south trending normal faults. These movements resulted in the formation of more than 150 fault-block mountain ranges rising abruptly above the adjacent basins. The essential character of today's Basin and Range province was now established.*

---

*Additional discussion of this topic is found in the section entitled "Fault-Block Mountains" in Chapter Fourteen and by examining Figure 14.11.

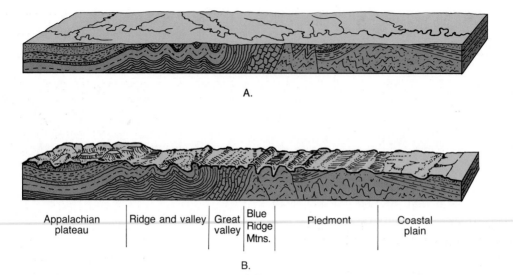

A.

B.

| Appalachian plateau | Ridge and valley | Great valley | Blue Ridge Mtns. | Piedmont | Coastal plain |

FIGURE 16.11 Recent upwarping and erosion in the Appalachians has produced the present topography. **A.** Eroded Appalachians. **B.** Uplifting and rejuvenation of the Appalachians. (Drawn by Erwin Raisz, from "Stream Sculpture on the Atlantic Slope," by Douglas Johnson, © 1931, used with permission of Columbia University Press)

As the Basin and Range province was experiencing block faulting, the entire western interior of the continent was gradually uplifted. This uplift re-elevated the Rockies and rejuvenated many of the West's major rivers. As the rivers became entrenched, many spectacular gorges were formed, including the Grand Canyons of the Colorado and Snake rivers and the Black Canyon of the Gunnison. The present topography of the Rocky Mountains is in large measure the result of this late Tertiary uplift and the subsequent excavation of the early Tertiary basin deposits by rejuvenated streams.

Volcanic activity was common in the West during much of Cenozoic time. Beginning in the Miocene epoch, great volumes of fluid basaltic lava were extruded as fissure eruptions in portions of present-day Washington, Oregon, and Idaho. These eruptions, which built the extensive (1.3 million square kilometers) Columbia Plateau, continued into the Pliocene, with some vents remaining active through the Pleistocene (Figure 16.12). Immediately west of the Columbia Plateau, the nature of volcanic activity was quite different. Here, magmas with a higher silica content erupted explosively, creating a chain of stratovolcanoes from northern California to the Canadian border (Figure 16.13). Although these volcanoes were most active during the Pliocene and Pleis-

tocene epochs, some (including Mount St. Helens) have remained active to the present.

A final episode of folding occurred in the West in late Tertiary time. This episode of orogenesis created the Coast Ranges that stretch along the Pacific Coast from California to British Columbia. Meanwhile, the Sierra Nevada became fault-block mountains as they were uplifted along their eastern flank, creating the imposing mountain front we know today.

As the Tertiary period drew to a close, the effects of mountain building, volcanic activity, isostatic adjustments, and extensive erosion and sedimentation had created a physical landscape very similar to the configuration of today. All that remained was the final two-millon-year episode called the Quaternary period. During this last phase of earth history, in which modern man evolved, the action of glacial ice and other erosional agents added the finishing touches.*

Cenozoic Life

The record of life during the Cenozoic era is more familiar than the records of preceding eras. Mam-

---

*A thorough discussion of glaciers and glaciation in the geologic past is found in Chapter Eight.

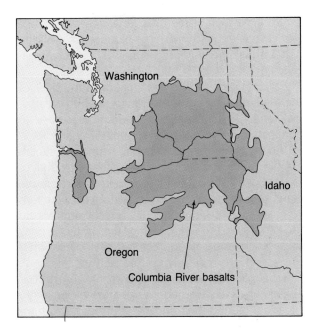

FIGURE 16.12 Area covered by the Columbia River basalts. (After U.S. Geological Survey)

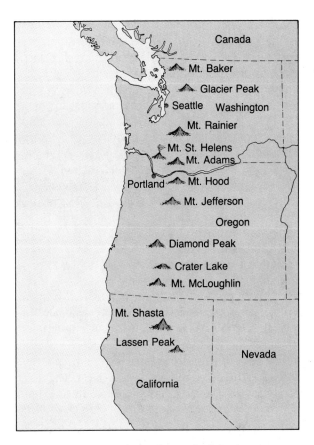

FIGURE 16.13 Locations of several of the larger composite cones that comprise the Cascade Range.

mals replaced the reptiles that had dominated the land during Mesozoic time. Moreover, as the Cenozoic era opened, angiosperms (flowering plants with covered seeds) had replaced gymnosperms as the dominant land plants. Marine invertebrates also took on a modern look. Although mollusks, especially pelecypods (clams, oysters) and gastropods (snails), are the most familiar and abundant macroscopic* fossils, microscopic foraminifera became especially important. Today foraminifera are among the most intensely studied of all fossils, because their widespread occurrence makes them invaluable in correlating Tertiary sediments. These strata are very important to the modern world, for they yield more oil than rocks of any other age.

The Cenozoic is often called the "age of mammals," because these animals dominated land life. It could also be called the "age of flowering plants," for the angiosperms enjoyed a similar status in the plant world. Due to advances in seed fertilization and dispersal, angiosperms experienced a period of rapid development and expansion as the Mesozoic drew to a close. Thus, as the Cenozoic era began, angiosperms were already the dominant land plants.

Development of the flowering plants, in turn, strongly influenced the evolution of both birds and mammals. Birds that feed on seeds and fruits, for example, evolved rapidly during the Cenozoic in close association with the flowering plants. During the middle Tertiary, grasses developed rapidly and spread over the plains. This fostered the emergence of herbivorous (plant-eating) mammals that were mainly grazers. In turn, the development and spread of grazing mammals established the setting for the evolution of the carnivorous mammals which preyed upon them.

An important evolutionary event was the appearance of primitive mammals in late Triassic time, about the same time as the dinosaurs emerged. Yet throughout the period of dinosaur dominance, mammals remained in the background as small and inconspicuous animals. By the close of the Mesozoic era, dinosaurs and other reptiles no longer dominated the land. It was only after these large reptiles

---

*Macroscopic means large enough to be examined with the unaided eye; that is, without instrumentation.

became extinct that mammals came into their own as the dominant land animals. This transition is a major example in the fossil record of the replacement of one large group by another.

Mammals are distinct from reptiles in several respects. Among these differences, mammalian young are born live and mammals maintain a constant body temperature; that is, they are "warm blooded."* This latter adaption allowed mammals to lead more active and diversified lives than reptiles because they could survive in cold regions and search for food during any season or time of day. Other mammalian adaptations included the development of insulating body hair and a more efficient heart and lungs.

With the demise of the Mesozoic reptiles, Cenozoic mammals diversified rapidly. The many forms that exist today evolved from small primitive mammals that were characterized by short legs, flat five-toed feet, and small brains. Their development and specialization took four principle directions: (1) Increase in size; (2) Increase in brain capacity;

---

*One minor group of mammals, the monotremes, still lays eggs. The two species in this group, the duck-billed platypus and the spiny anteater, are found only in Australia. Moreover, although modern reptiles are "cold blooded," some paleontologists believe that dinosaurs may have been "warm blooded."

(3) Specialization of teeth to better accommodate a particular diet; and (4) Specialization of limbs to better equip the animal for life in a particular environment.

Following the reptilian extinctions at the close of the Mesozoic, two groups of mammals, the marsupials and the placentals, evolved and expanded to dominate the Cenozoic. The groups differ principally in their modes of reproduction. Young marsupials are born live, but at a very early stage of development. After birth, the tiny and immature young crawl into the mother's external stomach pouch to complete their development. Placental mammals, on the other hand, develop within the mother's body for a much longer period, so that birth occurs after the young are relatively mature and independent.

Today marsupials are found primarily in Australia, where they went through a separate evolutionary expansion during the Cenozoic largely isolated from placental mammals. In South America both primitive marsupials and placentals reached the continent before the landmass became completely isolated. Evolution and specialization of both groups continued undisturbed for approximately 40 million years, until the close of the Pliocene epoch when the Central American land bridge was established. Then an invasion of advanced carnivores from North America

FIGURE 16.14 Most Pleistocene mammals in this mural, including the mastodon, mammoth, and giant bison in the center, are now extinct. (Peabody Museum of Natural History, Yale University)

brought the extinction of many hoofed mammals that had persisted for millions of years. The marsupials, except for opossums, also could not compete and became extinct. Both Australia and South America provide excellent examples of how isolation caused by the separation of continents increased the diversity of animals in the world.

As we have seen, during the Cenozoic era mammals diversified quite rapidly. One tendency was for some groups to become very large. For example, by the Oligocene epoch a hornless rhinoceros that stood nearly 5 meters high had evolved. It is the largest land mammal yet known to have existed. As time approached the present, many other types evolved to a large size as well; in fact, more so than now exist. Many of these large forms were common as recently as 11,000 years ago. However, a wave of late Pleistocene extinctions rapidly eliminated these animals from the landscape. In North America, huge relatives of the elephant, the mastodon and mammoth, became extinct (Figure 16.14). In addition, sabertoothed cats, giant beavers, large ground sloths, horses, camels, giant bison, and others died out. In Europe, late Pleistocene extinctions included wooly rhinos, large cave bears, and the Irish elk. The reason for this recent wave of large animal extinctions puzzles scientists. Since these animals had survived several major glacial advances and interglacial periods, it is difficult to ascribe these extinctions to climatic change. Some scientists believe that early man hastened the decline of these mammals by selectively hunting large forms. Although this hypothesis is advanced more than any other, it is not yet accepted by all who have studied the problem.

REVIEW QUESTIONS

1   Briefly describe the events believed to have led to the formation of the solar system.

2   What is the major source of free oxygen in the earth's atmosphere?

3   Explain why Precambrian history is more difficult to decipher than more recent geological history.

4   Match the following words and phrases to the most appropriate time span. Select among the following: Precambrian, early Paleozoic, late Paleozoic, Mesozoic, Cenozoic.
   a   Pangaea came into existence.
   b   Bacteria and blue-green algae preserved in chert.
   c   The era that encompasses the least amount of time.
   d   Shields.
   e   "Age of dinosaurs."
   f   Formation of the original northern Appalachian mountains.
   g   Mastodons and mammoths.
   h   Extensive deposits of rock salt.
   i   Triassic, Jurassic, and Cretaceous.
   j   Coal swamps extended across North America, Europe, and Siberia.
   k   Gulf Coast oil deposits form.
   l   Most of the world's major iron ore deposits.
   m   Massive sand dunes covered large portions of the Colorado Plateau region.
   n   The "age of fishes" occurred during this span.
   o   Cambrian, Ordovician, and Silurian.
   p   Pangaea began to break apart.
   q   "Age of mammals."
   r   Animals with hard parts first appear in abundance.
   s   Gymnosperms were the dominant trees.
   t   Columbia Plateau formed.
   u   Stromatolites are among its more common fossils.
   v   "Golden age of trilobites" occurred during this span.
   w   Fault-block mountains form in the Basin and Range region.

5   Briefly discuss two proposals that attempt to explain why several groups developed hard parts at the beginning of the Cambrian period. Do these proposals appear to provide a satisfactory explanation? Why or why not?

6   List some differences between amphibians and reptiles. Between reptiles and mammals.

7   Describe one hypothesis that attempts to explain the extinction of the dinosaurs. Cite a major objection to this proposal.

8   Contrast the eastern and western margins of North America during the Cenozoic era in terms of their relationships to plate boundaries.

APPENDIX A
METRIC AND ENGLISH
UNITS COMPARED

UNITS

1 kilometer (km)	= 1000 meters (m)
1 meter (m)	= 100 centimeters (cm)
1 centimeter (cm)	= 0.39 inch (in.)
1 mile (mi)	= 5280 feet (ft)
1 foot (ft)	= 12 inches (in.)
1 inch (in.)	= 2.54 centimeters (cm)
1 square mile (mi^2)	= 640 acres (a)
1 kilogram (kg)	= 1000 grams (g)
1 pound (lb)	= 16 ounces (oz)
1 fathom	= 6 feet (ft)

CONVERSIONS

When you want to convert:	Multiply by:	To find:
Length		
inches	2.54	centimeters
centimeters	0.39	inches
feet	0.30	meters
meters	3.28	feet
yards	0.91	meters
meters	1.09	yards
miles	1.61	kilometers
kilometers	0.62	miles
Area		
square inches	6.45	square centimeters
square centimeters	0.15	square inches
square feet	0.09	square meters
square meters	10.76	square feet
square miles	2.59	square kilometers
square kilometers	0.39	square miles

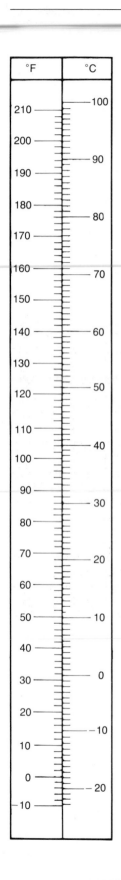

CONVERSIONS

When you want to convert:	Multiply by:	To find:

Volume

cubic inches	16.38	cubic centimeters
cubic centimeters	0.06	cubic inches
cubic feet	0.028	cubic meters
cubic meters	35.3	cubic feet
cubic miles	4.17	cubic kilometers
cubic kilometers	0.24	cubic miles
liters	1.06	quarts
liters	0.26	gallons
gallons	3.78	liters

Masses and Weights

ounces	20.33	grams
grams	0.035	ounces
pounds	0.45	kilograms
kilograms	2.205	pounds

Temperature

When you want to convert degrees Fahrenheit (°F) to degrees Celsius (°C), subtract 32 degrees and divide by 1.8.

When you want to convert degrees Celsius (°C) to degrees Fahrenheit (°F), multiply by 1.8 and add 32 degrees.

When you want to convert degrees Celsius (°C) to kelvins (K), delete the degree symbol and add 273.

When you want to convert kelvins (K) to degrees Celsius (°C), add the degree symbol and subtract 273.

APPENDIX B

PERIODIC TABLE OF

THE ELEMENTS

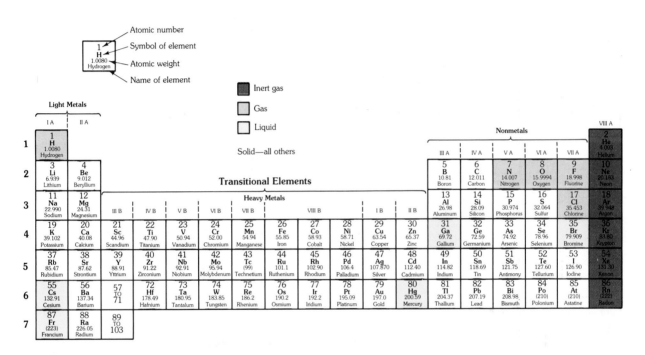

Atomic number

Symbol of element

Atomic weight

Name of element

1
H
1.0080
Hydrogen

Inert gas

Gas

Liquid

Solid—all others

Light Metals

Transitional Elements

Heavy Metals

Nonmetals

	I A	II A												III A	IV A	V A	VI A	VII A	VIII A
1	1 H 1.0080 Hydrogen																		2 He 4.003 Helium
2	3 Li 6.939 Lithium	4 Be 9.012 Beryllium												5 B 10.81 Boron	6 C 12.011 Carbon	7 N 14.007 Nitrogen	8 O 15.9994 Oxygen	9 F 18.998 Fluorine	10 Ne 20.183 Neon
3	11 Na 22.990 Sodium	12 Mg 24.31 Magnesium	III B	IV B	V B	VI B	VII B		VIII B			I B	II B	13 Al 26.98 Aluminum	14 Si 28.09 Silicon	15 P 30.974 Phosphorus	16 S 32.064 Sulfur	17 Cl 35.453 Chlorine	18 Ar 39.948 Argon
4	19 K 39.102 Potassium	20 Ca 40.08 Calcium	21 Sc 44.96 Scandium	22 Ti 47.90 Titanium	23 V 50.94 Vanadium	24 Cr 52.00 Chromium	25 Mn 54.94 Manganese	26 Fe 55.85 Iron	27 Co 58.93 Cobalt	28 Ni 58.71 Nickel	29 Cu 63.54 Copper	30 Zn 65.37 Zinc	31 Ga 69.72 Gallium	32 Ge 72.59 Germanium	33 As 74.92 Arsenic	34 Se 78.96 Selenium	35 Br 79.909 Bromine	36 Kr 83.80 Krypton
5	37 Rb 85.47 Rubidium	38 Sr 87.62 Strontium	39 Y 88.91 Yttrium	40 Zr 91.22 Zirconium	41 Nb 92.91 Niobium	42 Mo 95.94 Molybdenum	43 Tc (99) Technetium	44 Ru 101.1 Ruthenium	45 Rh 102.90 Rhodium	46 Pd 106.4 Palladium	47 Ag 107.870 Silver	48 Cd 112.40 Cadmium	49 In 114.82 Indium	50 Sn 118.69 Tin	51 Sb 121.75 Antimony	52 Te 127.60 Tellurium	53 I 126.90 Iodine	54 Xe 131.30 Xenon
6	55 Cs 132.91 Cesium	56 Ba 137.34 Barium	57 TO 71	72 Hf 178.49 Hafnium	73 Ta 180.95 Tantalum	74 W 183.85 Tungsten	75 Re 186.2 Rhenium	76 Os 190.2 Osmium	77 Ir 192.2 Iridium	78 Pt 195.09 Platinum	79 Au 197.0 Gold	80 Hg 200.59 Mercury	81 Tl 204.37 Thallium	82 Pb 207.19 Lead	83 Bi 208.98 Bismuth	84 Po (210) Polonium	85 At (210) Astatine	86 Rn (222) Radon
7	87 Fr (223) Francium	88 Ra 226.05 Radium	89 TO 103															

Lanthanide series

57 La 138.91 Lanthanum	58 Ce 140.12 Cerium	59 Pr 140.91 Praseodymium	60 Nd 144.24 Neodymium	61 Pm (147) Promethium	62 Sm 150.35 Samarium	63 Eu 151.96 Europium	64 Gd 157.25 Gadolinium	65 Tb 158.92 Terbium	66 Dy 162.50 Dysprosium	67 Ho 164.93 Holmium	68 Er 167.26 Erbium	69 Tm 168.93 Thullium	70 Yb 173.04 Ytterbium	71 Lu 174.97 Lutetium

Actinide series

89 Ac (227) Actinium	90 Th 232.04 Thorium	91 Pa (231) Protactinium	92 U 238.03 Uranium	93 Np (237) Neptunium	94 Pu (242) Plutonium	95 Am (243) Americium	96 Cm (247) Curium	97 Bk (249) Berkelium	98 Cf (251) Californium	99 Es (254) Einsteinium	100 Fm (253) Fermium	101 Md (256) Mendelevium	102 No (254) Nobelium	103 Lw (257) Lawrencium

APPENDIX C
COMMON MINERALS OF
THE EARTH'S CRUST

Mineral or Group Name	Composition	Cleavage/ Fracture	Color	Hardness	Other Properties/Comments
Albite See *Plagioclase feldspar*					
Amphibole (common member: hornblende)	Complex family of hydrous, Ca, Na, Mg, Fe, Al silicates	Two at 60 and 120 degrees	Deep green to black	5–6	Forms elongated crystals. Commonly found in igneous and metamorphic rocks.
Anorthite See *Plagioclase feldspar*					
Augite See *Pyroxene*					
Bauxite	Mixture of weathered clay minerals	Irregular fracture	Varied, reddish-brown common	Variable	Earthy luster commonly contains small spheres. Ore of aluminum.
Biotite	$K(Mg,Fe)_3(AlSi_3O_{10})(OH)_2$	Perfect cleavage in one direction	Black to dark brown	2–2.5	Splits into thin, flexible sheets. Common mica found in igneous and metamorphic rocks.
Bornite	Cu_5FeS_4	Uneven fracture	Brownish bronze on a fresh surface	3	Tarnishes to a variegated purple blue; hence, called peacock ore. High specific gravity (5). Ore of copper.
Calcite	$CaCO_3$ Calcium carbonate	Three perfect cleavages at 75 degrees	White or colorless	2.5–3	Common in sedimentary rocks. When transparent exhibits double refraction. Reacts with weak acid.
Chalcedony	SiO_2 Silicon dioxide	Conchoidal fracture	White when pure. Often multicolored	5–6.5	Microcrystalline form of quartz. Multicolored. Called agates when banded. Opal is an amorphous variety.

Mineral or Group Name	Composition	Cleavage/ Fracture	Color	Hardness	Other Properties/Comments
Chalcopyrite	$CuFeS_2$	Irregular fracture	Brass yellow	3.5–4	Usually massive. Specific gravity 4–4.5. Ore of copper.
Chlorite	$(Mg,Fe)_5(Al,Fe)_2Si_3O_{10}(OH)_8$	One direction of cleavage	Light to dark green	2–2.5	Occurs as mass of flaky scales. Common in metamorphic rocks.
Cinnabar	HgS	One direction, but not generally observed	Scarlet red	2.5	Occurs in masses mixed with other materials. Often dull earthy luster. Important ore of mercury.
Clay minerals (common member: kaolinite)	Complex group of hydrous aluminum silicates	Irregular fracture	Buff to brownish gray	1–2.5	Found in earthy masses as a main constituent of soil. Also abundant in shales and other sedimentary rocks.
Corundum	Al_2O_3	Two good cleavages with striations	Variable; red, blue, yellow, and green	9	Important gemstone. Red variety called ruby; blue variety is sapphire. Also used as an abrasive.
Dolomite	$CaMg(CO_3)_2$	Three good cleavages at 75 degrees	Variable; white when pure	3.5–4	Similar to calcite, but will effervesce with acid only when powdered. Common in sedimentary rocks.
Epidote	Complex Ca, Fe, and Al silicate	One good cleavage, one poor	Yellow-green to dark green	6–7	Commonly occurs as small elongated crystals in metamorphic rocks.

Feldspar   See *Orthoclase feldspar* and *Plagioclase feldspar*

Fluorite	CaF_2	Perfect cleavage in 4 directions	Colorless; violet, green, or yellow	4	Commonly found with ores of metals.
Galena	PbS	Three cleavages at right angles	Silver gray	2.5	Shiny metallic mineral with high specific gravity (7.6). Ore of lead.
Garnet	Complex family of silicate minerals containing Ca, Mg, Fe, Mn, Al, Ti, Cr	Uneven to conchoidal fracture	Various colors; commonly deep red to brown	6.5–7.5	Forms 12- or 24-sided crystals commonly found in metamorphic rocks.
Graphite	C	One direction of cleavage	Steel gray	1–2	Occurs in scaly, foliated masses. Used as a lubricant. Greasy feel.

Mineral or Group Name	Composition	Cleavage/ Fracture	Color	Hardness	Other Properties/Comments
Gypsum	$CaSO_4 \cdot 2H_2O$	Cleavage good in one direction, poor in two others	Colorless to white	2	Occurs as tabular crystals, or fibrous or finely crystalline masses. Common in sedimentary layers. Used for plaster.
Halite	$NaCl$	Three cleavages at right angles	Colorless to white	2.5	Common table salt. Occurs as granular masses. Common sedimentary mineral.
Hematite	Fe_2O_3	Uneven fracture	Reddish brown to steel gray	5.5–6.5	Occurs as earthy masses. High specific gravity (4.8–5.5). Important ore of iron.
Hornblende See *Amphibole*					
Kaolinite See *Clay minerals*					
Kyanite	Al_2SiO_5	One good direction of cleavage	White to light blue	5–7	Forms long, bladed, or tabular crystals. Common mineral in metamorphic rocks.
Labradorite See *Plagioclase feldspar*					
Limonite (goethite)	Mixture of hydrous iron oxides	Uneven fracture	Yellowish to brown	1–5.5	Earthy masses. Forms from the alteration of other iron-rich minerals. Gives rock surfaces and soils a yellow color.
Magnetite	Fe_3O_4	Uneven fracture	Black	5.5–6.5	Submetallic to metallic luster. Magnetic. High specific gravity (5). Generally occurs in granular masses. Ore of iron.
Malachite	$Cu_2CO_3(OH)_2$	Uneven fracture	Bright green	3.5–4	Effervesces in acid. Ore of copper.
Mica See *Biotite* and *Muscovite*					
Muscovite	$KAl_3Si_3O_{10}(OH)_2$	Perfect cleavage in one direction	Colorless to light gray	2–2.5	Splits into thin elastic sheets. Transparent in thin sheets. Common in all rock types.
Olivine	$(Mg,Fe)_2SiO_4$	Conchoidal fracture	Olive to dark green	6.5–7	Occurs as granular masses or some grains in dark-colored igneous rocks.
Orthoclase feldspar (K feldspar)	$KAlSi_3O_8$	Two cleavages at right angles	White to gray. Frequently salmon pink	6	Forms elongated crystals in igneous rocks. Also, commonly found in sedimentary and metamorphic rocks.

Mineral or Group Name	Composition	Cleavage/Fracture	Color	Hardness	Other Properties/Comments
Plagioclase feldspar	$NaAlSi_3O_8$ (albite) $CaAl_2Si_2O_8$ (anorthite)	Two cleavages at nearly right angles	White to gray	6	Forms elongated crystals in igneous rocks. Also commonly found in sedimentary and metamorphic rocks. Striations on some cleavage planes.
Pyrite	FeS_2	Uneven fracture	Brass yellow	6–6.5	Occurs as granular masses or well-formed cubic crystals. High specific gravity (4.8–5.2). Called "fool's gold."
Pyroxene (common member: augite)	Complex family of Mg, Fe, Ca, Na, and Al silicate	Good cleavage in two directions at nearly right angles	Green to black	5–6	Occurs as individual grains in igneous and metamorphic rocks.
Quartz	SiO_2	Conchoidal fracture	Colorless when pure	7	Common in all rock types. Often lightly colored, including gray, pink, yellow, and violet.
Serpentine	$Mg_3Si_2O_5(OH)_4$	Uneven fracture	Light to dark green	2.5–5	Fibrous variety is asbestos. Occurs most often in metamorphic rocks.
Sillimanite	Al_2SiO_5	One direction of cleavage	White to gray	6–7	High grade metamorphic mineral.
Sphalerite	ZnS	Six directions of cleavage	Yellow to brown	3.5–4	Moderate specific gravity (4.1–4.3). Smell of sulfur when powdered. Ore of zinc.
Staurolite	$FeAl_4(SiO_4)_2(OH)_2$	Cleavage not prominent	Brown to reddish brown	7	Elongated crystals, occasionally twinned to form a cross-shaped crystal. Commonly found in metamorphic rocks.
Sulfur	S	Irregular fracture	Yellow	1.5–2.5	Bright yellow mineral most often associated with sedimentary deposits, in coal, and near volcanoes.
Talc	$Mg_3(Si_4O_{10})(OH)_2$	Good cleavage in one direction	White to light green	1–1.5	Soapy feel. Found in foliated masses consisting of thin flakes or scales. Most often associated with metamorphic rocks.
Wollastonite	$CaSiO_3$	Two perfect cleavages	Colorless to white	4.5–5	Forms fibrous or bladed crystals. Common in contact metamorphic rocks.

APPENDIX D

TOPOGRAPHIC MAPS

A map is a representation on a flat surface of all or a part of the earth's surface drawn to a specific scale. Maps are often the most effective means for showing the locations of both natural and manmade features, their sizes, and their relationships to one another. Like photographs, maps readily display information that would be impractical to express in words.

While most maps show only the two horizontal dimensions, geologists, as well as other map users, often require that the third dimension, elevation, be shown on maps. Maps that show the shape of the land are called **topographic maps**. Although various techniques may be used to depict elevations, the most accurate method involves the use of contour lines.

Contour Lines

A **contour line** is a line on a map representing a corresponding imaginary line on the ground that has the same elevation above sea level along its entire length. While many map symbols are pictographs, resembling the objects they represent, a contour line is an abstraction that has no counterpart in nature. It is, however, an accurate and effective device for representing the third dimension on paper.

Some useful facts and rules concerning contour lines are listed as follows. This information should be studied in conjunction with Figure D.1.

1. Contour lines bend upstream or upvalley. The contours form Vs that point upstream, and in the upstream direction the successive contours represent higher elevations. For example, if you were standing on a stream bank and wished to get to the point at the same elevation directly opposite you on the other bank, without stepping up or down, you would need to walk upstream along the contour at that elevation to where it crosses the stream bed, cross the stream, and then walk back downstream along the same contour.

2. Contours near the upper parts of hills form closures. The top of a hill is higher than the highest closed contour.

3. Hollows (depressions) without outlets are shown by closed, hatched contours. Hatched contours are contours with short lines on the inside pointing downslope.

4. Contours are widely spaced on gentle slopes.

5. Contours are closely spaced on steep slopes.

6. Evenly spaced contours indicate a uniform slope.

7. Contours usually do not cross or intersect each other, except in the rare case of an overhanging cliff.

8. All contours eventually close, either on a map or beyond its margins.

9. A single high contour never occurs between two lower ones, and vice versa. In other words, a change in slope direction is always determined by the repetition of the same elevation either as two different contours of the same value or as the same contour crossed twice.

10. Spot elevations between contours are given at many places, such as road intersections, hill summits, and lake surfaces. Spot elevations differ from control elevation stations, such as bench marks, in not being permanently established by permanent markers.

Relief

Relief refers to the difference in elevation between any two points. Maximum relief refers to the difference in elevation between the highest and lowest points in the area being considered. Relief determines the **contour interval**, which is the difference in elevation between succeeding contour lines that is used on topographic maps. Where relief is low, a small contour interval, such as 10 or 20 feet, may be used. In flat areas, such as wide river valleys or broad, flat uplands, a contour interval of 5 feet is often used. In rugged mountainous terrain, where relief is many hundreds of feet, contour intervals as large as 50 or 100 feet are used.

Scale

Map **scale** expresses the relationship between distance or area on the map to the true distance or area on the earth's surface. This is generally expressed as a ratio or fraction, such as 1:24,000 or 1/24,000. The numerator, usually 1,

APPENDIX D

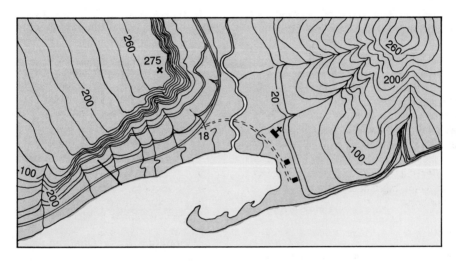

FIGURE D.1 Perspective view of an area and a contour map of the same area. These illustrations show how features are depicted on a topographic map. The upper illustration is a perspective view of a river valley and the adjoining hills. The river flows into a bay, which is partly enclosed by hooked sandbar. On either side of the valley are terraces through which streams have cut gullies. The hill on the right has a smoothly eroded form and gradual slopes, whereas the one on the left rises abruptly in a sharp precipice, from which it slopes gently, and forms an inclined plateau traversed by a few shallow gullies. A road provides access to a church and the two houses situated across the river from a highway that follows the seacoast and curves up the river valley. The lower illustration shows the same features represented by symbols on a topographic map. The contour interval (vertical distance between adjacent contours) is 20 feet. (After U.S. Geological Survey)

represents map distance, and the denominator, a large number, represents ground distance. Thus, 1:24,000 means that a distance of 1 unit on the map represents a distance of 24,000 such units on the surface of the earth. It does not matter what the units are.

Often, the graphic or bar scale is more useful than the fractional scale, because it is easier to use for measuring distances between points. The graphic scale (Figure D.2) consists of a bar divided into equal segments, which represent equal distances on the map. One segment on the left

side of the bar is usually divided into smaller units to permit more accurate estimates of fractional units.

Topographic maps, which are also referred to as quadrangles, are generally classified according to publication scale. Each series is intended to fulfill a specific type of map need. To select a map with the proper scale for a particular use, remember that large-scale maps show more detail and small-scale maps show less detail. The sizes and scales of topographic maps published by the U.S. Geological Survey are shown in Table D.1.

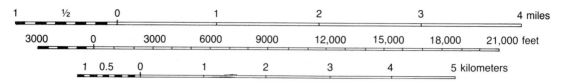

FIGURE D.2 Graphic scale.

Color and Symbol

Each color and symbol used on a U.S. Geological Survey topographic map has significance. Common topographic map symbols are shown in Figure D.3. The meaning of each color is as follows:

Blue—water features
Black—works of man, such as homes, schools, churches, roads, and so forth
Brown—contour lines
Green—woodlands, orchards, and so forth
Red—urban areas, important roads, public land subdivision lines

TABLE D.1 National topographic maps.

Series	Scale	1 Inch Represents	Standard Quadrangle Size (latitude-longitude)	Quadrangle Area (square miles)	Paper Size E-W N-S Width Length (inches)
7½-minute	1:24,000	2000 feet	7½′ × 7½′	49–70	22 × 27*
Puerto Rico 7½-minute	1:20,000	about 1667 feet	7½′ × 7½′	71	29½ × 32½
15-minute	1:62,500	nearly 1 mile	15′ × 15′	197–282	17 × 21*
Alaska 1:63,360	1:63,360	1 mile	15′ × 20′ − 36′	207–281	18 × 21**
U.S. 1:250,000	1:250,000	nearly 4 miles	1° × 2°†	4580–8669	34 × 22‡
U.S. 1:1,000,000	1:1,000,000	nearly 16 miles	4° × 6°†	73,734–102,759	27 × 27

Source: U.S. Geological Survey.
*South of latitude 31 degrees, 7½-minute sheets are 23 × 27 inches; 15-minute sheets are 18 × 21 inches.
**South of latitude 62 degrees, sheets are 17 × 21 inches.
†Maps of Alaska and Hawaii vary from these standards.
‡North of latitude 42 degrees, sheets are 29 × 22 inches; Alaska sheets are 30 × 23 inches.

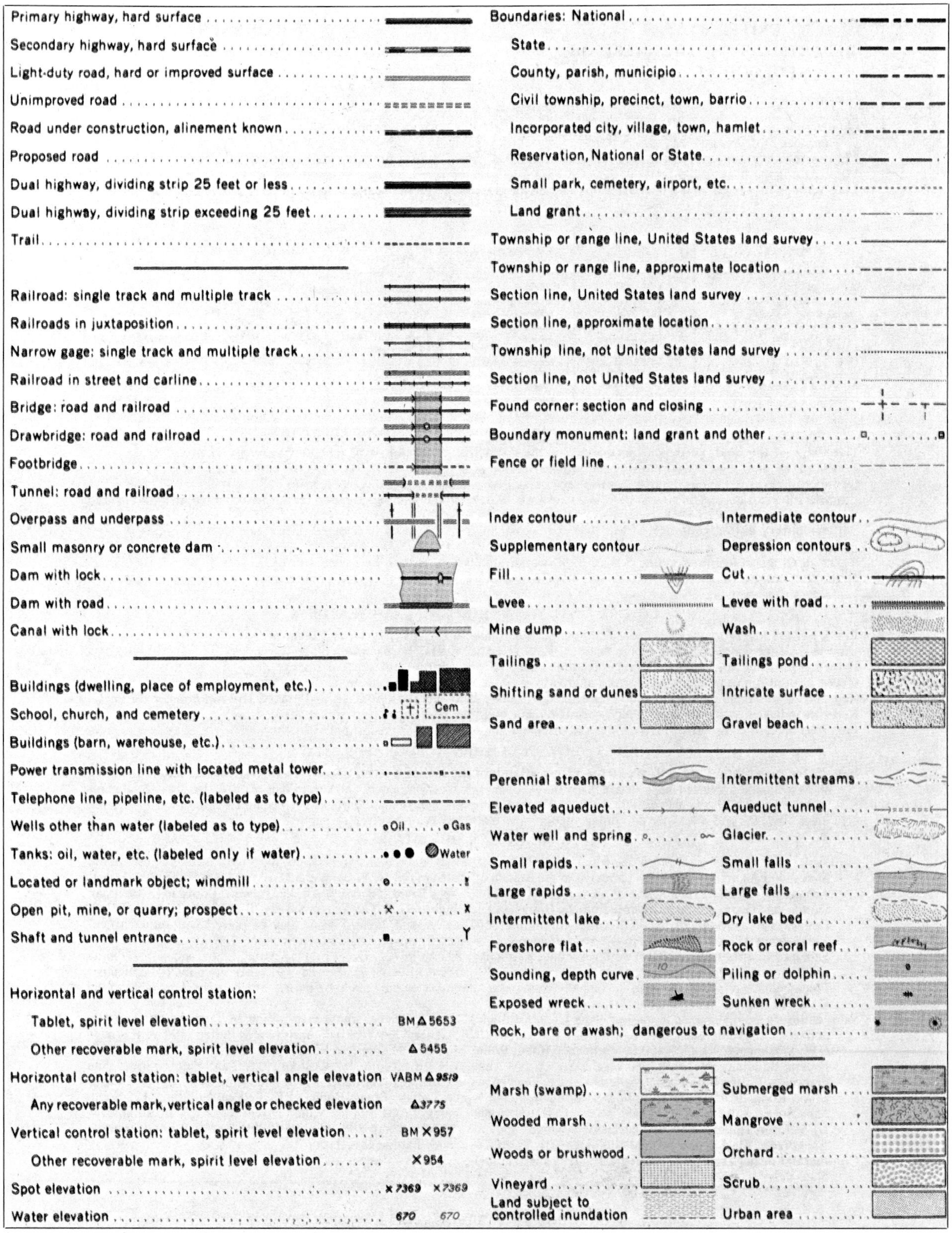

Primary highway, hard surface		Boundaries: National	
Secondary highway, hard surface		State	
Light-duty road, hard or improved surface		County, parish, municipio	
Unimproved road		Civil township, precinct, town, barrio	
Road under construction, alinement known		Incorporated city, village, town, hamlet	
Proposed road		Reservation, National or State	
Dual highway, dividing strip 25 feet or less		Small park, cemetery, airport, etc.	
Dual highway, dividing strip exceeding 25 feet		Land grant	
Trail		Township or range line, United States land survey	

Township or range line, approximate location

Section line, United States land survey

Section line, approximate location

Railroad: single track and multiple track		Township line, not United States land survey	
Railroads in juxtaposition		Section line, not United States land survey	
Narrow gage: single track and multiple track		Found corner: section and closing	
Railroad in street and carline		Boundary monument: land grant and other	
Bridge: road and railroad		Fence or field line	
Drawbridge: road and railroad			
Footbridge			
Tunnel: road and railroad		Index contour	Intermediate contour
Overpass and underpass		Supplementary contour	Depression contours
Small masonry or concrete dam		Fill	Cut
Dam with lock		Levee	Levee with road
Dam with road		Mine dump	Wash
Canal with lock		Tailings	Tailings pond
		Shifting sand or dunes	Intricate surface
Buildings (dwelling, place of employment, etc.)		Sand area	Gravel beach
School, church, and cemetery	Cem		
Buildings (barn, warehouse, etc.)			
Power transmission line with located metal tower		Perennial streams	Intermittent streams
Telephone line, pipeline, etc. (labeled as to type)		Elevated aqueduct	Aqueduct tunnel
Wells other than water (labeled as to type)	Oil Gas	Water well and spring	Glacier
Tanks: oil, water, etc. (labeled only if water)	Water	Small rapids	Small falls
Located or landmark object; windmill		Large rapids	Large falls
Open pit, mine, or quarry; prospect		Intermittent lake	Dry lake bed
Shaft and tunnel entrance		Foreshore flat	Rock or coral reef
		Sounding, depth curve	Piling or dolphin
Horizontal and vertical control station:		Exposed wreck	Sunken wreck
Tablet, spirit level elevation	BM △ 5653	Rock, bare or awash; dangerous to navigation	
Other recoverable mark, spirit level elevation	△ 5455		
Horizontal control station: tablet, vertical angle elevation	VABM △ 9519	Marsh (swamp)	Submerged marsh
Any recoverable mark, vertical angle or checked elevation	△ 3775	Wooded marsh	Mangrove
Vertical control station: tablet, spirit level elevation	BM × 957	Woods or brushwood	Orchard
Other recoverable mark, spirit level elevation	× 954	Vineyard	Scrub
Spot elevation	× 7369 × 7369	Land subject to controlled inundation	Urban area
Water elevation	670 670		

FIGURE D.3 U.S. Geological Survey topographic map symbols. (Variations will be found on older maps)

329

GLOSSARY

Aa A type of lava flow that has a jagged, blocky surface.

Ablation A general term for the loss of ice and snow from a glacier.

Abrasion The grinding and scraping of a rock surface by the friction and impact of rock particles carried by water, wind, or ice.

Absolute dating Determination of the number of years since the occurrence of a given geologic event.

Abyssal plain Very level area of the deep-ocean floor, usually lying at the foot of the continental rise.

Active layer The zone above the permafrost that thaws in summer and refreezes in winter.

Aftershock A smaller earthquake that follows the main earthquake.

Alluvial fan A fan-shaped deposit of sediment formed when a stream's slope is abruptly reduced.

Alluvium Unconsolidated sediment deposited by a stream.

Alpine glacier A glacier confined to a mountain valley, which in most instances had previously been a stream valley.

Angle of repose The steepest angle at which loose material remains stationary without sliding downslope.

Angular unconformity An unconformity in which the older strata dip at an angle different from that of the younger beds.

Anthracite A hard, metamorphic form of coal that burns clean and hot.

Anticline A fold in sedimentary strata that resembles an arch.

Aphanitic A texture of igneous rocks in which the crystals are too small for individual minerals to be distinguished with the unaided eye.

Aquiclude An impermeable bed that hinders or prevents groundwater movement.

Aquifer Rock or sediment through which groundwater moves easily.

Arête A narrow, knifelike ridge separating two adjacent glaciated valleys.

Arkose A feldspar-rich sandstone.

Artesian well A well in which the water rises above the level where it was initially encountered.

Asthenosphere A subdivision of the mantle situated below the lithosphere. This zone of weak material exists below a depth of about 100 kilometers and in some regions extends as deep as 700 kilometers. The rock within this zone is easily deformed.

Astronomical theory A theory of climatic change first developed by the Yugoslavian astronomer Milankovitch. It is based upon changes in the shape of the earth's orbit, variations in the obliquity of the earth's axis, and the wobbling of the earth's axis.

Atmosphere The gaseous portion of a planet; the planet's envelope of air. One of the traditional subdivisions of the earth's physical environment.

Atoll A continuous or broken ring of coral reef surrounding a central lagoon.

Atom The smallest particle that exists as an element.

Atomic number The number of protons in the nucleus of an atom.

Atomic weight The average of the atomic masses of isotopes for a given element.

Aureole A zone or halo of contact metamorphism found in the country rock surrounding an igneous intrusion.

Azoic zone A well-known but incorrect theory formulated around 1850 by Edward Forbes stating that no life existed in the ocean below about 550 meters.

Back swamp A poorly drained area on a floodplain resulting when natural levees are present.

Bajada An apron of sediment along a mountain front created by the coalescence of alluvial fans.

Barchan dune A solitary sand dune shaped like a crescent with its tips pointing downwind.

Barrier island A low, elongate ridge of sand that parallels the coast.

Basal slip A mechanism of glacial movement in which the ice mass slides over the surface below.

Basalt A fine-grained igneous rock of mafic composition.

Base level The level below which a stream cannot erode.

Basin A circular downfolded structure.

Batholith A large mass of igneous rock that formed when magma was emplaced at depth, crystallized, and was subsequently exposed by erosion.

Baymouth bar A sandbar that completely crosses a bay, sealing it off from the main body of water.

Beach drift The transport of sediment in a zigzag pattern along a beach caused by the uprush of water from obliquely breaking waves.

Bedding plane A nearly flat surface separating two beds of sedimentary rock. Each bedding plane marks the end of one deposit and the beginning of another having different characteristics.

Bed load Sediment rolled along the bottom of a stream by moving water, or particles rolled along the ground surface by wind.

Belt of soil moisture A zone in which water is held as a film on the surface of soil particles and may be used by plants or withdrawn by evaporation. The uppermost subdivision of the zone of aeration.

Benioff zone The zone of inclined seismic activity that extends from a trench downward into the asthenosphere.

Biogenous sediment Sea-floor sediments consisting of material of marine-organic origin.

Bituminous coal The most common form of coal, often called soft, black coal.

Blowout (deflation hollow) A depression excavated by wind in easily eroded materials.

Body wave A seismic wave that travels through the earth's interior.

Bottomset bed A layer of fine sediment deposited beyond the advancing edge of a delta and then buried by continued delta growth.

Braided stream A stream consisting of numerous intertwining channels.

Breakwater A structure protecting a nearshore area from breaking waves.

Breccia A sedimentary rock composed of angular fragments that were lithified.

Cactolith A quasi-horizontal chonolith composed of anastomosing ductoliths, whose distal ends curl like a harpolith, thin like a sphenolith, or bulge discordantly like an akmolith or ethmolith.

Caldera A large depression typically caused by collapse or ejection of the summit area of a volcano.

Caliche A hard layer, rich in calcium carbonate, that forms beneath the *B* horizon in soils of arid regions.

Calving Wastage of a glacier that occurs when large pieces of ice break off into water.

Capacity The total amount of sediment a stream is able to transport.

Capillary fringe A relatively narrow zone at the base of the zone of aeration. Here water rises from the water table in tiny threadlike openings between grains of soil or sediment.

Catastrophism The concept that the earth was shaped by catastrophic events of a short-term nature.

Cavern A naturally formed underground chamber or series of chambers most commonly produced by solution activity in limestone.

Cenozoic era A time span on the geologic calendar beginning about 65 million years ago following the Mesozoic era.

Chemical weathering The processes by which the internal structure of a mineral is altered by the removal and/or addition of elements.

Cinder cone A rather small volcano built primarily of pyroclastics ejected from a single vent.

Cirque An amphitheater-shaped basin at the head of a glaciated valley produced by frost wedging and plucking.

Clastic A sedimentary rock texture consisting of broken fragments of pre-existing rock.

Cleavage The tendency of a mineral to break along planes of weak bonding.

Col A pass between mountain valleys where the headwalls of two cirques intersect.

Column A feature found in caves that is formed when a stalactite and stalagmite join.

Columnar joints A pattern of cracks that forms during cooling of molten rock to generate columns.

Competence A measure of the largest particle a stream can transport; a factor dependent on velocity.

Composite cone A volcano composed of both lava flows and pyroclastic material.

Compound A substance formed by the chemical combination of two or more elements in definite proportions and usually having properties different from those of its constituent elements.

Concordant A term used to describe intrusive igneous masses that form parallel to the bedding of the surrounding rock.

Cone of depression A cone-shaped depression immediately surrounding a well.

Conformable layers Rock layers that were deposited without interruption.

Conglomerate A sedimentary rock composed of rounded gravel-sized particles.

Contact metamorphism Changes in rock caused by the heat from a nearby magma body.

Continental drift A hypothesis, credited largely to Alfred Wegener, that suggested all present continents once existed as a single supercontinent. Further, beginning about 200 million years ago, the supercontinent began breaking into smaller continents which then "drifted" to their present positions.

Continental glacier A massive accumulation of ice that covers extensive land areas and whose flow is not usually controlled by the underlying topography.

Continental margin That portion of the sea floor adjacent to the continents. It may include the continental shelf, continental slope, and continental rise.

Continental rise The gently sloping surface at the base of the continental slope.

Continental shelf The gently sloping submerged portion of the continental margin extending from the shoreline to the continental slope.

Continental slope The steep gradient that leads to the deep-ocean floor and marks the seaward edge of the continental shelf.

Convergent boundary A boundary in which two plates move together, causing one of the slabs of lithosphere to be consumed into the mantle as it descends beneath an overriding plate.

Correlation Establishing the equivalence of rocks of similar age in different areas.

Covalent bond A chemical bond produced by the sharing of electrons.

Crater The depression at the summit of a volcano, or that which is produced by a meteorite impact.

Creep The slow downhill movement of soil and regolith.

Crevasse A deep crack in the brittle surface of a glacier.

Cross-cutting A principle of relative dating. A rock or fault is younger than any rock (or fault) through which it cuts.

Crust The very thin outermost layer of the earth.

Crystal An orderly arrangement of atoms.

Crystal form The external appearance of a mineral as determined by its internal arrangement of atoms.

Crystallization The formation and growth of a crystalline solid from a liquid or gas.

Curie point The temperature above which a material loses its magnetization.

Cut bank The area of active erosion on the outside of a meander.

Cutoff A short channel segment created when a river erodes through the narrow neck of land between meanders.

Daughter product An isotope resulting from radioactive decay.

Deep-focus earthquake An earthquake focus at a depth of more than 300 kilometers.

Deep-ocean basin The portion of sea floor that lies between the continental margin and the oceanic ridge system. This region comprises almost 30 percent of the earth's surface.

Deflation The lifting and removal of loose material by wind.

Delta An accumulation of sediment formed where a stream enters a lake or ocean.

Dendritic pattern A stream system that resembles the pattern of a branching tree.

Density The weight per unit volume of a particular material.

Desert pavement A layer of coarse pebbles and gravel created when wind removed the finer material.

Detrital sedimentary rocks Rocks that form from the accumulation of materials that originate and are transported as solid particles derived from both mechanical and chemical weathering.

Dike A tabular-shaped intrusive igneous feature that cuts through the surrounding rock.

Dip The angle at which a rock layer is inclined from the horizontal. The direction of dip is at a right angle to the strike.

Dip-slip fault A fault in which the movement is parallel to the dip of the fault.

Discharge The quantity of water in a stream that passes a given point in a period of time.

Disconformity A type of unconformity in which the beds above and below are parallel.

Discontinuity A sudden change with depth in one or more of the physical properties of the material making up the earth's interior. The boundary between two dissimilar materials in the earth's interior as determined by the behavior of seismic waves.

Discordant A term used to describe plutons that cut across existing rock structures, such as bedding planes.

Dissolved load That portion of a stream's load carried in solution.

Distributary A section of a stream that leaves the main flow.

Diurnal tide A tide characterized by a single high and low water height each tidal day.

Divergent boundary A boundary in which two plates move apart, resulting in upwelling of material from the mantle to create new sea floor.

Divide An imaginary line that separates the drainage of two streams; often found along a ridge.

Dome A roughly circular upfolded structure.

Drainage basin The land area that contributes water to a stream.

Drawdown The difference in height between the bottom of a cone of depression and the original height of the water table.

Drift The general term for any glacial deposit.

Drumlin A streamlined asymmetrical hill composed of glacial till. The steep side of the hill faces the direction from which the ice advanced.

Dry climate A climate in which yearly precipitation is less than the potential loss of water by evaporation.

Dune A hill or ridge of wind-deposited sand.

Earthflow The downslope movement of water-saturated, clay-rich sediment. Most characteristic of humid regions.

Earthquake Vibration of the earth produced by the rapid release of energy.

Ebb tide The movement of tidal current away from the shore.

Echo sounder An instrument used to determine the depth of water by measuring the time interval between emission of a sound signal and the return of its echo from the bottom.

Effluent stream A stream channel that intersects the water table. Consequently, groundwater feeds into the stream.

Elastic deformation Nonpermanent deformation in which rock returns to its original shape when the stress is released.

Elastic rebound The sudden release of stored strain in rocks that results in movement along a fault.

Electron A negatively charged subatomic particle that has a negligible mass and is found outside an atom's nucleus.

Element A substance that cannot be decomposed into simpler substances by ordinary chemical or physical means.

Emergent coast A coast where land formerly below sea level has been exposed either by crustal uplift or a drop in sea level or both.

End moraine A ridge of till marking a former position of the front of a glacier.

Entrenched meander A meander cut into bedrock when uplifting rejuvenated a meandering stream.

Epicenter The location on the earth's surface that lies directly above the focus of an earthquake.

Epoch A unit of the geologic calendar that is a subdivision of a period.

Era A major division on the geologic calendar; eras are divided into shorter units called periods.

Erosion The incorporation and transportation of material by a mobile agent, such as water, wind, or ice.

Esker Sinuous ridge composed largely of sand and gravel deposited by a stream flowing in a tunnel beneath a glacier near its terminus.

Estuary A funnel-shaped inlet of the sea that formed when a rise in sea level or subsidence of land caused the mouth of a river to be flooded.

Evaporite A sedimentary rock formed of material deposited from solution by evaporation of the water.

Evapotranspiration The combined effect of evaporation and transpiration.

Exotic stream A permanent stream that traverses a desert and has its source in well-watered areas outside the desert.

Extrusive Igneous activity that occurs at the earth's surface.

Fall A type of movement common to mass wasting processes that refers to the free falling of detached individual pieces of any size.

Fault A break in a rock mass along which movement has occurred.

Fault-block mountain A mountain formed by the displacement of rock along a fault.

Faunal succession Fossil organisms succeed one another in a definite and determinable order, and any time period can be recognized by its fossil content.

Fetch The distance that the wind has traveled across the open water.

Fiord A steep-sided inlet of the sea formed when a glacial trough was partially submerged.

Firn Granular recrystallized snow. A transitional stage between snow and glacial ice.

Fissure eruption An eruption in which lava is extruded from narrow fractures or cracks in the crust.

Flood basalts Flows of basaltic lava that issue from numerous cracks or fissures and commonly cover extensive areas to thicknesses of hundreds of meters.

Floodplain The flat, low-lying portion of a stream valley subject to periodic inundation.

Flood tide The tidal current associated with the increase in the height of the tide.

Flow A type of movement common to mass wasting processes in which water-saturated material moves downslope as a viscous fluid.

Fluorescence The absorption of ultraviolet light, which is re-emitted as visible light.

Focus (earthquake) The zone within the earth where rock displacement produces an earthquake.

Foliated A texture of metamorphic rocks that gives the rock a layered appearance.

Foreset bed An inclined bed deposited along the front of a delta.

Foreshocks Small earthquakes that often precede a major earthquake.

Fossil The remains or traces of organisms preserved from the geologic past.

Fractional crystallization The process that separates magma into components having varied compositions and melting points.

Frost wedging The mechanical breakup of rock caused by the expansion of freezing water in cracks and crevices.

Fumarole A vent in a volcanic area from which fumes or gases escape.

Geology The science that examines the earth, its form and composition, and the changes which it has undergone and is undergoing.

Geosyncline A large linear downwarp in the earth's crust in which thousands of meters of sediment have accumulated.

Geothermal energy Natural steam used for power generation.

Geothermal gradient The gradual increase in temperature with depth in the crust. The average is 30°C per kilometer in the upper crust.

Geyser A fountain of hot water ejected periodically from the ground.

Glacial erratic An ice-transported boulder that was not derived from the bedrock near its present site.

Glacial striations Scratches and grooves on bedrock caused by glacial abrasion.

Glacial trough A mountain valley that has been widened, deepened, and straightened by a glacier.

Glacier A thick mass of ice originating on land from the compaction and recrystallization of snow that shows evidence of past or present flow.

Glass (volcanic) Natural glass produced when molten lava cools too rapidly to permit crystallization. Volcanic glass is a solid composed of unordered atoms.

Glassy A term used to describe the texture of certain igneous rocks, such as obsidian, that contain no crystals.

Gondwanaland The southern portion of Pangaea consisting of South America, Africa, Australia, India, and Antarctica.

Graben A valley formed by the downward displacement of a fault-bounded block.

Graded bed A sediment layer characterized by a decrease in sediment size from bottom to top.

Graded stream A stream that has the correct channel characteristics to maintain exactly the velocity required to transport the material supplied to it.

Gradient The slope of a stream; generally measured in feet per mile.

Granitization The process of converting country rock into granite. The process is thought to occur when hot, ion-rich fluids migrate through a rock and chemically alter its composition.

Groin A short wall built at a right angle to the seashore to trap moving sand.

Groundmass The matrix of smaller crystals within an igneous rock that has porphyritic texture.

Ground moraine An undulating layer of till deposited as the ice front retreats.

Groundwater Water in the zone of saturation.

Guyot A submerged flat-topped seamount.

Half-life The time required for one-half of the atoms of a radioactive substance to decay.

Hanging valley A tributary valley that enters a glacial trough at a considerable height above the floor of the trough.

Hardness A mineral's resistance to scratching and abrasion.

Head The vertical distance between the recharge and discharge points of a water table.

Historical geology A major division of geology that deals with the origin of the earth and its development through time. Usually involves the study of fossils and their sequence in rock beds.

Hogback A narrow, sharp-crested ridge formed by the upturned edge of a steeply dipping bed of resistant rock.

Horn A pyramid-like peak formed by glacial action in three or more cirques surrounding a mountain summit.

Horst An elongate, uplifted block of crust bounded by faults.

Hot spot A proposed concentration of heat in the mantle capable of producing magma which, in turn, extrudes onto the earth's surface. The intraplate volcanism that produced the Hawaiian Islands is one example.

Hot spring A spring in which the water is 6–9°C (10–15°F) warmer than the mean annual air temperature of its locality.

Humus Organic matter in soil produced by the decomposition of plants and animals.

Hydrogenous sediment Sea-floor sediments consisting of minerals that crystallize from seawater. The principal example is manganese nodules.

Hydrologic cycle The unending circulation of the earth's water supply. The cycle is powered by energy from the sun and is characterized by continuous exchanges of water among the oceans, the atmosphere, and the continents.

Hydrolysis A chemical weathering process in which minerals are altered by chemically reacting with water and acids.

Hydrosphere The water portion of our planet; one of the traditional subdivisions of the earth's physical environment.

Hydrothermal solution The hot, watery solution that escapes from a mass of magma during the latter stages of crystallization. Such solutions may alter the surrounding country rock and are frequently the source of significant ore deposits.

Ice-contact deposit An accumulation of stratified drift deposited in contact with a supporting mass of ice.

Igneous rock A rock formed by the crystallization of molten magma.

Immature soil A soil lacking horizons.

Index fossil A fossil that is associated with a particular span of geologic time.

Index mineral A mineral that is a good indicator of the metamorphic environment in which it formed. Used to distinguish different zones of regional metamorphism.

Infiltration The movement of surface water into rock or soil through cracks and pore spaces.

Infiltration capacity The maximum rate at which soil can absorb water.

Influent stream A stream channel that is above the water table level. Water seeps downward from the channel to the zone of saturation to produce an upward bulge in the water table.

Inner core The solid innermost layer of the earth, about 1216 kilometers (754 miles) in radius.

Inselberg An isolated mountain remnant characteristic of the late stage of erosion in a mountainous arid region.

Interior drainage A discontinuous pattern of intermittent streams that do not flow to the ocean.

Intermediate focus An earthquake focus at a depth of between 60 and 300 kilometers.

Intrusive rock Igneous rock that formed below the earth's surface.

Ion An atom or molecule that possesses an electrical charge.

Ionic bond A chemical bond between two oppositely charged ions formed by the transfer of valence electrons from one atom to the other.

Island arc A chain of volcanic islands generally located a few hundred kilometers from a trench where active subduction of one oceanic slab beneath another is occurring.

Isostasy The concept that the earth's crust is "floating" in gravitational balance upon the material of the mantle.

Isotopes Varieties of the same element that have different mass numbers; their nuclei contain the same number of protons but different numbers of neutrons.

Jetties A pair of structures extending into the ocean at the entrance to a harbor or river that are built for the purpose of protecting against storm waves and sediment deposition.

Joint A fracture in rock along which there has been no movement.

Kame A steep-sided hill composed of sand and gravel originating when sediment collected in openings in stagnant glacial ice.

Kame terrace A narrow, terrace-like mass of stratified drift deposited between a glacier and an adjacent valley wall.

Karst A topography consisting of numerous depressions called sinkholes.

Kettle holes Depressions created when blocks of ice become lodged in glacial deposits and subsequently melt.

Laccolith A massive, concordant igneous body intruded between pre-existing strata.

Laminar flow The movement of water particles in straightline paths that are parallel to the channel. The water particles move downstream without mixing.

Lateral moraine A ridge of till along the sides of an alpine glacier composed primarily of debris that fell to the glacier from the valley walls.

Laterite A red, highly leached soil type found in the tropics that is rich in oxides of iron and aluminum.

Laurasia The northern portion of Pangaea consisting of North America and Eurasia.

Lava Magma that reaches the earth's surface.

Lava dome A bulbous mass associated with an old-age volcano, produced when thick lava is slowly squeezed from the vent. Lava domes may act as plugs to deflect subsequent gaseous eruptions.

Law of superposition In any undeformed sequence of sedimentary rocks or surface-deposited igneous materials, each layer is older than the one above it and younger than the one below.

Lithification The process, generally cementation and/or compaction, of converting sediments to solid rock.

Lithogenous sediment Sea-floor sediment consisting primarily of mineral grains that were weathered from continental rocks and transported to the ocean.

Lithosphere The rigid outer layer of the earth, including the crust and upper mantle.

Loess Deposits of windblown silt, lacking visible layers, generally buff colored, and capable of maintaining a nearly vertical cliff.

Longitudinal dunes Long ridges of sand oriented parallel to the prevailing wind; these dunes form where sand supplies are limited.

Longitudinal profile A cross section of a stream channel along its descending course from the head to the mouth.

Longshore current A nearshore current that flows parallel to the shore.

Luster The appearance or quality of light reflected from the surface of a mineral.

Magma A body of molten rock found at depth, including any dissolved gases and crystals.

Magnetometer A sensitive instrument used to measure the intensity of the earth's magnetic field at various points.

Magnitude (earthquake) The total amount of energy released during an earthquake.

Manganese nodules A type of hydrogenous sediment scattered on the ocean floor, consisting mainly of manganese and iron, and usually containing small amounts of copper, nickel, and cobalt.

Mantle The 2885-kilometer (1789-mile) thick layer of the earth located below the crust.

Mass number The sum of the number of neutrons and protons in the nucleus of an atom.

Mass wasting The downslope movement of rock, regolith, and soil under the direct influence of gravity.

Meander A looplike bend in the course of a stream.

Meander scar A floodplain feature created when an oxbow lake becomes filled with sediment.

Mechanical weathering The physical disintegration of rock, resulting in smaller fragments.

Medial moraine A ridge of till formed when lateral moraines from two coalescing alpine glaciers join.

Mélange A highly deformed mixture of rock material formed in areas of plate convergence.

Melt The liquid portion of magma excluding the solid crystals.

Mercalli intensity scale A 12-point scale originally developed to evaluate earthquake intensity based upon the amount of damage to various types of structures.

Mesozoic era A time span on the geologic calendar between the Paleozoic and Cenozoic eras—from about 225 to 65 million years ago.

Metallic bond A chemical bond present in all metals that may be characterized as an extreme type of electron sharing in which the electrons move freely from atom to atom.

Metamorphic rock Rock formed by the alteration of pre-existing rock deep within the earth (but still in the solid state) by heat, pressure, and/or chemically active fluids.

Metamorphism The changes in mineral composition and texture of a rock subjected to high temperature and pressure within the earth.

Mid-ocean ridge A continuous mountainous ridge on the floor of all the major ocean basins and varying in width from 500–5000 kilometers (300–3000 miles). The rifts at the crests of these ridges represent divergent plate boundaries.

Migmatite A rock exhibiting both igneous and metamorphic rock characteristics. Such rocks may form when light-colored silicate minerals melt and then crystallize, while the dark silicate minerals remain solid.

Mineral A naturally occurring, inorganic crystalline material with a unique chemical structure.

Mohorovičić discontinuity (Moho) The boundary separating the crust and the mantle, discernible by an increase in seismic velocity.

Mohs scale A series of ten minerals used as a standard in determining hardness.

Monocline A one-limbed flexure in strata. The strata are usually flat lying or very gently dipping on both sides of the monocline.

Mud crack A feature in some sedimentary rocks that forms when wet mud dries out, shrinks, and cracks.

Mudflow The flowage of debris containing a large amount of water; most characteristic of canyons and gullies in dry, mountainous regions.

Natural levees The elevated landforms composed of alluvium that parallel some streams and act to confine their waters, except during floodstage.

Neap tide The lowest tidal range, occurring near the times of the first and third quarters of the moon.

Neutron A subatomic particle found in the nucleus of an atom. The neutron is electrically neutral with a mass approximately equal to that of a proton.

Nonclastic A term for the texture of sedimentary rocks in which the minerals form a pattern of interlocking crystals.

Nonfoliated Metamorphic rocks that do not exhibit foliation.

Normal fault A fault in which the rock above the fault plane has moved down relative to the rock below.

Normal polarity A magnetic field the same as that which presently exists.

Nucleus The small, heavy core of an atom that contains all of its positive charge and most of its mass.

Nuée ardente Incandescent volcanic debris buoyed up by hot gases that moves downslope in an avalanche fashion.

Oblique-slip fault A fault having both vertical and horizontal movement.

Octet rule Atoms combine in order that each may have the electron arrangement of a noble gas; that is, the outer energy level contains eight electrons.

Ophiolite complex The sequence of rocks that make up the oceanic crust. The three-layer sequence includes an upper layer of pillow basalts, a middle zone of sheeted dikes, and a lower layer of gabbro.

Orogenesis The processes that collectively result in the formation of mountains.

Outer core A layer beneath the mantle about 2270 kilometers (1407 miles) thick which has the properties of a liquid.

Outwash Sediments deposited by glacial meltwater.

Outwash plain A relatively flat, gently sloping plain consisting of materials deposited by meltwater streams in front of the margin of an ice sheet.

Oxbow lake A curved lake produced when a stream cuts off a meander.

Oxidation The removal of one or more electrons from an atom or ion. So named because elements commonly combine with oxygen.

Pahoehoe A lava flow with a smooth-to-ropy surface.

Paleomagnetism The natural remnant magnetism in rock bodies. The permanent magnetization acquired by rock which can be used to determine the location of the magnetic poles and the latitude of the rock at the time it became magnetized.

Paleontology The systematic study of fossils and the history of life on earth.

Paleozoic era A time span on the geologic calendar between the Precambrian and Mesozoic eras—from about 570 million to 225 million years ago.

Pangaea The proposed supercontinent which 200 million years ago began to break apart and form the present landmasses.

Parabolic dune A sand dune similar in shape to a barchan dune except that its tip points into the wind. These dunes often form along coasts that have strong onshore winds, abundant sand, and vegetation that partly covers the sand.

Parasitic cone A volcanic cone which forms on the flank of a larger volcano.

Parent material The material upon which a soil develops.

Partial melting The process by which most igneous rocks melt. Since individual minerals have different melting points, most igneous rocks melt over a temperature range of a few hundred degrees. If the liquid is squeezed out after some melting has occurred, a melt with a higher silica content results.

Pater noster lakes A chain of small lakes in a glacial trough that occupy basins created by glacial erosion.

Pedalfer Soil of humid regions characterized by the accumulation of iron oxides and aluminum-rich clays in the B horizon.

Pediment A sloping bedrock surface fringing a mountain base in an arid region, formed when erosion causes the mountain front to retreat.

Pedocal Soil associated with drier regions and characterized by an accumulation of calcium carbonate in the upper horizons.

Pegmatite A very coarse-grained igneous rock (typically granite) commonly found as a dike associated with a large mass of plutonic rock that has smaller crystals. Crystallization in a water-rich environment is believed to be responsible for the very large crystals.

Peneplain In the idealized cycle of landscape evolution in a humid region, an undulating plain near base level associated with old age.

Perched water table A localized zone of saturation above the main water table created by an impermeable layer (aquiclude).

Peridotite An igneous rock of ultramafic composition thought to be abundant in the upper mantle.

Period A basic unit of the geologic calendar that is a subdivision of an era. Periods may be divided into smaller units called epochs.

Permafrost Any permanently frozen subsoil. Usually found in the subarctic and arctic regions.

Permeability A measure of a material's ability to transmit water.

Phaneritic An igneous rock texture in which the crystals are roughly equal in size and large enough so that individual minerals can be identified with the unaided eye.

Phenocryst Conspicuously large crystals in a porphyry that are imbedded in a matrix of finer-grained crystals (the groundmass).

Physical geology A major division of geology that examines the materials of the earth and seeks to understand the processes and forces acting beneath and upon the earth's surface.

Pillow lava Basaltic lava that solidifies in an underwater environment and develops a structure that resembles a pile of pillows.

Plastic deformation Permanent deformation that results in a change in size and shape through folding or flowing.

Plastic flow A type of glacial movement that occurs within the glacier, below a depth of approximately 50 meters, in which the ice is not fractured.

Plate One of numerous rigid sections of the lithosphere that moves as a unit over the material of the asthenosphere.

Plate tectonics The theory which proposes that the earth's outer shell consists of individual plates which interact in various ways and thereby produce earthquakes, volcanoes, mountains, and the crust itself.

Playa The flat central area of an undrained desert basin.

Playa lake A temporary lake in a playa.

Pleistocene epoch An epoch of the Quaternary period beginning about 2.5 million years ago and ending about 10,000 years ago. Best known as a time of extensive continental glaciation.

Plucking The process by which pieces of bedrock are lifted out of place by a glacier.

Pluton A structure that results from the emplacement and crystallization of magma beneath the surface of the earth.

Pluvial lake A lake formed during a period of increased rainfall. For example, this occurred in many nonglaciated areas during periods of ice advance elsewhere.

Point bar A crescent-shaped accumulation of sand and gravel deposited on the inside of a meander.

Polar wandering hypothesis As the result of paleomagnetic studies in the 1950s, researchers proposed that either the magnetic poles migrated greatly through time or the continents had gradually shifted their positions.

Polymorphs Two or more minerals having the same chemical composition but different crystalline structures. Exemplified by the diamond and graphite forms of carbon.

Porosity The volume of open spaces in rock or soil.

Porphyritic An igneous rock texture characterized by two distinctively different crystal sizes. The larger crystals are called phenocrysts and the matrix of smaller crystals is termed the groundmass.

Porphyry An igneous rock with a porphyritic texture.

Pothole A depression formed in a stream channel by the abrasive action of the water's sediment load.

Precambrian All geologic time prior to the Paleozoic era.

Principle of faunal succession Fossil organisms succeed one another in a definite and determinable order, and any time period can be recognized by its fossil content.

Principle of original horizontality Layers of sediment are generally deposited in a horizontal or nearly horizontal position.

Proton A positively charged subatomic particle found in the nucleus of an atom.

P wave The fastest earthquake wave, which travels by compression and expansion of the medium.

Pyroclastic An igneous rock texture resulting from the consolidation of individual rock fragments that are ejected during a violent eruption.

Pyroclastic material The volcanic rock ejected during an eruption. Pyroclastics include ash, bombs, and blocks.

Radial drainage A system of streams running in all directions away from a central elevated structure, such as a volcano.

Radioactivity The spontaneous decay of certain unstable atomic nuclei.

Radiocarbon (carbon-14) The radioactive isotope of carbon produced continuously in the atmosphere and used in dating events as far back as 75,000 years.

Radiometric dating The procedure of calculating the absolute ages of rocks and minerals that contain certain radioactive isotopes.

Recessional moraine An end moraine formed as the ice front stagnated during glacial retreat.

Rectangular pattern A drainage pattern characterized by numerous right angle bends that develops on jointed or fractured bedrock.

Refraction A change in direction of waves as they enter shallow water. The portion of the wave in shallow water is slowed, which causes the wave to bend and align with the underwater contours.

Regional metamorphism Metamorphism associated with large-scale mountain building.

Regolith The layer of rock and mineral fragments that nearly everywhere covers the earth's land surface.

Rejuvenation A change in relation to base level, often caused by regional uplift, that causes the forces of erosion to intensify.

Relative dating Rocks are placed in their proper sequence or order. Only the chronological order of events is determined.

Residual soil Soil developed directly from the weathering of the bedrock below.

Reverse fault A fault in which the material above the fault plane moves up in relation to the material below.

Reverse polarity A magnetic field opposite to that which presently exists.

Richter scale A scale of earthquake magnitude based on the motion of a seismograph.

Rift A region of the earth's crust along which divergence is taking place.

Ripple marks Small waves of sand that develop on the surface of a sediment layer by the action of moving water or air.

Roche moutonnée An asymmetrical knob of bedrock formed when glacial abrasion smoothes the gentle slope facing the advancing ice sheet and plucking steepens the opposite side as the ice overrides the knob.

Rock A consolidated mixture of minerals.

Rock avalanche The very rapid downslope movement of rock and debris. These rapid movements may be aided by a layer of air trapped beneath the debris, and they have been known to reach speeds in excess of 200 kilometers per hour.

Rock cleavage The tendency of rock to split along parallel, closely spaced surfaces. These surfaces are often highly inclined to the bedding planes in the rock.

Rock flour Ground-up rock produced by the grinding effect of a glacier.

Rockslide The rapid slide of a mass of rock downslope along planes of weakness.

Runoff Water that flows over the land rather than infiltrating into the ground.

Saltation Transportation of sediment through a series of leaps or bounces.

Salt flat A white crust on the ground produced when water evaporates and leaves its dissolved materials behind.

Schistosity A type of foliation characteristic of coarser-grained metamorphic rocks. Such rocks have a parallel arrangement of platy minerals such as the micas.

Scoria Hardened lava which has retained the vesicles produced by escaping gases.

Sea arch An arch formed by wave erosion when caves on opposite sides of a headland unite.

Sea-flooring spreading The hypothesis first proposed in the 1960s by Harry Hess which suggested that new oceanic crust is produced at the crests of mid-ocean ridges, which are the sites of divergence.

Seamount An isolated volcanic peak that rises at least 1000 meters (3300 feet) above the deep-ocean floor.

Sea stack An isolated mass of rock standing just offshore, produced by wave erosion of a headland.

Sediment Unconsolidated particles created by the weathering and erosion of rock, by chemical precipitation from solution in water, or from the secretions of organisms, and transported by water, wind, or glaciers.

Sedimentary rock Rock formed from the weathered products of pre-existing rocks that have been transported, deposited, and lithified.

Seiche The rhythmic sloshing of water in lakes, reservoirs, and other smaller enclosed basins. Some seiches are initiated by earthquake activity.

Seismic sea wave A rapidly moving ocean wave generated by earthquake activity which is capable of inflicting heavy damage in coastal regions.

Seismogram The record made by a seismograph.

Seismograph An instrument that records earthquake waves.

Seismology The study of earthquakes and seismic waves.

Settling velocity The speed at which a particle falls through a still fluid. The size, shape, and specific gravity of particles influence settling velocity.

Shadow zone The zone between 105 and 140 degrees distance from an earthquake epicenter which direct waves do not penetrate because of refraction by the earth's core.

Shallow-focus earthquake An earthquake focus at a depth of less than 60 kilometers.

Shear Stress that causes two adjacent parts of a body to slide past one another.

Sheeting A mechanical weathering process characterized by the splitting off of slablike sheets of rock.

Shelf break The point at which a rapid steepening of the gradient occurs, marking the outer edge of the continental shelf and the beginning of the continental slope.

Shield A large, relatively flat expanse of ancient metamorphic rock within the stable continental interior.

Shield volcano A broad, gently sloping volcano built from fluid basaltic lavas.

Silicate Any one of numerous minerals that have the silicon-oxygen tetrahedron as their basic structure.

Silicon-oxygen tetrahedron A structure composed of four oxygen atoms surrounding a silicon atom that constitutes the basic building block of silicate minerals.

Sill A tabular igneous body that was intruded parallel to the layering of pre-existing rock.

Sinkhole A depression produced in a region where soluble rock has been removed by groundwater.

Slide A movement common to mass wasting processes in which the material moving downslope remains fairly coherent and moves along a well-defined surface.

Slip face The steep, leeward surface of a sand dune which maintains a slope of about 34 degrees.

Slump The downward slipping of a mass of rock or unconsolidated material moving as a unit along a curved surface.

Snowfield An area where snow persists throughout the year.

Snowline Lower limit of perennial snow.

Soil A combination of mineral and organic matter, water, and air; that portion of the regolith that supports plant growth.

Soil horizon A layer of soil that has identifiable characteristics produced by chemical weathering and other soil-forming processes.

Soil profile A vertical section through a soil showing its succession of horizons and the underlying parent material.

Solifluction Slow, downslope flow of water-saturated materials common to permafrost areas.

Solum The O, A, and B horizons in a soil profile. Living roots and other plant and animal life are largely confined to this zone.

Solution The change of matter from the solid or gaseous state into the liquid state by its combination with a liquid.

Specific gravity The ratio of a substance's weight to the weight of an equal volume of water.

Speleothem A collective term for the dripstone features found in caverns.

Spheroidal weathering Any weathering process that tends to produce a spherical shape from an initially blocky shape.

Spit An elongate ridge of sand that projects from the land into the mouth of an adjacent bay.

Spring A flow of groundwater that emerges naturally at the ground surface.

Spring tide The highest tidal range. Occurs near the times of the new and full moons.

Stalactite The iciclelike structure that hangs from the ceiling of a cavern.

Stalagmite The columnlike form that grows upward from the floor of a cavern.

Stock A pluton similar to but smaller than a batholith.

Strata Parallel layers of sedimentary rock.

Stratovolcano See *Composite cone.*

Streak The color of a mineral in powdered form.

Stream A general term to denote the flow of water within any natural channel. Thus, a small creek and a large river are both streams.

Stress The force per unit area acting on any surface within a solid. Also known as *directed pressure.*

Striations The multitude of fine parallel lines found on some cleavage faces of plagioclase feldspars, but that are not present on orthoclase feldspar.

Striations (glacial) Scratches or grooves in a bedrock surface caused by the grinding action of a glacier and its load of sediment.

Strike The compass direction of the line of intersection created by a dipping bed or fault and a horizontal surface. Strike is always perpendicular to the direction of dip.

Strike-slip fault A fault along which the movement is horizontal.

Subduction The process of thrusting oceanic lithosphere into the mantle along a convergent zone.

Submarine canyon A seaward extension of a valley that was cut on the continental shelf during a time when sea level was lower, or a canyon carved into the outer continental shelf, slope, and rise by turbidity currents.

Submergent coast A coast whose form is largely the result of the partial drowning of a former land surface either due to a rise of sea level or subsidence of the crust, or both.

Subsoil A term applied to the *B* horizon of a soil profile.

Surf A collective term for breakers; also the wave activity in the area between the shoreline and the outer limit of breakers.

Surface soil The upper portion of a soil profile consisting of the *O* and *A* horizons.

Surface waves Seismic waves that travel along the outer layer of the earth.

Surge A period of rapid glacial advance. Surges are typically sporadic and short lived.

Suspended load The fine sediment carried within the body of flowing water or air.

S wave An earthquake wave, slower than a P wave, that travels only in solids.

Swells Wind-generated waves that have moved into an area of weaker winds or calm.

Syncline A linear downfold in sedimentary strata; the opposite of anticline.

Talus An accumulation of rock debris at the base of a cliff.

Tarn A small lake in a cirque.

Tectonics The study of the large-scale processes that collectively deform the earth's crust.

Temporary (local) base level The level of a lake, resistant rock layer, or any other base level that stands above sea level.

Terminal moraine The end moraine marking the farthest advance of a glacier.

Terrace A flat, benchlike structure produced by a stream, which was left elevated as the stream cut downward.

Texture The size, shape, and distribution of the particles that collectively constitute a rock.

Thrust fault A low-angle reverse fault.

Tidal current The alternating horizontal movement of water associated with the rise and fall of the tide.

Tidal flat A marshy or muddy area that is alternately covered and uncovered by the rise and fall of the tide.

Tide Periodic change in the elevation of the ocean surface.

Till Unsorted sediment deposited directly by a glacier.

Tillite A rock formed when glacial till is lithified.

Tombolo A ridge of sand that connects an island to the mainland or to another island.

Topset bed An essentially horizontal sedimentary layer deposited on top of a delta during floodstage.

Transform fault boundary A boundary in which two plates slide past one another without creating or destroying lithosphere.

Transpiration The release of water vapor to the atmosphere by plants.

Transported soil Soils that form on unconsolidated deposits.

Transverse dunes A series of long ridges oriented at right angles to the prevailing wind; these dunes form where vegetation is sparse and sand is very plentiful.

Travertine A form of limestone ($CaCO_3$) that is deposited by hot springs or as a cave deposit.

Trellis drainage A system of streams in which nearly parallel tributaries occupy valleys cut in folded strata.

Trench An elongate depression in the sea floor produced by bending of oceanic crust during subduction.

Truncated spurs Triangular-shaped cliffs produced when spurs of land that extend into a valley are removed by the great erosional force of an alpine glacier.

Tsunami The Japanese word for a seismic sea wave.

Turbidite Turbidity current deposit characterized by graded bedding.

Turbidity current A downslope movement of dense, sediment-laden water created when sand and mud on the continental shelf and slope are dislodged and thrown into suspension.

Turbulent flow The movement of water in an erratic fashion often characterized by swirling, whirlpool-like eddies. Most streamflow is of this type.

Ultimate base level Sea level; the lowest level to which stream erosion could lower the land.

Unconformity A surface that represents a break in the rock record, caused by erosion or nondeposition.

Uniformitarianism The concept that the processes that have shaped the earth in the geologic past are essentially the same as those operating today.

Valence electron The electrons involved in the bonding process; the electrons occupying the highest principal energy level of an atom.

Valley train A relatively narrow body of stratified drift deposited on a valley floor by meltwater streams that issue from the terminus of an alpine glacier.

Ventifact A cobble or pebble polished and shaped by the sandblasting effect of wind.

Vesicles Spherical or elongated openings on the outer portion of a lava flow that were created by escaping gases.

Vesicular A term applied to igneous rocks that contain small cavities called vesicles, which are formed when gases escape from lava.

Viscosity A measure of a fluid's resistance to flow.

Volcanic arc Mountains formed in part by igneous activity associated with the subduction of oceanic lithosphere beneath a continent. Examples include the Andes and the Cascades.

Volcanic bomb A streamlined pyroclastic fragment ejected from a volcano while molten.

Volcanic neck An isolated, steep-sided, erosional remnant consisting of lava that once occupied the vent of a volcano.

Volcano A mountain formed from lava and/or pyroclastics.

Wash A desert stream course that is typically dry except for brief periods immediately following rainfall.

Water gap A pass through a ridge or mountain in which a stream flows.

Water table The upper level of the saturated zone of goundwater.

Wave-cut cliff A seaward-facing cliff along a steep shoreline formed by wave erosion at its base and mass wasting.

Wave-cut platform A bench or shelf along a shore at sea level, cut by wave erosion.

Wave height The vertical distance between the trough and crest of a wave.

Wave length The horizontal distance separating successive crests or troughs.

Wave period The time interval between the passage of successive crests at a stationary point.

Weathering The disintegration and decomposition of rock at or near the surface of the earth.

Welded tuff A pyroclastic deposit composed of particles fused together by the combination of heat still contained in the deposit after it has come to rest and the weight of overlying material.

Well An opening bored into the zone of saturation.

Wind gap An abandoned water gap. These gorges typically result from stream piracy.

Xenolith An inclusion of unmelted country rock in an igneous pluton.

Xerophyte A plant highly tolerant of drought.

Yazoo tributary A tributary that flows parallel to the main stream because a natural levee is present.

Zone of accumulation The part of a glacier characterized by snow accumulation and ice formation. The outer limit of this zone is the snowline.

Zone of aeration Area above the water table where openings in soil, sediment, and rock are not saturated but filled mainly with air.

Zone of fracture The upper portion of a glacier consisting of brittle ice.

Zone of saturation Zone where all open spaces in sediment and rock are completely filled with water.

INDEX